Proteomics

METHODS EXPRESS

The **METHODS EXPRESS** series

Series editor: B. David Hames

Faculty of Biological Sciences, University of Leeds, Leeds LS2 9JT, UK

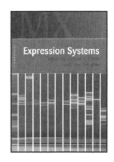

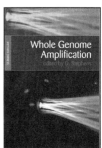

Coming Soon   **For full details see: www.scionpublishing.com/mx**

Animal Cell Culture

Biosensors

Genomics

GPCR

In Situ Hybridization

Plant Cell Culture

Protein Arrays

Proteomics

METHODS EXPRESS

edited by **C.D. O'Connor**
*Centre for Proteomic Research, School of Biological Sciences,
University of Southampton, Southampton, UK*

and **B.D. Hames**
*Faculty of Biological Sciences,
University of Leeds, Leeds, UK*

Scion

© Scion Publishing Ltd, 2008

First published 2008

All rights reserved. No part of this book may be reproduced or transmitted, in any form or by any means, without permission.

A CIP catalogue record for this book is available from the British Library.

ISBN: 978 1 904842 13 2 (paperback)
ISBN: 978 1 904842 20 0 (hardback)

Scion Publishing Limited
Bloxham Mill, Barford Road, Bloxham, Oxfordshire OX15 4FF
www.scionpublishing.com

Important note from the Publisher

The information contained within this book was obtained by Scion Publishing Limited from sources believed by us to be reliable. However, while every effort has been made to ensure its accuracy, no responsibility for loss or injury whatsoever occasioned to any person acting or refraining from action as a result of information contained herein can be accepted by the authors or publishers.

Typeset by Phoenix Photosetting, Chatham, Kent, UK
Printed by Gutenberg Press, Malta

Contents

Contributors	x
Preface	xiii
Abbreviations	xvi
Color section	xix

Chapter 1. Sample preparation and subcellular fractionation approaches: purification of membranes and their microdomains for mass spectrometry analysis
Yan Li, Phil Oh, and Jan E. Schnitzer

1. Introduction	1
2. Methods and approaches	2
2.1 Isolation of plasma membrane and its microdomains	2
2.2 Principles of the technology	3
2.3 Methodology	3
3. Troubleshooting	13
4. Identification of proteins from purified membranes by liquid chromatography tandem MS	14
4.1 Mapping of the proteome of the rat lung microvascular endothelial cell surface by multi-dimensional protein identification technology	14
4.2 Target proteins identified by two-dimensional gel electrophoresis combined with LC-MS/MS	14
5. References	15

Chapter 2. An isotope-coding strategy for quantitative proteomics
Xian Chen

1. Introduction	17
2. Methods and approaches	18
2.1 Amino acid-coded mass tagging	18
2.2 High-throughput MS for AACT-based quantitative proteomic analysis	23

    2.3 AACT/epitope dual-tagging strategy for pathway scale profiling of protein–protein interactions regulating gene expression   27
     2.4 Mapping 'real-time' phosphorylation sites of signal proteins involved in signal transduction   29
3. Conclusions   30
4. Troubleshooting   30
5. References   31

Chapter 3. Gel-based approaches
Stuart J. Cordwell, Ben Crossett, and Melanie Y. White

1. Introduction   33
2. Methods and approaches   34
     2.1 Sample preparation   34
     2.2 Protein pre-fractionation   36
     2.3 Isoelectric focusing   41
     2.4 Difference in-gel electrophoresis   42
     2.5 Reduction, alkylation, and detergent exchange (equilibration)   45
     2.6 SDS-PAGE   46
     2.7 Staining 2DE gels   47
3. Troubleshooting   53
     3.1 Interfering molecules during IEF   53
     3.2 'Difficult' proteins in 2DE   54
     3.3 Alternatives to 2DE gel-based approaches   56
4. Conclusions   56
5. References   57

Chapter 4. Peptide sorting by reverse-phase diagonal chromatography
Kris Gevaert and Joël Vandekerckhove

1. Introduction   61
2. Methods and approaches   62
     2.1 Principles of combined fractional diagonal chromatography   62
     2.2 General applications of COFRADIC for gel-free proteomics   65
     2.3 Methodology   66
3. Troubleshooting   79
4. References   81

Chapter 5. Mass spectrometry strategies for protein identification
David R. Goodlett and Garry L. Corthals

1. Introduction   83
2. Sample preparation   85
     2.1 Digestion   85
     2.2 Sample clean-up prior to LC-MS   87
3. MS analysis   89

	3.1	Microcapillary LC-MS/MS	89
	3.2	Data-dependent MS/MS allowing automated ion selection	90
	3.3	Iterative gas-phase fractionation increasing proteome coverage	93
4.	Database searching		94
	4.1	Matching peptide fragmentation patterns to amino acid sequence	95
5.	Concluding remarks		97
6.	References		98

Chapter 6. Desorption electrospray ionization: proteomics studies by a method that bridges ESI and MALDI
Zoltán Takáts, Justin M. Wiseman, Demian R. Ifa, and R. Graham Cooks

1.	Introduction		99
	1.1	DESI instrumentation	100
	1.2	Ion formation	102
2.	Methods and approaches		104
	2.1	Analysis of intact proteins	105
	2.2	Analysis of tryptic digests	106
	2.3	Coupling DESI with separation methods	110
	2.4	Reactive DESI	112
	2.5	*In situ* proteomics	113
3.	Troubleshooting		116
4.	References		117

Chapter 7. Analysis of cellular protein complexes by affinity purification and mass spectrometry
Tilmann Bürckstümmer and Keiryn L. Bennett

1.	Introduction		119
2.	Methods and approaches		120
	2.1	TAP compared with other experimental approaches	120
	2.2	Methodology	121
3.	Troubleshooting		132
4.	References		133

Chapter 8. Clinical proteomic profiling and disease signatures
Rosamonde E. Banks, David A. Cairns, David N. Perkins, and Jennifer H. Barrett

1.	Introduction		135
2.	Methods and approaches		137
	2.1	Principles of MALDI and SELDI	138
	2.2	Choice of sample type	139
	2.3	Study design, pre-analytical and analytical issues	140
	2.4	Data processing	141

	2.5 Data analysis	142
	2.6 Validation and translation	143
	2.7 Methodology	144
3.	Troubleshooting	157
4.	References	157

Chapter 9. Characterization of post-translational modifications: undertaking the phosphoproteome
W. Andy Tao, Bernd Bodenmiller, and Ruedi Aebersold

1.	Introduction	161
2.	Methods and approaches	163
	2.1 Enrichment of phosphoproteins	163
	2.2 Enrichment of phosphopeptides	165
	2.3 MS data acquisition, phosphopeptide identification, and determination of sites of phosphorylation	173
	2.4 Quantitative phosphoproteomics	176
3.	Troubleshooting	178
4.	References	179

Chapter 10. Protein microarray technologies
Chien-Sheng Chen, Sheng-Ce Tao, and Heng Zhu

1.	Introduction	183
2.	Protein microarray fabrication	184
	2.1 Surface materials	184
	2.2 Protein printing	185
	2.3 Assay platforms	186
	2.4 Detection methods	187
3.	Analytical protein microarrays	190
4.	Functional protein microarrays	191
	4.1 Expression-ready open reading frame collections and high-throughput production of proteins	192
	4.2 Protein–protein and protein–lipid interactions	194
	4.3 Protein–DNA interactions	195
	4.4 Protein–drug interactions	195
	4.5 Protein–peptide interactions	196
	4.6 Protein–cell interactions	196
	4.7 Identification of kinase substrates on protein chips	198
	4.8 Protein glycosylation analysis	198
	4.9 Profiling immune responses	198
5.	Development of new protein microarray technologies	199
	5.1 Reverse-phase protein microarrays	199
	5.2 Making protein microarrays without large-scale cloning and expressions	200
6.	Outlook	202
7.	References	202

Chapter 11. Intelligent mining of complex data: challenging the proteomic bottleneck.
Dan Bach Kristensen and Alexandre Podtelejnikov

1. Introduction	207
2. Methods and approaches	209
2.1 Instrumentation	210
2.2 Database selection	212
2.3 Selection of a search engine	213
2.4 Data mining in shotgun proteomics	215
3. EPICENTER	216
3.1 Data organization and import	216
3.2 Automatic and manual peptide validation	218
3.3 Data mining – exploring datasets	220
3.4 Data mining – comparing datasets	222
3.5 Additional EPICENTER features	223
3.6 Conclusions	224
4. References	224

Chapter 12. Bioinformatic approaches in proteomics
Sandra Orchard and Henning Hermjakob

1. Introduction	227
2. Methods and approaches	228
2.1 *In silico* characterization of proteins	228
2.2 Data standardization	230
2.3 Methodology	232
3. Troubleshooting	242
4. References	243

Appendix

List of suppliers	245

Index

	249

Contributors

Aebersold, Ruedi Institute for Molecular Systems Biology, Federal Institute of Technology, Campus Hoenggerberg, HPT E78, Wolfgang Pauli Strasse 16, CH-8093, Zurich, Switzerland; and Faculty of Natural Sciences, University of Zurich, Zurich, Switzerland. E-mail: aebersold@imsb.biol.ethz.ch

Banks, Rosamonde E. Cancer Research UK Clinical Centre, St James's University Hospital, Beckett Street, Leeds LS9 7TF, UK. E-mail: r.banks@leeds.ac.uk

Barrett, Jennifer H. Cancer Research UK Clinical Centre, St James's University Hospital, Beckett Street, Leeds LS9 7TF, UK. E-mail: j.h.barrett@leeds.ac.uk

Bennett, Keiryn L. Research Center for Molecular Medicine of the Austrian Academy of Sciences (CeMM), Lazarettgasse 19/3, A-1090 Vienna, Austria. E-mail: kbennett@cemm.oeaw.ac.at

Bodenmiller, Bernd Institute for Molecular Systems Biology, Federal Institute of Technology, Campus Hoenggerberg, HPT E78, Wolfgang Pauli Strasse 16, CH-8093, Zurich, Switzerland. E-mail: bodenmiller@imsb.biol.ethz.ch

Bürckstümmer, Tilmann Research Center for Molecular Medicine of the Austrian Academy of Sciences (CeMM), Lazarettgasse 19/3, A-1090 Vienna, Austria. E-mail: tbuerckstuemmer@cemm.oeaw.ac.at

Cairns, David A. Cancer Research UK Clinical Centre, St James's University Hospital, Beckett Street, Leeds LS9 7TF, UK. E-mail: d.a.cairns@leeds.ac.uk

Chen, Chien-Sheng Department of Pharmacology and Molecular Sciences/High Throughput Biology Center, Johns Hopkins University School of Medicine, Baltimore, MD 21205, USA; and Department of Food Science, National Taiwan Ocean University, Keelung 20224, Taiwan.

Chen, Xian Department of Biochemistry and Biophysics, University of North Carolina at Chapel Hill, NC, USA. E-mail: xian_chen@med.unc.edu

Cooks, R. Graham Department of Chemistry, Purdue University, West Lafayette, IN 47907, USA. E-mail: cooks@purdue.edu

Cordwell, Stuart J. School of Molecular and Microbial Biosciences, The University of Sydney, Australia. E-mail: s.cordwell@mmb.usyd.edu.au

Corthals, Garry L. Protein Research Group, Turku Centre for Biotechnology, University of Turku and Åbo Akademi, FIN-20521 Turku, Finland. E-mail: garry.corthals@btk.fi

Crossett, Ben School of Molecular and Microbial Biosciences, The University of Sydney, New South Wales, Australia.

Gevaert, Kris Department of Biochemistry (GE07) and Medical Protein Research (VIB09), Faculty of Medicine and Health Sciences, Ghent University and Flanders Interuniversity Institute for Biotechnology, Ghent, Belgium. E-mail: kris.gevaert@ugent.be

Goodlett, David R. Department of Medicinal Chemistry, University of Washington, 1959 NE Pacific Street, Health Sciences Building, Seattle, WA 98195-7610, USA. E-mail: goodlett@u.washington.edu

Hermjakob, Henning EMBL Outstation — European Bioinformatics Institute, Wellcome Trust, Genome Campus, Hinxton, Cambridge, UK. E-mail: hhe@ebi.ac.uk

Ifa, Demian R. Department of Chemistry, Purdue University, West Lafayette, IN 47907, USA. E-mail: difa@purdue.edu

Kristensen, Dan Bach Maxygen, Agern Alle 1, DK-2970 Hoersholm, Denmark. E-mail: dbk@maxygen.dk

Li, Y. Sidney Kimmel Cancer Center, 10905 Road to the Cure, San Diego, CA 92121, USA. E-mail: yli@skcc.org

Oh, P. Sidney Kimmel Cancer Center, 10905 Road to the Cure, San Diego, CA 92121, USA. E-mail: poh@skcc.org

Orchard, Sandra EMBL Outstation — European Bioinformatics Institute, Wellcome Trust, Genome Campus, Hinxton, Cambridge, UK. E-mail: orchard@ebi.ac.uk

Perkins, David N. Cancer Research UK Clinical Centre, St James's University Hospital, Beckett Street, Leeds LS9 7TF, UK. E-mail: d.n.perkins@cruk.leeds.ac.uk

Podtelejnikov, Alexandre Proxeon, Staermosegaardsvej 6, DK-5230 Odense M, Denmark. E-mail: apodtelejnikov@proxeon.com

Schnitzer, Jan E. Sidney Kimmel Cancer Center, 10905 Road to the Cure, San Diego, CA 92121, USA. E-mail: jschnitzer@skcc.org

Takáts, Zoltán Department of Chemistry, Purdue University, West Lafayette, IN 47907, USA. Present address: Institute of Structural Chemistry, Chemical Research Centre of the Hungarian Academy of Sciences, Pusztaszeri ut 59-67, Budapest, Hungary. E-mail: takáts@kkk.org.hu

Tao, Sheng-Ce Department of Pharmacology and Molecular Sciences/High Throughput Biology Center, Johns Hopkins University School of Medicine, Baltimore, MD 21205, USA.

Tao, W. Andy Bindley Bioscience Center and Department of Biochemistry, Purdue University, West Lafayette, IN 47907, USA. E-mail: taow@purdue.edu

Vandekerckhove, Joël Department of Biochemistry (GE07) and Medical Protein Research (VIB09), Faculty of Medicine and Health Sciences, Ghent University and Flanders Interuniversity Institute for Biotechnology, Ghent, Belgium. E-mail: joel.vandekerckhove@ugent.be

White, Melanie Y. Minomic Pty Ltd, Frenchs Forest, New South Wales, Australia.

Wiseman, Justin M. Department of Chemistry, Purdue University, West Lafayette, IN 47907, USA. Present address: Prosolia, Inc., Indianapolis, IN, USA. E-mail: wiseman@prosolia.com

Zhu, Heng Department of Pharmacology and Molecular Sciences/High Throughput Biology Center, Johns Hopkins University School of Medicine, Baltimore, MD 21205, USA. E-mail: heng.zhu@jhmi.edu

Preface

Proteins are the ultimate effectors of essentially all cellular processes. They also form the vast majority of molecular targets for therapeutic drugs, serve as predictive biomarkers of disease and are the basis of most vaccines. In keeping with their central importance, studies on proteins have relevance to every major branch of the life sciences. Until recently, however, it has been difficult to investigate them systematically and on a large scale due to their extraordinarily diverse molecular properties and wide-ranging levels of expression. This situation is now changing, primarily due to advances in mass spectrometry (MS), which allow the detection and quantification of proteins from sub-picomole amounts of material, and the availability of genome sequences, which greatly facilitate protein identification via bioinformatic-based approaches. Collectively, these technologies are known as proteomics.

The vast molecular diversity of proteins has triggered the development of a proportionately large number of proteomic approaches. Whilst the availability of so many techniques is a testament to the ingenuity of researchers in this area, it also confuses novices, who need a core set of reliable protocols that are relatively painless to implement but sufficiently powerful to yield valuable new scientific information. This book provides such a set of procedures, with the emphasis on those with the potential to provide valuable scientific insights and that can be readily adapted to address the specific needs of individual researchers. Currently, most proteomic experiments require access to a suitable mass spectrometer, which could discourage researchers from entering this field. Like computers, however, these instruments are rapidly dropping in price while becoming more user-friendly. There have also been major advances in mass accuracy, resolution and sensitivity. Many universities and research institutes now run mass spectrometry facilities that allow external users to access proteomics without major capital investment. Thus, the barriers to taking up proteomics have never been lower.

The order of chapters in this book approximately reflects the workflows associated with proteomics. Thus, we start with the crucial topics of sample preparation and isotope labeling strategies before considering powerful methods to separate and characterize peptides, proteins and protein complexes. The chapters that follow focus on novel technologies, such as the MS-based analysis of living cells and tissues and the use of protein microarrays, as well as clinical approaches to identify protein and peptide biomarkers. Finally, the book considers important bioinformatic approaches to mine the often prodigious amounts of data that are generated, thereby producing biologically meaningful information.

Careful sample preparation is undoubtedly critical for success in proteomics, both to prevent artefacts and to ensure that as many relevant proteins as possible are available for analysis. Inevitably, it is impossible to devise a 'one size fits all' protocol due to the enormous variety of samples that can be investigated. Instead, Li et al. outline the major factors that influence sample quality and before using a specific example – the endothelial cell surface and its microdomains – to illustrate how difficult subcellular fractions can be purified.

A traditional entry point into proteomics has often been the use of two-dimensional gel electrophoresis for protein separations. This is to be expected as such an approach is a natural extension of methods that are widely used by life scientists. As demonstrated in the chapter by Cordwell et al., it remains an excellent way to obtain a quantitative snapshot of cellular processes at the protein level and is particularly effective at revealing protein isoforms, whose relative levels often reflect the physiological status of a cell as well as specific disease states. However, such approaches do not lend themselves to large-scale studies, and they tend to under-represent proteins with extreme physico-chemical properties. For these reasons, much of the rest of the book focuses on approaches that avoid the need to run two-dimensional gels, and hence circumvent several of the problems associated with traditional proteomic methodology. The use of such approaches, coupled with careful experimental design, can significantly increase sample throughput and the degree of coverage of the proteome, while facilitating accurate protein quantification. In this respect, the chapters by Chen, by Gevaert and Vandekerckhove, by Goodlett and Corthals, and by Bürckstümmer and Bennett are invaluable.

Most proteomic approaches are 'bottom-up' strategies in which highly enriched components are ultimately analyzed by mass spectrometry *in vacuo*. Recently, however, it has become possible to modify electrospray ionization MS so that the surfaces of living cells and tissues are directly interrogated at ambient temperature and pressure. This versatile 'top-down' approach is significant since, among other applications, it allows the molecular imaging of complex samples. For example, the investigator is able to superimpose the distribution of components with specific *m/z* values onto the corresponding optical image of a tissue section. As described by Takáts et al., the instrument modifications required are relatively simple and well within the scope of many laboratories using ESI-MS.

Many of the ultimate applications of proteomics lie in the clinic, e.g. in the discovery of biomarkers of diagnostic and prognostic value. While the number of protein biomarkers in routine clinical use remains stubbornly low, this is an area of intense activity and great promise. Equally, however, researchers need to appraise potential pitfalls in this area critically before embarking on such major projects. The chapter by Banks et al. carefully describes factors that are likely to be critical to obtain proteomic profiles and disease signatures of real diagnostic value. Many of the surface capture concepts described in the chapter can be extended to protein microarrays. As indicated by Chen et al., these highly adaptable tools are still very much under development. However, they will have a major impact on proteomic coverage and consequently should provide insights into a much larger number of cellular processes and disease conditions.

A major future priority is to develop methods that allow individual protein species bearing specific post-translation modifications (PTMs) to be arrayed, thereby increasing the resolution and depth of proteomic data. Comprehensive consideration of methods to characterize PTMs is outside the scope of this book, not least because over two hundred modifications fall into this category. However, the chapter by Tao *et al.* uses protein phosphorylation to illustrate how such modified proteins and peptides can be efficiently captured and quantitatively characterized. Like most proteomic methods, these techniques generate large amounts of data that need to be refined, manipulated and analyzed to extract biological insights. The final two chapters, by Kristensen and Podtelejnikov and by Orchard and Hermjakob, describe some fundamental approaches to handle this data and to mine it. Several themes emerge, notably the need to avoid swamping end-users with superfluous information and for readily accessible interfaces, the importance of standardization of proteomic data to increase confidence in the quality of the data generated, and the use of data repositories that can be accessed by other investigators.

One of the current grand challenges in proteomics is to achieve a complete description of an organism's components at the protein level and to quantify these components in different physiological states. Such important objectives will certainly be reached more rapidly and with greater accomplishment if proteomics is able to recruit a new generation of creative researchers with fresh ideas. It is therefore hoped that, in addition to helping investigators address their own scientific questions, the methods described here will help to stimulate further advances in proteomics and progress towards these exciting goals.

The editors wish to thank the authors for generously sharing their expertise and methods, including protocols previously only available to members of their labs, and for their forbearance during the production of this book. We also thank Jonathan Ray for valuable help and guidance.

C. David O'Connor
B. David Hames
October 2007

Abbreviations

2DE	two-dimensional electrophoresis
2DLC	two-dimensional liquid chromatography
2D-PAGE	two-dimensional polyacrylamide gel electrophoresis
AACT	amino acid-coded mass tagging
ACE	angiotensin-converting enzyme
a.m.u.	atomic mass unit
ANS	1-anilino-naphthalene-8-sulfonic acid
ASB-14	amidosulfobetaine-14
BSA	bovine serum albumin
CA	carrier ampholyte
CCD	charge-coupled device
CHAPS	3-((3-cholamidopropyl)dimethylammonio)-1-propanesulfonate
CID	collision-induced dissociation
CIP	calf intestinal alkaline phosphatase
COFRADIC	combined fractional diagonal chromatography
ε-COP	coatomer ε-subunit
CV	coefficient of variation
DESI	desorption electrospray ionization
DI	desorption ionization
DIGE	difference in-gel electrophoresis
DMF	dimethylformamide
DTT	dithiothreitol
ECM	extracellular matrix
EDC	N-(3-dimethylaminopropyl)-N'-ethylcarbodiimide
ELISA	enzyme-linked immunosorbent assay
eNOS	endothelial nitric synthase
EPICENTER	Experimental Peptide Identification Center
ESI	electrospray ionization
EST	expressed sequence tag
ETD	electron transfer dissociation
FAB	fast atom bombardment
FDR	false discovery rate
FFE	free-flow electrophoresis
FTICR	Fourier transform ion cyclotron resonance

GFP	green fluorescent protein
GO	gene ontology
GPF	gas-phase fractionation
GST	glutathione S-transferase
HPLC	high-performance liquid chromatography
HUPO	Human Proteome Organization
IAA	iodoacetamide
ICAT	isotope-coded affinity tag
IEF	isoelectric focusing
IFA	immunofluorescent assay
IMAC	immobilized metal affinity chromatography
IPG	immobilized pH gradient
IPI	International Protein Index
Lamp1	lysosomal membrane glycoprotein 1
LC	liquid chromatography
LD	laser desorption
MALDI	matrix-assisted laser desorption ionization
MBS	MES-buffered saline
MCAT	mass-coded abundance tagging
MCC	moth cytochrome c
MES	2-(N-morpholine) ethanesulfonic acid
MS	mass spectrometry
MS/MS	tandem mass spectrometry
M/Z	mass-to-change ratio
NCBI	National Center for Biotechnology Information
ORF	open reading frame
PAA	polyacrylic acid
PAGE	polyacrylamide gel electrophoresis
PBS	phosphate-buffered saline
PD	plasma desorption
pI	isoelectric point
PMMA	polymethyl methacrylate
PSI	Proteomics Standards Initiative
PTFE	polytetrafluoroethylene
PTM	post-translational modification
PVDF	polyvinylidene difluoride
QC	quality control
QTOF	quadrupole time of flight
RP	reverse phase
RT-PCR	reverse transcriptase polymerase chain reaction
S/N	signal-to-noise
SARS-CoV	severe acute respiratory syndrome coronavirus
SB3-10	sulfobetaine 3-10
SCX	strong cation exchange
SDS	sodium dodecyl sulfate
SELDI	surface-enhanced laser desorption/ionization

SH2	Src homology 2
SIL	stable isotope label
SILAC	stable isotope labeling in cell culture
SIMS	secondary ion mass spectrometry
SMIR	small-molecule inhibitors of rapamycin
SPA	sinapinic acid
SPR	surface plasmon resonance
TAP	tandem affinity purification
TBP	tributylphosphine
TCA	trichloroacetic acid
TCEP	Tris(2-carboxyethyl)phosphine
TEMED	tetramethylethylenediamine
TEV	tobacco etch virus
TFA	trifluoroacetic acid
TIC	total ion current
TLC	thin-layer chromatography
TNBS	trinitrobenzene sulfonic acid
TOF	time of flight
TOR	target of rapamycin
TPP	Trans-Proteomic Pipeline
UHQ	ultrahigh quality

Color section

Chapter 2. An isotope-coding strategy for quantitative proteomics

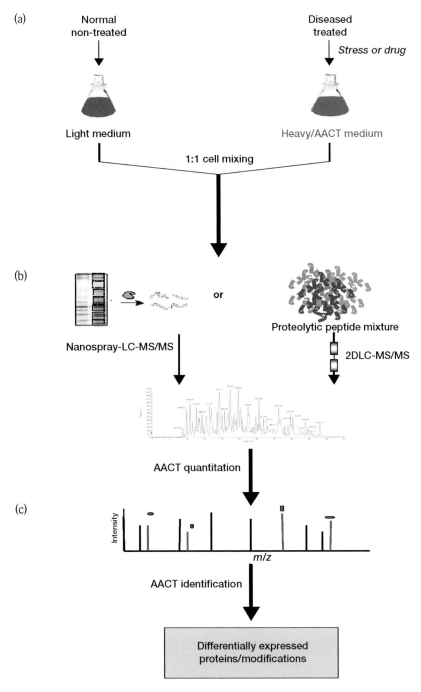

Figure 1. Overall experimental design of AACT-assisted quantitative proteomics (see page 19).
(a) In vivo/in vitro cell culturing; (b) sample preparation; (c) MS measurements and data analyses. Symbols (●, ○, ■, ─) are defined in the Fig. 3 legend.

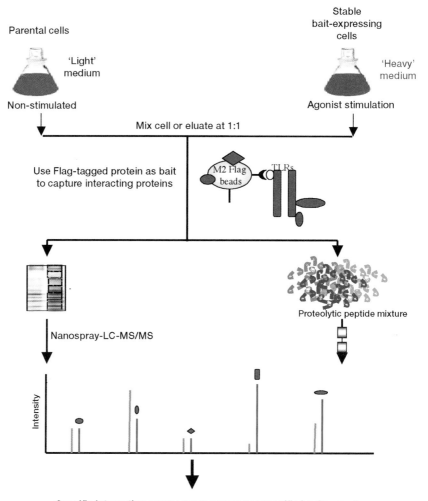

Figure 3. A dual-tagging proteomic platform for the characterization of protein complexes (see page 27).
(a) Epitope-tagging, *in vivo* cell culturing and *in situ* stimulation; (b) affinity purification; and (c) MS-based quantitative identification of specific protein interacting partners. Background binding proteins are identified based on a 1 : 1 ratio between labeled and unlabeled states (e.g. peaks ● and ◆). Peaks that show a divergence from this ratio will represent interacting proteins (peaks ◐, ◘, and ━). TLRs, toll-like receptors.

Chapter 3. Gel-based approaches

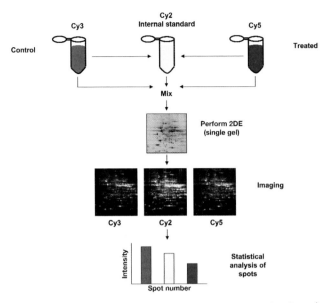

Figure 3. Schematic diagram of 2DE performed using DIGE technology (see page 43).
Two samples can be compared by labeling with Cy3 and Cy5 dyes, whilst a pooled internal standard is labeled with Cy2. Fluorescent images are then taken at the respective dye wavelengths and the images overlapped to quantify changes in protein abundance. The pooled standard is used to normalize the resulting image data.

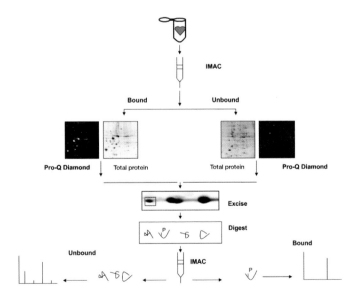

Figure 5. Large-scale 2DE gel analysis of phosphorylated proteins from rabbit myocardium (see page 51).
Affinity chromatography (IMAC) was used to bind phosphorylated proteins. Both bound and unbound proteins were then separated by 2DE and the resulting gels multiplex-stained with Pro-Q Diamond and SYPRO Ruby. Proteins of interest can then be excised, digested with trypsin, and subjected to IMAC. Bound and unbound peptides are then analyzed using MS.

Chapter 5. Mass spectrometry strategies for protein identification

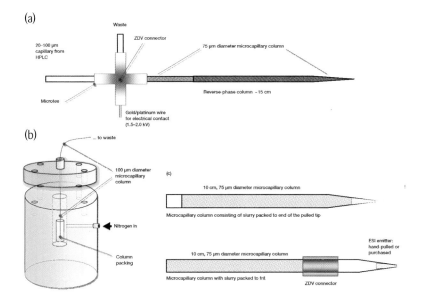

Figure 2. Construction of in-house capillary columns using a pressure cell and connection to an HPLC system using a zero dead volume cross-connection (see page 89).
(a) A zero dead volume union 'cross' showing connection to a capillary column (right), the waste flow (top), the flow from HPLC (left), and an electrode (bottom). (b) A pressure cell (Brechbuehler, Inc.) for packing capillary columns or loading samples onto a capillary column. (c) Examples of packed capillary columns.

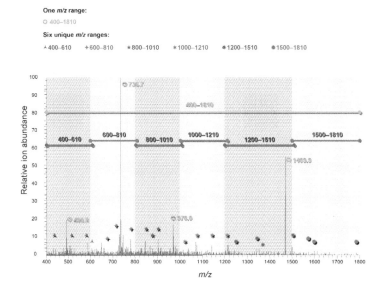

Figure 3. Gas-phase fractionation (see page 91).
Comparison of the increase in ions subjected to CID among six randomly chosen m/z ranges used for gas-phase fractionation and one normal wide m/z range.

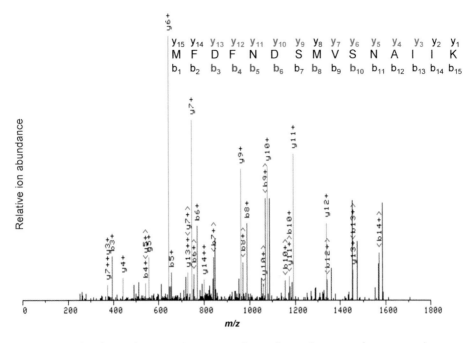

Figure 4. Example of a tandem mass spectrum of peptide ion fragments (see page 92).
The mass spectrum shows fragment ions generated after CID of a specific parental peptide ion. The spectrum was generated using a linear ion-trap mass spectrometer (ThermoElectron) controlled by data-dependent rules during LC introduction for ESI-MS/MS. Bracketed ions (<>) indicate that they have lost either H_2O or NH_3.

Table 1. List of all b and y ions for the peptide with amino acid sequence MFDFNDSMVSNAIIK[a] (see page 96)

b+ ions		Amino acid	y+ and y^{2+} ions		
m/z value of b+ ion	Ion number		m/z value of y+ ions	m/z value of y^{2+} ions	Ion number
132.2005	1	M	–	–	15
279.3771	2	F	1601.8145	801.4112	14
394.4657	3	D	1454.6379	727.8229	13
541.6422	4	F	1339.5493	670.2786	12
655.7461	5	N	1192.3728	596.6904	11
770.8347	6	D	1078.2689	539.6384	10
857.9129	7	S	963.1803	482.0941	9
989.1054	8	M	876.1021	438.555	8
1088.238	9	V	744.9096	372.9588	7
1175.3162	10	S	645.777	323.3925	6
1289.42	11	N	558.6988	279.8534	5
1360.4988	12	A	444.595	222.8015	4
1473.6583	13	I	373.7162	187.2621	3
1586.8177	14	I	260.3567	130.6823	2
–	15	K	147.1973	74.1026	1

[a]The b and y ions detected experimentally are highlighted in red and blue, respectively, and correspond to those shown in Fig. 4.

Chapter 6. Desorption electrospray ionization: proteomics studies by a method that bridges ESI and MALDI

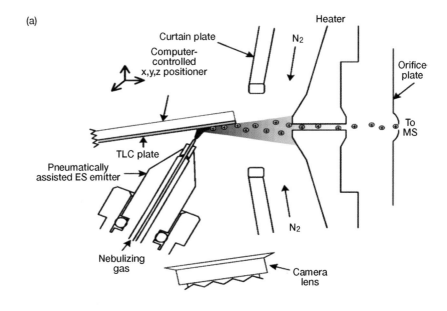

Figure 9. Example of a TLC/DESI-MS experiment (see page 110).
(a) Schematic illustration of the TLC/DESI-MS experimental set-up. (b) Color photograph of the DESI emitter and the TLC plate as viewed through the camera monitor during a TLC/DESI-MS experiment. Separate bands of rhodamine 6G (orange band), rhodamine B (pink band), and rhodamine 123 (yellow band) are observed on the RP C8 TLC plate.

Chapter 7. Analysis of cellular protein complexes by affinity purification and mass spectrometry

Expression of TAP-tagged entry point in target cells

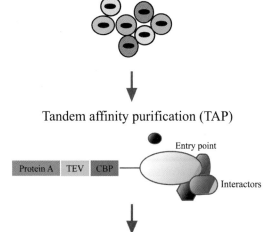

Tandem affinity purification (TAP)

Mass spectrometric protein identification by LC-MS/MS

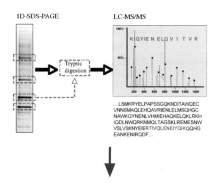

Physical network

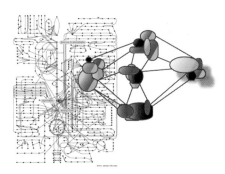

Figure 1. Overview of the TAP procedure (see page 122).
The gene of interest is expressed as a TAP-tagged fusion protein in the target cell line. TAP-tagged proteins are purified by TAP and co-purified proteins are identified by SDS-PAGE and subsequent MS analysis. Information gathered by multiple TAP is integrated to generate a physical map of the pathway of interest. CBP, calmodulin-binding protein.

Chapter 8. Clinical proteomic profiling and disease signatures

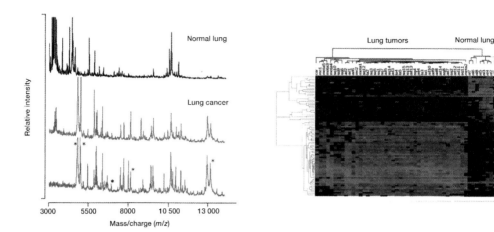

Figure 1. Detection of optimum discriminatory biomarker sets in lung tumors (see page 136).
(a) Representative MALDI-TOF-MS spectra obtained from frozen tissue sections showing the relative intensities of the peaks of different molecular masses (m/z values; asterisks indicate examples of discriminatory peaks). (b) Hierarchical clustering analysis of 42 lung tumors and eight normal lung tissues in the training cohort based on 82 MS signals. Each row represents an individual signal and each column an individual sample with the dendrogram at the top showing the similarity in protein expression profiles of the samples. Substantially raised expression is indicated in red. Reproduced with permission from (1).

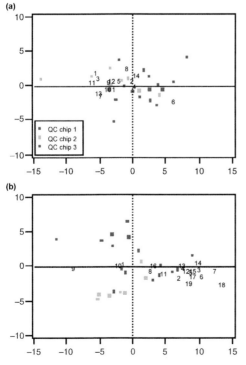

Figure 5. Examples of the use of plots of the first two principal components in QC (see page 153).
These examples are based on the method outlined by Coombes et al. (13). Routinely, three chips with a pooled QC sample are run at the beginning of the experiment (indicated by colored squares). Principal components are calculated on the correlation matrix of the identified peak intensities for these 24 samples. Subsequently, one spot on each chip contains the same pooled QC sample (indicated by 1 for QC from chip 1, 2 for QC spot from chip 2, etc.). Peak intensities are projected into the principal component space of the original 24 samples, the Mahalanobis distance from the center of the space is calculated, and the QC spot is assessed by comparing this with χ^2_2 distribution. (a) An example of acceptable QC. None of the sample chip QC spots are very far from the origin and the spread of values is generally within or very close to those found on the QC chips. As the spot from chip 6 is at the edge of the space occupied by the QC samples, the spectra from this chip should be examined further. (b) An example of QC flagging unacceptable results. The QC spots from sample chips are on the edge of the space occupied by the original QC spots, indicating that the spectra from these chips ought to be investigated further. The QC spots from chips 7 and 18 are identified as being unusual, as they have statistically significant large values of Mahalanobis distance, i.e. these QC spots cannot be said to be the same as those from the first three QC chips. Subsequently, a machine fault was identified. Note also the marked chip effects between the three QC chips, also observed to a much lesser extent in (a).

COLOR SECTION xxvii

Significant: 34
Median number of false positives: 1.93
False discovery rate (%): 5.67

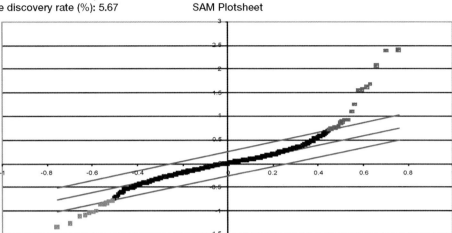

Figure 6. Comparison of MS profiles of two sample groups using the significance analysis of microarrays (SAM) method (94) (see page 155).

The x-axis shows the expected score (in this case a two-sample t-statistic) for each of 510 peaks, ordered from lowest to highest, under the null hypothesis of no difference between groups. This is obtained by randomly permuting the group labels (e.g. case–control status), calculating the t-statistic for each peak on this permuted data set, and ordering these 510 statistics. The mean of each ordered statistic is then obtained on the basis of 1000 such permuted samples. The y-axis shows the observed scores for the 510 peaks, again ordered by magnitude. Deviation greater than a certain distance from the line y=x indicates a significant result. In Microsoft EXCEL, a slider tool allows the adjustment of the width of the lines parallel to y=x. This allows the interactive selection of the number of significant results with an acceptable FDR, which is indicated in the top left-hand corner of the graph.

Chapter 10. Protein microarray technologies

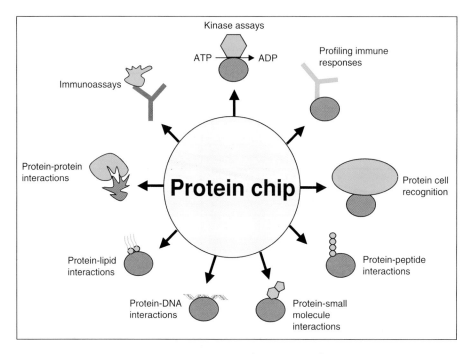

Figure 2. Image of a typical protein microarray (see page 188).
A yeast protein microarray was probed with an anti-GST antibody, followed by a Cy3-conjugated secondary antibody to visualize the immobilized proteins. An enlarged image of one of the 48 blocks is shown below the protein chip.

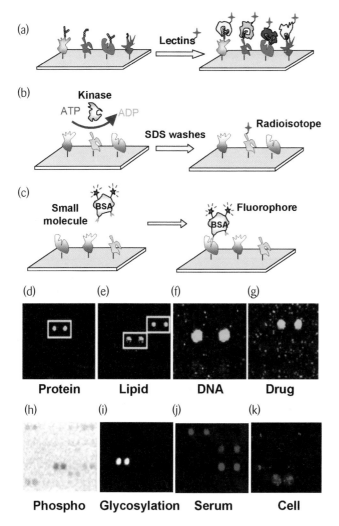

Figure 3. Protocols and examples of different assays using functional protein microarrays (see page 194). Experimental protocols for three assays for screening glycosylated proteins using a lectin probe (a), in vitro kinase substrates (b), and drug targets (c) are illustrated. Chip images of various biochemical assays for protein–protein (d), protein–lipid (e), protein–DNA (f), protein–drug (g) and protein–cell (k) interactions are also shown. In addition, the protein chip approach may be applied to monitor immune responses in patients (j) or post-translational modifications of proteins, such as phosphorylation (h) and glycosylation (i).

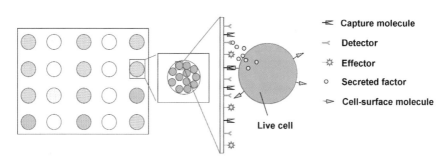

Figure 4. Analytical protein microarrays for live-cell studies (see page 197).
A schematic of an analytical protein microarray for profiling of live cells is shown. On such types of chip, affinity reagents, such as capture molecules, detectors, and effectors, can be immobilized, with the ability to recognize cell-surface molecules, such as surface glycans, antigens, and receptors, or molecules secreted by cells.

Chapter 11. Intelligent mining of complex data: challenging the proteomic bottleneck

(a)

(b)

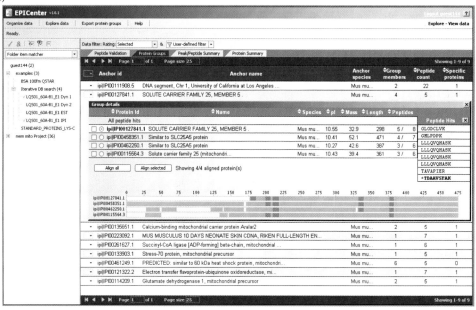

Figure 2. Data mining in EPICENTER (see page 217).
The user-definable data tree is shown to the left in (a) and (b). In this example, two data folders (LQS01_60A-B1_E1 IPI and LQS01_60A-B1_E1 Dyn1) have been selected in the path guest144/examples/Iterative DB search. To the right of the data tree, one of the four result views ('Peptide Validation', 'Protein Groups', 'Peak/Peptide Summary', and 'Protein Summary') is displayed, in this case the 'Protein Groups' view. Each line in the 'Protein Groups' view in (a) represents a protein group. To access the protein group, click on the anchor Id or the name, which will open a 'Group details' window (b). This window contains the accession number, protein name, species, pI, average mass, length, and number of peptides (number of unique peptides/number of matching precursor ions) for all of the protein members of the group. Unambiguously identified proteins are indicated with bold protein Ids, and the anchor protein is selected by clicking its radio button to the left of the protein Id. To the left in the group details views, the peptide hits are shown for the highlighted group member. Unique peptide hits (TDAAVSFAK) are shown in bold and modified peptides are indicated with an asterisk (TDAAVSFAK is N-terminally acetylated). In addition, the 'Group details' view shows a CLUSTALW alignment of all group members at the bottom of the window, with the sequenced peptides highlighted in red boxes.

	Peak ⇕ 350.224 / ⇕ 16.76	Database	Search	Rating	Δ mass	Quant ratio	Ions match/exp	y/b ion score	Proline score	Total NSP	Engine score
VVQPTR		IPI_mouse 60A MitoMem IPI L...			0.02 (2+)	-	6/46	41.6/1.0	40.0	0	24.53
VVQVVR		IPI_mouse 60A MitoMem IPI L...			0.01 (2+)	-	6/36	42.4/1.0	0.0	0	24.53

	Peak ⇕ 350.224 / ⇕ 32.08	Database	Search	Rating	Δ mass	Quant ratio	Ions match/exp	y/b ion score	Proline score	Total NSP	Engine score
IQLLGR		IPI_mouse 60A MitoMem IPI L...			0.01 (2+)	-	9/38	98.7/9.4	0.0	9	33.54
LAGLIGR		IPI_mouse 60A MitoMem IPI L...			0.01 (2+)	-	8/36	75.3/8.0	0.0	0	31.57
LAGLLGR		IPI_mouse 60A MitoMem IPI L...			0.01 (2+)	-	8/36	75.3/8.0	0.0	0	31.57
IQLGIR		IPI_mouse 60A MitoMem IPI L...			0.01 (2+)	-	5/38	37.9/1.0	0.0	0	25.73
LQLVAR		IPI_mouse 60A MitoMem IPI L...			0.01 (2+)	-	5/38	37.9/1.0	0.0	0	25.73
IKLAVR		IPI_mouse 60A MitoMem IPI L...			0.05 (2+)	-	5/38	37.7/1.0	0.0	0	25.73
LKIGLR		IPI_mouse 60A MitoMem IPI L...			0.05 (2+)	-	5/38	37.7/1.0	0.0	0	25.73
IAGIGLR		IPI_mouse 60A MitoMem IPI L...			0.01 (2+)	-	5/36	11.0/1.0	0.0	0	25.73
LQLGLR		IPI_mouse 60A MitoMem IPI L...			0.01 (2+)	-	5/38	37.9/1.0	0.0	0	25.73
LKLGLR		IPI_mouse 60A MitoMem IPI L...			0.05 (2+)	-	5/38	37.7/1.0	0.0	0	25.73

	Peak ⇕ 350.226 / ⇕ 25.17	Database	Search	Rating	Δ mass	Quant ratio	Ions match/exp	y/b ion score	Proline score	Total NSP	Engine score
LIGLQR		IPI_mouse 60A MitoMem IPI L...			0.01 (2+)	-	7/32	42.5/1.0	0.0	0	25.51
LLGLQR		IPI_mouse 60A MitoMem IPI L...			0.01 (2+)	-	7/32	42.5/1.0	0.0	0	25.51
ILGLKR		IPI_mouse 60A MitoMem IPI L...			0.04 (2+)	-	7/32	41.6/1.0	0.0	0	25.51
LLGLKR		IPI_mouse 60A MitoMem IPI L...			0.04 (2+)	-	7/32	41.6/1.0	0.0	0	25.51
LLLGKR		IPI_mouse 60A MitoMem IPI L...			0.04 (2+)	-	6/32	41.6/1.0	0.0	0	23.58
LLGKLR		IPI_mouse 60A MitoMem IPI L...			0.04 (2+)	-	6/34	41.6/1.0	0.0	0	22.42
*QNGAIGR		IPI_mouse 60A MitoMem IPI L...			0.10 (2+)	-	6/48	34.2/1.0	0.0	0	21.55
ILIANR		IPI_mouse 60A MitoMem IPI L...			0.01 (2+)	-	5/32	42.5/1.0	0.0	20	20.39

Figure 3. Peptide validation (see page 219).
After the automatic validation, the 'Peptide Validation' view can then be used for manually inspecting and rating peptide hits. For instance, the rating filter can be set to 'Potential' to evaluate only precursor ions with uncertain peptide assignments. The 'Peptide Validation' view shows a list of precursor ions together with one or more suggested peptide hits, with a maximum of one selected hit per precursor ion. In this example, three precursor ions with 2, 10, and 8 suggested peptide hits, respectively, are shown. In the rating column, the peptide rating (selected, rejected, or potential) can be changed manually. In this case, a peptide hit has been assigned manually to the top precursor ion for illustrative purposes. Modified peptides are indicated in red and with an asterisk; in this case, a pyroglutamic acid (Q) is seen for a suggested peptide for the lower precursor ion.

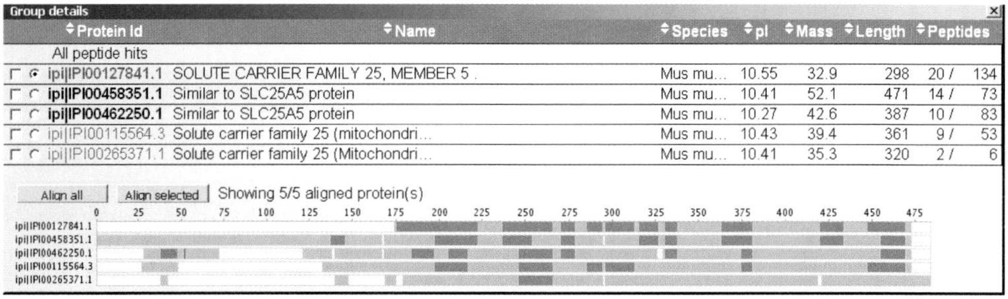

Figure 4. Detection of protein isoforms in the 'Group details' view (see page 221).
Five protein members of the solute carrier family 25 are found in this protein group, and EPICENTER has automatically selected the anchor protein based on the maximum number of peptide hits (the radio button next to the protein Id is selected). At the same time, there are two other protein entries (protein Ids highlighted in bold) with unique peptide hits that are not found in the other group members. This clearly indicates the presence of three distinct variants of the solute carrier family 25. At the bottom, a CLUSTALW alignment is shown for the five group members.

CHAPTER 1

Sample preparation and subcellular fractionation approaches: purification of membranes and their microdomains for mass spectrometry analysis

Yan Li, Phil Oh, and Jan E. Schnitzer

1. INTRODUCTION

Proteomics has the ability to view a large number of proteins in a given system and to identify specific proteins that are modified in a disease state or in response to drug treatment. However, it has been difficult to characterize the global proteome because of the enormous complexity and dynamic range of biological samples. Although there are only about 25 000 genes in the human genome, the number of distinct proteins in a tissue may reach 10^5 or 10^6 due, in part, to alternative splicing and many different types of post-translational modification (1, 2). This diversity occurs over a wide range of protein concentrations (exceeding sometimes ten orders of magnitude) (3, 4). It is estimated that the dynamic range of protein concentration within plasma is likely to be $>10^{10}$ and $\geq 10^6$ within the cells. The current mass spectrometry (MS)-based techniques are thought to cover a range of four to six orders of magnitude under the best experimental conditions (2, 5). Bearing in mind that biologically relevant proteins or biomarkers may be present at lower abundance, the biggest challenge in the proteomics field is finding a way of isolating and identifying proteins of interest, especially low abundance, highly modulated proteins. To overcome these difficulties, various sample separation strategies, such as protein-enrichment by affinity purification, depletion of specific highly abundant proteins from desired samples, subcellular fractionation, and incorporation of proteins into functional, biochemical, and biological groups, have been developed in recent years to obtain protein profiling data with high resolution and high detection sensitivity.

Our research has focused on identifying and characterizing proteins expressed on the luminal cell surface of the vascular endothelium. The vascular endothelium,

Proteomics: *Methods Express* (C.D. O'Connor and B.D. Hames, eds)
© Scion Publishing, 2008

directly exposed to the circulation, forms the critical barrier of the vascular system in living organs. In addition to regulating the exchange of small solutes to macromolecules between the intravascular compartment and extravascular space, its involvement in the development of many diseases has been well documented. It is likely that tissue- and disease-specific proteins on the surface of endothelial cell membranes play an essential role in transporting selected molecules from the circulating blood to the underlying tissue cells (6–10). Moreover, a distinctive feature of the surface of many vascular endothelia is its abundance of noncoated plasmalemmal vesicles (11, 12) known as caveolae. Caveolae provide a primary pathway for the transport of blood molecules across endothelium through the shuttling of these molecules from blood into and across the endothelium (11, 13–16). Specific proteins have been identified on the surface of endothelial cells that mediate selective transport of important blood proteins via caveolae found in certain endothelia (9, 11, 13, 16–18). However, little is known about which proteins are uniquely expressed on the surface of the endothelium of different tissues or diseases. Identifying such proteins in various tissues or diseases may provide better, more effective tissue- or disease-specific imaging and therapies (19–21).

In view of the vast diversity of biological material that can be analyzed by proteomics, it is not possible to provide sample preparation protocols that are universally applicable. Instead, we describe a specific method for the purification of plasma membranes and their microdomains from one particular source – luminal endothelial cells – in the hope that the approach can be adapted for use with other cellular systems. The method provides membrane fractions useful not only for identifying specifically expressed proteins but also for purifying any identified protein of interest (22). It enables us to reveal proteins that were not detected previously by reducing the protein diversity of the tissue sample by three to four orders of magnitude.

2. METHODS AND APPROACHES

2.1 Isolation of plasma membrane and its microdomains

The vascular endothelium is a very small constituent of tissue, even in highly vascularized organs. Thus, more specialized techniques beyond classic plasma membrane isolation are required to detect the tissue/disease-specific proteins on the surface of the endothelia *in vivo*. Isolation of endothelial cells from tissues removes the endothelial cells from their environment, resulting in significant changes in morphology and protein expression (3). In the 1990s, a new nanotechnology was developed to subfractionate tissues for the selective isolation of luminal endothelial cell plasma membranes and their specific microdomains, such as caveolae (23, 24). Briefly, after flushing out the blood, the vasculature is perfused with a solution of positively charged colloidal silica beads that adhere to the negatively charged proteins found on the luminal endothelial cell surface, thus altering the density of the luminal endothelial membranes for selective membrane isolation. More than 90% of the microvasculature *in situ* was coated with silica and approximately 80% of the silica-coated membrane was isolated from the starting material. Also, the isolated luminal endothelial plasma membranes derived from the microvasculature of rat lung were highly enriched in endothelial cells and plasma membrane markers with little or no detectable markers for other cell and intracellular organelles (24–26). Caveolae separated from the purified silica-coated plasma membranes can be isolated away from the other microdomains to ≥95% purity (24, 27). Electron microscopy of the purified microdomains shows a relatively homogenous population of vesicles of caveolae. They are enriched in seven caveolar markers: caveolin 1 (Cav-1), Ca^{2+}-ATPase, vesicle-associated membrane protein (VAMP), endothelial nitric oxide synthase (eNOS), dynamin, inositol 1,4,5-trisphosphate receptor, and cholera toxin-binding ganglioside G_{M1} (24–26), whereas angiotensin-converting enzyme (ACE), band 4.1, and β-actin are markedly depleted from the caveolae compared with the silica-coated endothelial plasma membranes.

2.2 Principles of the technology

The technology is based on the principle of ionic interaction of the naturally polyanionic cell surface with cationic colloidal silica particles to allow them to coat the endothelial cell surface upon intravascular perfusion *in situ*. Negatively charged cell membranes interact with positively charged colloidal silica particles to form a stable and positively charged cell surface. To neutralize the residual cationic sites of the silica particles, the positively charged cell surface is then reacted with negatively charged polymers. The cross-linking of the silica particles with the polymer creates a strong adherent silica membrane pellicle. The density of cell plasma membrane is increased by the silica coating, thereby enabling the silica-coated luminal endothelial surface membrane sheets produced by tissue or cell homogenization to be separated easily from all other cell and tissue components by ultracentrifugation through a light-density-media gradient. The silica coating forms a firm adherent layer over the noninvaginated plasma membrane and the silica nanoparticles are large enough to avoid entering into the caveolae whilst uniformly coating all other parts of the plasma membrane. Caveolae are physically separated from the membranes by homogenization. Sucrose density centrifugation is then used to isolate the caveolae in a low-buoyancy fraction away from the re-pelleted silica-coated membranes stripped of caveolae. These procedures produce purified membranes enriched 20-fold in the proper markers, whilst being 20-fold depleted in markers of other organelle compartments (all relative to the proceeding fraction). *Fig. 1* shows a flow diagram of the purification process.

2.3 Methodology

The silica-coating technology can be applied to purify plasma membranes directly from tissues (see *Protocols 1* and *2*) and from cell suspensions and monolayers of cultured cells (see *Protocol 4*). The resulting membranes can be used as a source of microdomains such as caveolae (see *Protocol 3*) and membrane proteins can be recovered from the silica beads (see *Protocol 5*) for proteomic analysis by MS. Where tissues are used, blood must be flushed away from the entire vasculature of the desired tissues by *in situ* intravascular perfusion prior to coating with silica.

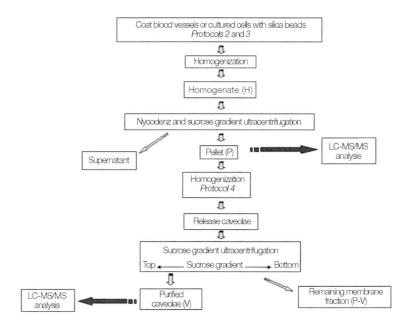

Figure 1. Procedure for the isolation and purification of endothelial membrane and its caveolae.
The silica-coated membrane purification strategy can be performed on both tissue vasculatures and cultured cells. Clearance of blood from the tissue vasculatures by buffer perfusion (*Protocol 1*) is required prior to silica coating. Isolated membrane pellet (P) and purified caveolae (V) are used in downstream applications.

Protocol 1 describes the perfusion of rat lung ready for the isolation of endothelial cell luminal plasma membranes and caveolae.

Protocol 1

Perfusion of rat lung vasculature for the purification of silica-coated endothelial cell plasma membranes and caveolae

Equipment and Reagents
- 3-0 Silk suture (Ethicon Inc.)
- Perfusion tubing (Tygon tubing, 0.0625 inch internal diameter; Fisher Scientific)
- Ketamine (Wyeth)
- Xylazine (Vedco Inc.)
- Heparin
- Mammalian Ringer's solution (114 mM sodium chloride, 4.5 mM potassium chloride, 1 mM magnesium sulfate, 11 mM glucose, 1 mM sodium phosphate (dibasic), 25 mM sodium bicarbonate; adjusted pH to 7.4 with hydrochloric acid)[a]
- MES-buffered saline (MBS) (20 mM 2-(*N*-morpholine) ethanesulfonic acid (MES), 135 mM sodium chloride; adjusted pH to 6.0 with sodium hydroxide)[b]

- Colloidal silica solution (prepare by diluting a 30% positively charged colloidal silica stock solution to 1% with MBS, pH 6.0)
- 0.1% Polyacrylic acid (PAA) (Polysciences Inc.) (dilute a 25% stock solution to 0.1% with MBS, pH 6.0)
- Sucrose/HEPES (250 mM sucrose, 25 mM HEPES, 20 mM potassium chloride; adjusted pH to 7.4 with sodium hydroxide)
- 100× Protease inhibitors (EMD Biosciences Inc.)

Method

1. Anesthetize the animals with a mixture of ketamine at 60 mg/kg and xylazine at 1.6 mg/kg.
2. After tracheotomy and thoracotomy, inflate the lungs to approximately three-quarters of tidal volume using a small animal respirator.
3. Remove the pericardium and inject 200 units of heparin into the right ventricle as an anti-coagulant.
4. Make a small cut in the right ventricle, insert the perfusion tubing through this cut, and secure it into the pulmonary artery with a 3-0 silk suture.
5. Cut the left atrium to allow vascular fluid outflow.
6. Flush the vessels at a pressure of 6–8 mmHg (4–5 ml/min) with mammalian Ringer's solution for 5 min. Reduce the temperature to 10°C after 1.5 min by cooling the perfusion tube in an ice-cold water bath[c].
7. Bring the pH of the vascular environment to 6.0 by perfusion with MBS for 1.5 min[d].
8. Inject 3–5 ml of air into the lung with a syringe to keep it inflated.
9. Coat the luminal surface of the endothelium of the vasculature with a 1% colloidal silica solution for 1.5 min.
10. Flush the vessels with MBS for 1.5 min to remove any unbound silica particles.
11. Perfuse 0.1% PAA for 1.5 min to cross-link the silica particles on the luminal endothelial cell surface and quench any positive charges on the colloidal silica that are not interacted with endothelial cell-surface molecules.
12. Flush the vessels with 8 ml of sucrose/HEPES containing 1× protease inhibitors to remove PAA and to prevent protein degradation by proteases released during tissue disruption[e].

Notes

[a]Use ddH$_2$O for all reagents. All solutions must be filtered with at least a 0.45 µm filter and can be pre-prepared and stored at 4°C. Check the pH of solutions carefully, especially those associated with the cationic silica particles. Any alterations in pH may affect the binding of the silica to the luminal plasma membrane.

[b]Phosphate solutions are incompatible with solutions containing colloidal silica, as the phosphates will cause the silica to precipitate.

[c]The tissue must remain ice-cold throughout the procedure after the initial 1.5 min flush with mammalian Ringer's solution.

[d]Avoid getting air bubbles in the line of the perfusion apparatus, as they will interfere with the vessel lumen coating procedure and decrease the amount of isolated material. The concentration of the perfusion solution and the flow rate of the perfusion are two important factors to ensure uniform protein profiles between experiments.

[e]Coated silica beads are strongly attached to the plasma membranes and greatly increase their density. The silica-coated endothelial cell plasma membrane can be separated from all other components in the tissue homogenates by ultracentrifugation through high-density media.

CHAPTER 1: SAMPLE PREPARATION AND SUBCELLULAR FRACTIONATION APPROACHES

The next step after perfusion of the tissue is purification of the endothelial cell luminal plasma membranes and caveolae. This is described in *Protocol 2*.

Protocol 2

Purification of endothelial cell luminal plasma membranes (P fraction) and caveolae (V fraction) from rat lungs

Equipment and Reagents
- Perfused rat lung vasculature (from *Protocol 1*)
- Aluminum block
- Type C Teflon pestle/glass homogenizer with a high speed motor (Thomas Scientific)
- MBS (pH 6.0) (see *Protocol 1*)
- Sucrose/HEPES (see *Protocol 1*)
- 100× Protease inhibitors (see *Protocol 1*)
- 30 and 53 μm Nitex filters (Sefar America Inc.)
- Ultracentrifuge (Beckman L8-80, Sorvall Pro 80, or Optima MaxE Ultracentrifuge) and Beckman SW55 and SW28 rotor tubes
- 100% (v/w) Nycodenz (Accurate Chemical & Scientific Corp.)
- 60% Sucrose (385.95 g of sucrose and 10 ml of 1 M potassium chloride, made up to 500 ml with ddH$_2$O)
- 70% Nycodenz/sucrose/HEPES (7 ml of 100% Nycodenz, 1 ml of 60% sucrose, 1 ml of 250 mM HEPES, pH 7.4, and 200 μl of 1 M potassium chloride, made up to 10 ml with ddH$_2$O)
- 65% Nycodenz/sucrose/HEPES (6.5 ml of 100% Nycodenz, 1 ml of 60% sucrose, 1 ml of 250 mM HEPES, pH 7.4, and 200 μl of 1 M potassium chloride, made up to 10 ml with ddH$_2$O)
- 60% Nycodenz/sucrose/HEPES (6 ml of 100% Nycodenz, 1 ml of 60% sucrose, 1 ml of 250 mM HEPES, pH 7.4, and 200 μl of 1 M potassium chloride, made up to 10 ml with ddH$_2$O)
- 55% Nycodenz/sucrose/HEPES (5.5 ml of 100% Nycodenz, 1 ml of 60% sucrose, 1 ml of 250 mM HEPES, pH 7.4, and 200 μl of 1 M potassium chloride, made up to 10 ml with ddH$_2$O)
- 80% Nycodenz (8 ml of 100% Nycodenz and 200 μl of 1 M potassium chloride, made up to 10 ml with ddH$_2$O)
- 75% Nycodenz (7.5 ml of 100% Nycodenz and 200 μl of 1 M potassium chloride, made up to 10 ml with ddH$_2$O)
- 70% Nycodenz (7 ml of 100% Nycodenz and 200 μl of 1 M potassium chloride, made up to 10 ml with ddH$_2$O)
- 65% Nycodenz (6.5 ml of 100% Nycodenz and 200 μl of 1 M potassium chloride, made up to 10 ml with ddH$_2$O)
- 60% Nycodenz (6 ml of 100% Nycodenz and 200 μl of 1 M potassium chloride, made up to 10 ml with ddH$_2$O)

Method
1. Excise the perfused rat lung vasculature from the animal (from *Protocol 1*) and mince the excised tissue using a new razor blade on an aluminum block pre-chilled by embedding it in ice[a].

2. Transfer the minced tissue into a type C homogenizer containing 20 ml of ice-cold sucrose/HEPES containing 1× protease inhibitors.
3. Homogenize the tissue at 4°C for 12 strokes at 1800 r.p.m.
4. Filter the tissue homogenate, first with a 53 mm Nitex filter, then with a 30 mm Nitex filter.
5. Save 200 µl of the filtered solution, label as 'H' (for homogenate), and store at −20°C.
6. Add an equal volume of 100% Nycodenz to the homogenate (H) and mix well to form a 50% Nycodenz/sample solution.
7. In a 30 ml SW28 centrifuge tube, gently place 3 ml of 70, 65, 60, and 55% Nycodenz/sucrose/HEPES in sequence to form a 55–70% Nycodenz/sucrose/HEPES gradient.
8. Add 20 ml of the 50% Nycodenz sample solution (from step 6) on the top of the gradient and carefully swirl the solution five to ten times by holding the tube at a 45° angle to make the gradient linear.
9. Separate the silica-coated membranes from other cellular components by spinning at 15 000 r.p.m. for 30 min at 4°C in an SW55 rotor.
10. Remove the supernatant and keep the pellet.
11. Wash the pellet twice by centrifugation with 1 ml of MBS (pH 6.0) containing 1× protease inhibitors.
12. Resuspend the pellet in 1 ml of MBS (pH 6.0) containing 1× protease inhibitors and mix with an equal volume of 100% Nycodenz.
13. In an SW55 centrifuge tube, gently place 350 µl of 80, 75, 70, 65, and 60% Nycodenz in sequence to form a 60–80% Nycodenz gradient.
14. Carefully add 2 ml of the 50% Nycodenz/sample solution (from step 12) on the top of the gradient and carefully swirl the solution five to ten times by holding the tube at a 45° angle to make the gradient linear.
15. Spin at 30 000 r.p.m. for 30 min at 4°C to remove the other remaining cell components.
16. Remove the supernatant, resuspend the pellet in 1 ml of MBS (pH 6.0) and label as 'P' for pellet[b]. Store at −20 or −80°C.

Notes

[a] All steps should be performed at 4°C.
[b] We have shown previously by electron microscopy that the sedimented pellet contains silica-coated luminal endothelial cell plasma membranes and its caveolae with little or no contamination from other tissue components (23). A good membrane preparation generally has a 20-fold enrichment of several endothelial cell-surface markers, for example caveolin, ACE, eNOS, and β-actin, compared with the starting material (H fraction). The marker proteins of the other cellular compartments should be largely depleted in the P fraction. These proteins include: GTP-binding nuclear protein Ran, protein disulfide isomerase A4 (ERP72), coatomer (β-COP and ε-COP), lysosomal membrane glycoprotein 1 (Lamp1), and early endosomal antigen 1 (EEA1), which are nuclear, endoplasmic reticulum, Golgi, lysosomal, and endosomal markers, respectively (see *Fig. 2*).

The silica-coated membranes (P fraction) prepared as described in *Protocol 2* have many associated caveolae that are attached on the cytoplasmic side of the plasma membrane opposite the silica coating. The attached caveolae can be separated by homogenization and purified by sucrose density gradient flotation (see *Protocol 3*). Homogenization is effective not only in removing the caveolae from the

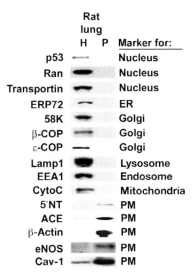

Figure 2. Immunoblot analysis of proteins from rat normal lung.
Five micrograms of total tissue homogenate (H) and endothelial cell plasma membranes (P) were loaded onto an SDS-polyacrylaimde gel (8–16%T, 10 × 10 cm; Cambrex Bio Science), transferred, and immunoblotted with antibodies to the indicated proteins. Ran, GTP-binding nuclear protein Ran; ERP72, protein disulfide isomerase A4; 58K, Golgi marker p58; β-COP, coatomer β-subunit; ε-COP, coatomer ε-subunit; Lamp1, lysosomal membrane glycoprotein-1; EEA1, early endosomal antigen 1; CytoC, cytochrome c; 5′NT, 5′ nucleotidase; ACE, angiotensin-converting enzyme; eNOS, endothelial nitric oxide synthase; Cav-1, caveolin 1.

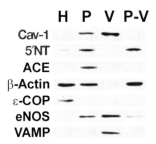

Figure 3. Immunoblot analysis of proteins from various membrane subfractions of rat lungs.
Rat lung endothelial cell luminal plasma membranes were isolated using silica-coated beads prior to purification of caveolae. Proteins (2 μg) from each fraction were loaded onto an SDS-polyacrylaimde gel (8–16%T, 10 × 10 cm; Cambrex Bio Science), transferred, and immunoblotted with antibodies to the indicated proteins. H, total tissue homogenate; P, silica-coated membrane pellet; V, purified caveolae; P-V, silica-coated membrane pellet after removal of V. Cav-1, caveolin-1; 5′NT, 5′ nucleotidase; ACE, angiotensin-converting enzyme; ε-COP, coatomer ε-subunit; eNOS, endothelial nitric oxide synthase; VAMP, vesicle-associated membrane protein.

P fraction (>80% loss in caveolin signal in the re-sedimented silica-coated membranes) but also in yielding a caveolin-enriched, low-density vesicular fraction after continuous sucrose density centrifugation. *Fig. 3* shows an immunoblot analysis under equal protein loads. Relative to the P fraction, the low-density fraction released by shearing (V fraction) is enriched more than tenfold in caveolin, whereas other proteins, such as the cytoskeletal protein β-actin, ACE, and the GPI-anchored protein 5′ nucleotidase (5′NT) are enriched in the P fraction with little or no signal in the caveolae (V) fraction. Markers with little or no signal in the P fraction, such as ε-COP, were not enriched in caveolae. Thus, the V fraction has a molecular composition that is quite distinct from the P fraction.

Protocol 3

Purification of caveolae from plasma membranes

Equipment and Reagents
- Endothelial cell luminal plasma membranes (P fraction from *Protocol 2*)
- 10% Triton X-100 in PBS (pH 7.4)
- Type AA Teflon pestle/glass homogenizer with a high-speed motor (Thomas Scientific)
- Ultracentrifuge (Beckman L8-80, Sorvall Pro 80, or Optima MaxE Ultracentrifuge), MLS-50 rotor or equivalent, TLA-55 rotor or equivalent, and 5 ml SW55 tubes
- 60% Sucrose/20 mM KCl (385.95 g of sucrose and 10 ml of 1 M potassium chloride, made up to 500 ml with ddH$_2$O)
- 35% Sucrose/20 mM KCl (5.2 ml of 60% sucrose and 200 μl of 1 M potassium chloride, made up to 10 ml with ddH$_2$O)
- 30% Sucrose/20 mM KCl (4.4 ml of 60% sucrose and 200 μl of 1 M potassium chloride, made up to 10 ml with ddH$_2$O)
- 25% Sucrose/20 mM KCl (3.8 ml of 60% sucrose and 200 μl of 1 M potassium chloride, made up to 10 ml with ddH$_2$O)
- 20% Sucrose/20 mM KCl (2.8 ml of 60% sucrose and 200 μl of 1 M potassium chloride, made up to 10 ml with ddH$_2$O)
- 15% Sucrose/20 mM KCl (2.1 ml of 60% sucrose and 200 μl of 1 M potassium chloride, made up to 10 ml with ddH$_2$O)
- 10% Sucrose/20 mM KCl (1.3 ml of 60% sucrose and 200 μl of 1 M potassium chloride, made up to 10 ml with ddH$_2$O)
- 5% Sucrose/20 mM KCl (0.7 ml of 60% sucrose and 200 μl of 1 M potassium chloride, made up to 10 ml with ddH$_2$O)
- 20 mM KCl
- MBS (pH 6.0) (see *Protocol 1*)

Method
1. Add 100 μl of 10% Triton X-100 to 900 μl of endothelial luminal plasma membranes (P fraction from *Protocol 2*) and mix well[a].

2. Mix the Triton X-containing P solution on a nutator for 10 min at 4°C.

3. Transfer the solution into a type AA homogenizer and homogenize for 20 strokes at 1800 r.p.m.

4. Transfer the homogenate to a 5 ml SW55 rotor centrifuge tube and add 1.5 ml of 60% sucrose/20 mM KCl to make a final 40% sucrose solution.

5. Make a discontinuous 35–0% sucrose/20 mM KCl gradient in the sample tube by carefully adding in sequence 350 µl each of 35, 30, 25, 20, 15, 10, and 5% sucrose/20 mM KCl and 20 mM KCl.
6. Separate the caveolae from the other membrane components by centrifugation at 50 000 r.p.m. for 4 h at 4°C in an MLS-50 rotor.
7. Collect the band between the 10 and 15% sucrose gradient and add MBS (pH 6.0) (two to three times the volume of the collected solution)[b]. Also keep the pellet (for use in step 10 below).
8. Remove the sucrose by centrifugation at 55 000 r.p.m. for 20 min at 4°C in a TLA-55 rotor.
9. Resuspend the pellet in 100 ml of MBS (pH 6.0), label as the caveolae ('V') fraction and store at −20°C.
10. Resuspend the pellet at the bottom of the centrifuge tube from step 7 in 1 ml of MBS (pH 6.0). Store this 'P-V' fraction at −20°C.

Notes

[a] Keep all solutions at 4°C, especially when working with Triton X-100. Triton X-100 facilitates the shearing of caveolae away from the plasma membranes. However, any increase in temperature may solubilize a wide range of proteins, which may alter the protein profile of the sample.

[b] A membrane band (V fraction) is easily detected at a density of 15–20% sucrose and contains a homogeneous population of caveolae as shown by electron microscopy (see *Fig. 4*).

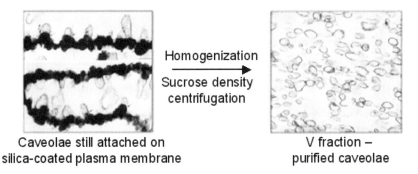

Figure 4. Electron micrograph of caveolae purified from silica-coated luminal endothelial membranes of rat lung vasculatures.
Electron microscopy shows a homogeneous population of small vesicles in the purified V fraction from the silica-coated membranes. These vesicles retain the characteristic flask shape of caveolae with diameters of <90 nm.

Protocol 4 describes the preparation of endothelial cell membranes (P fraction) from cultured cells using the silica-coating approach.

Protocol 4

Purification of endothelial cell plasma membranes (P fraction) from cultured cells

Equipment and Reagents
- Cultured cells
- Type AA Teflon pestle/glass homogenizer with high-speed motor (Thomas Scientific)
- T75 flasks (BD Falcon cell culture flasks; Becton Dickinson)
- Mammalian Ringer's solution (see *Protocol 1*)
- Colloidal silica solution (self-generated; see *Protocol 1*)
- MBS (pH 6.0) (see *Protocol 1*)
- 0.1% PAA (see *Protocol 1*)
- Sucrose/HEPES (see *Protocol 1*)
- 100× Protease inhibitors (see *Protocol 1*)
- 100% Nycodenz (see *Protocol 2*)
- 70, 65, 60, and 55% Nycodenz/sucrose/HEPES (see *Protocol 2*)
- Ultracentrifuge (Beckman L8-80, Sorvall Pro 80, or Optima MaxE Ultracentrifuge), plus SW55 rotor or equivalent, and SW55 tubes

Method
1. Rinse the cell monolayer with mammalian Ringer's solution and then wash three times with ice-cold MBS (pH 6.0). This is enough to bring the pH down to 6.0[a].
2. Add 1% ice-cold colloidal silica solution and incubate for 10 min to bind the silica nanoparticles to the surface of the cell monolayer.
3. Wash the cell monolayer three times with ice-cold MBS (pH 6.0) to remove unbound silica particles.
4. Add ice-cold 0.1% PAA and incubate for 10 min in order to cross-link the silica beads to the cell surface and to quench any noninteracting cationic sites on the silica beads.
5. Wash the cell monolayer three times with ice-cold MBS (pH 6.0) to remove the unbound PAA.
6. Add 5 ml of ice-cold sucrose/HEPES containing 1× protease inhibitors before scraping the cells from the culture dish[b].
7. Transfer the solution with released cells to a 15 ml centrifuge tube. Spin at 1000 *g* for 5 min at 4°C.
8. Remove the supernatant and resuspend the pellet in 1 ml of ice-cold sucrose/HEPES containing 1× protease inhibitors.
9. Transfer the resuspension into a type AA homogenizer. Homogenize for 20 strokes at 1800 r.p.m.
10. Keep 100 μl of the homogenate in a fresh microcentrifuge tube labeled as the 'H' fraction and store at −20°C.
11. Add an equal volume of 100% Nycodenz to the remaining homogenate and mix well.
12. In an SW55 centrifuge tube, gently place 350 μl each of 70, 65, 60, and 55% Nycodenz/sucrose/HEPES in sequence to form a 55–70% continuous Nycodenz/sucrose/HEPES gradient.
13. Add the 50% Nycodenz/sample solution (from step 11) on the top of the gradient gently and carefully swirl the solution five to ten times by holding the tube at a 45° angle to make the gradient linear.

14. Place the tube in an SW55 rotor. Balance the tube with a tube containing sucrose/HEPES containing 1× protease inhibitors.
15. Centrifuge at 30 000 r.p.m. for 30 min at 4°C.
16. Remove the supernatant. Resuspend the pellet in 1 ml of MBS (pH 6.0) and label as the plasma membrane ('P') fraction. Store at -20°C.

> **Notes**
> [a] All procedures should be performed at 4°C.
> [b] The protocol can be followed for one to six T75 flasks if the volumes are adjusted proportionately for larger or smaller quantities.

Membrane proteins attached to the silica beads can readily be recovered by incubating the beads with a strong detergent such as sodium dodecyl sulfate (SDS) (see Protocol 5). The membrane proteins can be separated by SDS-polyacrylamide gel electrophoresis with 3% SDS in the sample loading buffer. The separated protein bands can then be excised, digested in situ with trypsin, and the resulting peptides analyzed by MS. On the other hand, if the samples are prepared for two-dimensional gel electrophoresis or multiple-dimensional protein identification technology analysis, it will be necessary to remove the coated beads from the purified membranes, as well as the SDS used for detaching membrane proteins from the beads. A protein precipitation method can be used for this final protein clean-up and concentration step (28).

Protocol 5

Recovery of membrane proteins from the silica beads

Equipment and Reagents
- Purified membrane P fraction
- Microcentrifuge
- Methanol
- Chloroform
- MBS (pH 6.0) (see Protocol 1)
- 3× Lysis buffer (0.5 M Tris base, pH 6.8, 9 mM EDTA, 3.6% β-mercaptoethanol, 9% SDS)

Method
1. Add 30 µl of 3× lysis buffer to a 1.5 ml microcentrifuge tube containing 60 µl of the purified membrane P fraction in MBS to make a final concentration of 1× lysis buffer[a].
2. Mix well and incubate the sample mixture at 95°C for 5 min to remove the proteins from the silica beads.
3. Centrifuge the sample mixture at 15 000 r.p.m. for 5 min to pellet the unbound silica beads.
4. Place the supernatant in a fresh tube and discard the pellet.
5. Add 270 µl of methanol, 180 µl of chloroform, and 180 µl of water to the sample tube. (The ratio between the volume of the sample and the precipitation solution is 1 : 3 : 2 : 2, sample : methanol : chloroform : water)[b].

6. Mix the tube contents completely and centrifuge at 8000 r.p.m. for 5 min.
7. Isolated proteins will be concentrated at the interface between the water and the solvents and will form a visible protein band. Remove the solution above the protein band carefully. This solution contains the majority of the detergents used for the extraction of the endothelial proteins from the silica beads.
8. Gently add 500 µl of methanol down the side of the tube and mix well by inverting the tube a few times.
9. Spin at 8000 r.p.m. for 5 min to pellet the protein.
10. Discard as much of the supernatant as possible and allow the residual solvent to evaporate from the pellet by leaving the tube on the bench with its lid on for a few minutes[c].
11. Dissolve the protein pellet in the buffer of choice for downstream analysis.

Notes

[a]Treated membrane proteins are not suited to long-term storage. Therefore, it is recommended that the precipitation step is carried out immediately prior to downstream analysis.
[b]Methanol/chloroform precipitation seems to be the method of choice in our hands. We achieve much lower protein recovery with trichloroacetic acid precipitation.
[c]Do not over-dry the precipitated proteins – this makes it difficult to resolubilize them.

3. TROUBLESHOOTING

A good plasma membrane preparation should be enriched 20-fold in cell-surface marker proteins such as caveolin and eNOS. It should also be depleted by 20-fold in protein markers expressed in intracellular organelles such as nuclei (Ran, transportin), Golgi (p58, β-COP, ε-COP), lysosomes (Lamp1), and mitochondria (cytochrome c). Any factors that potentially decrease the electrostatic interactions between the membrane and the silica beads, such as the salt concentration and a higher pH, will detach the plasma membrane sufficiently from the silica pellicle in the P fraction. This effect will result in the co-isolation of caveolae with the other membrane microdomains, as well as a reduction in the amount of purified P fraction. We have isolated caveolae from other plasmalemmal microdomains, such as lipid rafts, based on this effect (24).

Processing time appears to be very important as some signaling molecules can dissociate from caveolae, e.g. eNOS and G-proteins. This dissociation may occur throughout the fractionation process, for instance during homogenization and also centrifugation to isolate the plasma membrane and caveolae. The dissociation may be anticipated because many proteins, especially signaling proteins, are not integral to the membrane and translocate back and forth between the cytosol and the membrane. Consequently, it is important to monitor the mass balance of all proteins. This can be accomplished by immunoblot analysis of the proteins present in different fractions of the membrane preparations. In comparison with the proteins left behind in the P-V fraction, one can verify whether the signals of the P and V fractions are accounted for by the fractionation.

Detergent extraction studies performed on the P fraction revealed that caveolae were not solubilized effectively in Triton X-100 at 4°C or using 3-((3-

cholamidopropyl)dimethylammonio)-1-propanesulfonate, which enables us to purify caveolae from the other membrane components with little contamination. In contrast, caveolae proteins were significantly solubilized by SDS, NP-40, β-D-glucoside, and deoxycholate, along with other membrane microdomains. We have also developed a detergent-free method to isolate caveolae from these membranes as well as from other types of membrane (27).

Endothelial and most cell types we have studied effectively lose most of their caveolae when recovered from tissues and grown in cell culture. This reduction (up to 100-fold) makes it difficult to purify caveolae with sufficient yields and little contamination for analysis. The plasma membrane is about 1% of the total membrane in the cells. Caveolae can comprise 80% of the endothelial cell-surface membrane *in vivo*, whereas *in vitro* the density of caveolae is less than 2%. Despite this, there are reports of the isolation of caveolae in fractions containing 1% of the total cell proteins. However, subsequent detailed analysis has shown that such preparations are contaminated by components from several different subcellular compartments.

4. IDENTIFICATION OF PROTEINS FROM PURIFIED MEMBRANES BY LIQUID CHROMATOGRAPHY TANDEM MS (LC-MS/MS)

4.1 Mapping of the proteome of the rat lung microvascular endothelial cell surface by multi-dimensional protein identification technology

We have analysed the proteins expressed in the endothelial membrane isolated from both rat lung and cultured rat lung microvascular endothelial cells by separating peptides using multi-dimensional high-performance liquid chromatography (HPLC) and nanospray tandem MS (28). In this way, 450 unique proteins were identified in the rat lung P fraction. This study showed that 81% of the identified proteins were associated with the plasma membrane, including 31% of integral or lipid-anchored membrane proteins, 35% of inner peripheral membrane proteins, 8.6% of externally bound proteins, and 25% of junctional and/or cytoskeletal proteins. In addition to 73 known vascular endothelial cell-surface marker proteins being identified, we also detected and validated 27 new proteins that are expressed on the surface of rat lung endothelial cells (28). These data demonstrate the enrichment of plasma membrane and endothelial marker proteins that can be achieved and are consistent with our previous observation that the silica-coating technology selectively binds to the endothelial cell surface.

4.2 Target proteins identified by two-dimensional gel electrophoresis combined with LC-MS/MS

Proteins from endothelial plasma membranes isolated from major rat organs as well as tumor-bearing lungs have been analyzed using two-dimensional gel electrophoresis. Comparisons of two-dimensional maps of the P and H fractions from different organs showed that the complexity of the purified membranes was greatly reduced and that distinctive protein profiles were obtained (29). The two-dimensional maps obtained with highly purified preparations of vascular endothelial plasma membrane have allowed us to isolate a number of proteins that are expressed uniquely in certain organs and in tumor-bearing lungs. Combined with MS analysis, proteins associated with normal tissues, as well as multiple tumor-induced proteins, have been identified. One-dimensional gel electrophoresis coupled with LC-MS/MS analysis has yielded many new protein identifications. In vivo γ-scintigraphy imaging of intravenously injected ^{125}I-labelled monoclonal antibodies to specific proteins that have been identified has confirmed their tissue specificity and helped to validate the overall strategy of finding targets for directed delivery *in vivo* (29, 30). Further studies have shown that the injection of the ^{125}I-labeled AnnA1 antibody destroyed several different types of solid tumor as assessed by histopathology or intravital microscopy. It also improved survival with approximately 80% of test animals apparently in complete remission, whilst all control animals died within 1 week of treatment (30). To provide more comprehensive analysis, we are pursuing further separation of proteins by gel electrophoresis as a prelude to trypsin digestion and one- and two-dimensional HPLC-based MS analysis.

5. REFERENCES

1. Maguire B, Moran N, Cagney G & Fitzgerald DJ (2004) *Trends Cardiovasc. Med.* **14**, 207–220.
2. Jacobs J, Adkins J, Qian W-J, *et al.* (2005) *J. Proteome Res.* **4**, 1073–1085.
3. Phizicky E, Bastiaens P, Zhu H, *et al.* (2003) *Nature*, **422**, 208–215.
4. Hanash S (2003) *Nature*, **422**, 226–232.
5. Schrattenholz A (2004) *Drug Discov. Today: Technologies*, **1**, 1–8.
6. Schnitzer JE, Carley WW & Palade GE (1988) *Am. J. Physiol.* **254**, H425–H437.
7. Schnitzer JE (1992) *Am. J. Physiol.* **31**, H246–H254.
8. Schnitzer JE & Oh P (1992) *Am. J. Physiol.* **263**, H1872–H1879.
9. Schnitzer JE & Oh P (1994) *J. Biol. Chem.* **269**, 6072–6082.
10. Jeffries WA, Brandon MR, Hunt SA, Williams AF, Gatter KC & Mason DY (1984) *Nature*, **312**, 162–163.
11. Ghitescu L, Fixman A, Simionescu M & Simionescu N (1986) *J. Cell Biol.* **102**, 1304–1311.
12. Villaschi S, Johns L, Cirigliano M & Pietra GG (1986) *Microvasc. Res.* **32**, 190–199.
13. Schnitzer JE (1993) *Trends Cardiovasc. Med.* **3**, 124–130.
14. Palade GE (1960) *Anat. Rec.* **136**, 254.
15. Ghitescu L & Bendayan M (1992) *J. Cell Biol.* **117**, 745–755.
16. Milici AJ, Watrous NE, Stukenbrok H & Palade GE (1987) *J. Cell Biol.* **105**, 2603–2612.
17. Schnitzer JE, Oh P, Pinney E & Allard A (1994) *J. Cell Biol.* **127**, 1217–1232.
18. Schnitzer JE & Bravo J (1993) *J. Biol. Chem.* **268**, 7562–7570.
19. Carver LA & Schnitzer JE (2003) *Nat. Rev. Cancer*, **3**, 571–581.
20. Schnitzer JE (1998) *N. Engl. J. Med.* **339**, 472–474.
21. Schnitzer JE (2001) *Adv. Drug Deliv. Rev.* **49**, 265–280.
★★ 22. Oh P & Schnitzer JE (1998) In *Cell Biology: a Laboratory Handbook*, vol. 2, pp. 34–36. Edited by J. Celis. Academic Press, Orlando. – *This chapter describes related protocols for the isolation of endothelial plasma membrane and caveolae using the silica-coating technique.*
23. Jacobson BS, Schnitzer JE & Palade GE (1992) *Eur. J. Cell Biol.* **58**, 296–306.

★★ 24. Schnitzer JE, McIntosh DP, Dvorak AM, Liu J & Oh P (1995) *Science*, **269**, 1435–1439. – *The initial publication describing purification of caveolae from lipid rafts.*
★ 25. Schnitzer JE, Liu J & Oh P (1995) *J. Biol. Chem.* **270**, 14399–14404. – *The initial publication describing the purification of caveolae from the luminal plasmalemma of rat lung endothelium.*
★ 26. Schnitzer JE, Oh P, Jacobson BS & Dvorak AM (1995c) *Proc. Natl. Acad. Sci. U.S.A.* **92**, 1759–1763. – *The initial publication describing the isolation of caveolae from the luminal plasma membranes of rat lung endothelium.*
27. Oh P & Schnitzer JE (1999) *J. Biol. Chem.* **274**, 23144–23154.
★★ 28. Durr E, Yu J, Krasinska K, *et al.* (2004) *Nat. Biotech.* **22**, 985–992. – *Proteome studies of membrane proteins of the rat lung microvascular endothelial cell surface* in vivo *and in cultured cells.*
★★ 29. Oh P, Li Y, Yu J, *et al.* (2004) *Nature* **429**, 629–635. – *Proteome studies of the membrane proteins of microvascular endothelial cell surface of the major organs of* rat in vivo.
30. Oh P, Borgström P, Witkiewicz H, *et al.* (2007) *Nat. Biotechnol.* **25**, 327–337.

CHAPTER 2

An isotope-coding strategy for quantitative proteomics

Xian Chen

1. INTRODUCTION

Quantitative measurements of protein expression changes using mass spectrometry (MS) are not fully reliable because the MS spectral intensity of peptide signals may not correlate linearly with the quantity of peptide/protein due to uneven ionization efficiency of peptides with different sequences. Two-dimensional electrophoresis (2DE) has commonly been used for high-resolution separation of protein mixtures with protein quantitation performed based on the visual intensity of resolved protein spots (1). However, the poor solubility of hydrophobic and membrane proteins may prevent them being fully resolved in 2DE gels, leading to them being excluded in quantitative analysis. Furthermore, 2DE is a tedious procedure and has low reproducibility. In addition, the lower level of MS detection compared with the high sensitivity of protein staining methods can indirectly reduce the sensitivity of integrated 2DE/MS quantitative proteomic approaches overall.

Isotope-coding schemes have been introduced for global-scale quantitative studies of differentially expressed proteins/modifications to improve the efficiency, accuracy, reproducibility, throughput, and proteome coverage of MS-based quantitative proteomics approaches. There are two major strategies to quantitate cellular proteins through stable isotope labeling (2):

- Chemically introduction of stable isotope label (SIL) tags *in vitro*
- *In vivo* metabolic cell culture (3–5)

Chemical SIL approaches usually target a particular residue of tryptic peptides, thus reducing the complexity of a sample. The most representative methods are isotope-coded affinity tags (ICAT) (6) and mass-coded abundance tagging (MCAT) (7). A common disadvantage is the relatively low efficiency of these chemical modification reactions and the limited abundance of target residues. In contrast, during cell growth, for example, Oda et al. (3) used ^{15}N uniformly labeled medium to label all nitrogen atoms in the whole proteome and applied this strategy to quantitate protein expression and to identify modifications.

Our laboratory has developed a different isotope-tagging strategy using SIL amino acids as precursors that can be incorporated into cellular proteins in a residue-specific manner. These amino acid-coded mass tags then label each individual proteins or modifications for subsequent quantitation and identification (5, 8–13). Amino acid-coded mass tagging (AACT) – also known as stable isotope labeling in cell culture (SILAC) – can be used in a wide range of experimental situations. Residue-specific incorporation of SIL amino acids (essential amino acids in particular) has been observed for a variety of cells including *Escherichia coli*, yeast, and human cell lines (5, 8, 9, 13). The SIL amino acids used to monitor post-translational modification would depend upon the event being studied, for example, Tyr-d_2 and Ser-d_3 in protein phosphorylation (11). Frequently used SILs for AACT are:

- Deuterium-enriched amino acids including [5,5,5-d_3]leucine (Leu-d_3), [4,4,5,5-d_4]lysine (Lys-d_4), [methyl-d_3]methionine (Met-d_3), [3,3-d_2]tyrosine (Tyr-d_2), [2,3,3-d_3]serine (Ser-d_3), [2,2-d_2]glycine (Gly-d_2), etc. (2)
- ^{15}N-enriched amino acids including $^{13}C_6^{15}N_2$-lysine

This chapter describes the use of a quantitative proteomics strategy based on AACT, emphasizing its unique strengths in identifying differentially expressed or post-translationally modified proteins on a genomic scale in a high-throughput manner (5, 10). Importantly, this 'gel-independent, in-spectra' quantitative approach has recently been extended to identify and characterize protein–protein interactions through a new AACT-based strategy of dual-tagging quantitative proteomics. Taken together, the information about the quantitative changes of multiple regulated proteins and their interactions is a powerful tool for identifying novel targets for diagnosis and therapeutic intervention.

2. METHODS AND APPROACHES

2.1 Amino acid-coded mass tagging

AACT is generally applicable to large-scale analyses of simultaneous expression changes of multiple proteins in different cell states. In an AACT quantitative experiment (see *Fig. 1*, also available in the color section), one cell sample grown in a regular or 'light' medium is compared with another cell sample grown in a 'heavy' medium (i.e. containing stable-isotope-labeled or 'heavy' amino acids). *In vivo* metabolic labeling with stable isotopes is described in *Protocol 1*.

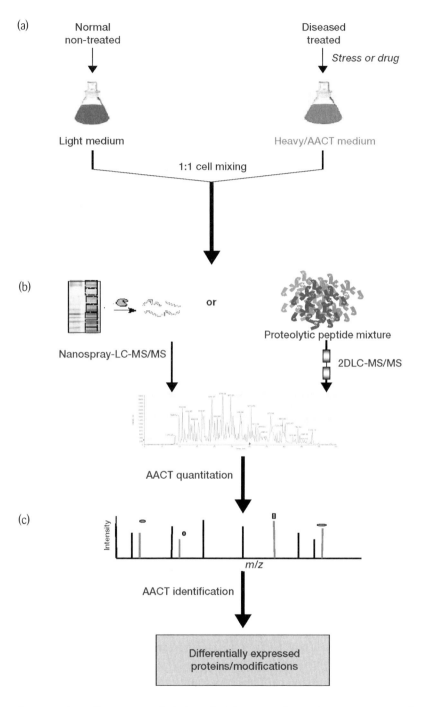

Figure 1. Overall experimental design of AACT-assisted quantitative proteomics (see page xix for color version).
(a) *In vivo/in vitro* cell culturing; (b) sample preparation; (c) MS measurements and data analyses. Symbols (●, ○, ■, ━) are defined in the *Fig. 3* legend.

Protocol 1

In vivo metabolic labeling

Equipment and Reagents
- Stable isotope-labeled/enriched (2) amino acid precursors (Cambridge Isotope Laboratories, Inc.)
- Unlabeled amino acids (Sigma)
- Lysis buffer (50 mM Tris/HCl, pH 8.0, 10 mM EDTA, 100 mM NaCl, 10 mM dithiothreitol (DTT) or 5% β-mercaptoethanol, 0.1% SDS, and protease inhibitor cocktail (1 tablet/50 ml; Sigma)
- Phosphate-buffered saline (PBS) (for analysis of mammalian cell lines)
- *For labeling of bacterial cells*: M9 minimal medium (Gibco-BRL) supplemented with 200 mg/l of the selected SIL amino acid (5), 20 mg/l of unlabeled cysteine, and 200 mg/l of each of the other unlabeled amino acids
- *For labeling of yeast cells*: a synthetic defined (SD) medium comprising 20 g of dextrose, 1.7 g of yeast nitrogen base without amino acids, 5.0 g of ammonium sulfate, and 100 mg each of adenine, histidine, tryptophan, and leucine in 1 l of H_2O. The SIL amino acids are added at a final concentration of 100 mg/l (13)
- *For labeling of mammalian cell lines*: α-MEM (Gibco-BRL) depleted of particular essential amino acids. This can be obtained from the UCSF Cell Culture Facility (University of California, San Francisco, CA 94143, USA). This AACT/α-MEM is supplemented with 10% dialyzed fetal bovine serum, 100 units/ml penicillin, 100 μg/ml streptomycin sulfate, and the selected isotope-enriched amino acid precursor such as Leu-d_3, Ser-d_3, or Tyr-d_2 completely substituting its unlabeled counterpart depleted in the α-MEM (9); for example, 88 mg/l of Lys-d_4 or Leu-d_3.

Method
Bacterial cells

1. Culture bacterial cells in M9 minimal medium supplemented with 200 mg/l of the selected SIL amino acid (5), 20 mg/l of unlabeled cysteine, and 200 mg/l of each of the other unlabeled amino acids.

2. Centrifuge the cells and wash the cell pellet twice with Milli-Q H_2O to remove excess medium.

3. Resuspend the cell pellet in lysis buffer and incubate for 10 min at 4°C.

Yeast cells

1. Inoculate the yeast cells into 10 ml of SD medium and grow overnight at 30°C.

2. Dilute the overnight yeast culture to an optical density (OD_{600}) of 0.02 in 100 ml of SD medium containing naturally occurring amino acids and likewise prepare 100 ml of cell culture with SIL amino acids[a].

3. Grow both the 'light' and the 'heavy' cultures under the same conditions, except that the heavy culture is exposed to the stimulus to be studied. For example, to study the response of exposure of a cellular proteome to a heavy metal (14), add cadmium ions in the form of $CdSO_4$ to a mid-exponentially growing *Saccharomyces pombe* culture ($OD_{600\ of}$ ~0.35) in the heavy medium at a defined concentration. Grow the control *S. pombe* cells in the light medium under the same conditions, but without Cd^{2+} treatment.

4. Measure the OD_{600} of the cell cultures in the light and heavy media at various time intervals.
5. At designated times after exposure of the heavy cell culture to the stimulus (e.g. Cd^{2+} addition), mix equivalent numbers of cells from each culture, based on the relative OD_{600}.
6. Centrifuge the cells and wash the cell pellet twice with Milli-Q H_2O to remove excess medium.
7. Resuspend the cell pellet in lysis buffer and prepare a cell lysate by vortexing with glass beads for 10 min at 4°C.
8. Centrifuge at 10 000 g for 10 min at 4°C and recover the supernatant (soluble proteins) for analysis.

Mammalian or human cell line

1. Grow human skin fibroblast cells[b] in α-MEM at 37°C with 5% CO_2 and 90% relative humidity (13, 15). Grow two cultures, one in heavy medium and one in light medium.
2. Subject the cells grown in heavy medium to the stimulus under study[c]. Grow the cells in the light medium without stress.
3. At about 80% confluency, mix equal numbers of cells from the light and heavy cultures and harvest these using trypsin/EDTA treatment.
4. Centrifuge the cells and wash the cell pellets with PBS.
5. Resuspend the cell pellet in lysis buffer for analysis.

Notes

[a] No growth difference is observed between cells in regular and labeled media, i.e. isotope-labeled amino acids do not affect cell growth.
[b] Human skin fibroblast cells are described here as an example; the exact conditions used will vary depending on the cell line under study.
[c] For example, to study the cellular response to heat shock or radiation, subject the heavy culture to incubation at 43°C for 8 h (9) or irradiate it with 500 Rads of γ-rays before recovering for 4 h under normal conditions (10).

Following cell mixing, protein extraction, and separation (see *Protocol 2*), quantitative MS (see *Protocol 3*) is used to recognize and measure the relative intensity ratios of different isotope forms of individual peptide sequences (16). Pairs of 'heavy' and 'light' isotope signals from the same peptides derived from an equal mixture of light and heavy cells, respectively, will provide 'in-spectra' markers in MS analysis for both quantitative measurements and concurrent protein identification (16). Because the isotopic ratios correlate with the relative peptide/protein abundance, particular differentially expressed/modified proteins can readily be distinguished from the huge number of unchanged proteins. Identification of the differentially expressed or modified proteins is carried out using peptide mass fingerprinting or tandem MS (MS/MS). Proteolytic peptide masses are typically in the range of 500 to 3000 atomic mass units (a.m.u.), where both electrospray ionization (ESI) and matrix-assisted laser desorption ionization time of flight (MALDI-TOF) mass spectrometers have sufficient resolution and sensitivity to distinguish AACT peptides by their characteristic mass-split patterns.

Protocol 2

Protein extraction, separation, and preparation for MS[a]

Equipment and Reagents
- Resuspension buffer (see *Protocol 1*)
- 10× SDS-PAGE loading buffer
- Equipment and reagents for one-dimensional (1D) SDS-PAGE or 2DE
- Rehydration buffer (7 M urea, 2 M thiourea, 4% CHAPS, 2% DTT, 2% IPG buffer, pH 4-7, 1 mM benzamidine, 1.5 mM EDTA/EGTA, 1 mM sodium vanadate, 1 µM microcytin-LR, 2 µg/ml pepstatin A, 10 µg/ml aprotinin, 20 µg/ml leupeptin)
- IPGphor II isoelectric focusing system (required for 2DE; Amersham Biosciences)
- Equilibration buffer (6 M urea, 2% SDS, 0.375 M Tris/HCl, pH 8.8, 20% glycerol)
- DTT
- Iodoacetamide
- Silver staining reagents or SYPRO Ruby protein gel stain (Invitrogen)
- Acetonitrile
- NH_4HCO_3
- Speed-vacuum centrifuge
- Trypsin (10 µg/ml)
- Acetic acid
- 0.1% Trifluoroacetic acid (TFA)[b]

Method
1. Centrifuge the cells and resuspend the cell pellet in resuspension buffer.
2. Lyse the cells using a method appropriate to the cell type, e.g. *E. coli* (5), yeast (10, 13), or human (9).
3. Either analyze the whole cell lysate or fractionate it into different subcellular compartments (17) before analysis.

1D-SDS-PAGE

1. Add 10× SDS-PAGE loading buffer to a final concentration of 1× and fully solubilize the cell fraction by boiling for 10 min.
2. Run 50-200 µg of total protein on a 4-20% gradient SDS-polyacrylamide gel (10 cm in length, 1 mm in thickness).
3. Visualize the protein bands by staining with Coomassie blue (G-250).
4. Excise the protein bands in 1-2 mm gel slices.
5. Chop each gel slice into 1 mm^3 cubes for destaining and digestion with trypsin (see below).

2DE

1. Add 50-200 µg of protein sample to rehydration buffer to give a final volume of 350 µl and rehydrate at a low voltage (20 V) for 12 h.
2. Carry out isoelectric focusing at an ambient temperature on an 18 cm Immobiline pH dry strip with a defined pH range. Hold the voltage at 500 V for 30 min and then at 1000 V for 2 h, and then apply a gradient from 1000 to 7000 V over 3 h. Finally, hold the voltage at 7000 V for another 4 h.
3. Incubate the strips for 10 min in equilibration buffer containing 130 mM DTT.

4. Incubate the strips in equilibration buffer containing 135 mM iodoacetamide.
5. Place each equilibrated strip onto a 20 cm × 20 cm 10% Duracryl gel and run the second dimension at 500 V for 5 h in a 2DE system.
6. Stain the gel using silver staining or SYPRO Ruby (18).
7. Excise the stained protein spots for destaining and trypsin digestion (see below).

Destaining and trypsin digestion

1. Destain the gel spots or bands with 50% (v/v) acetonitrile in 50 mM NH_4HCO_3.
2. Incubate with 100% acetonitrile for 20 min and dry the gel fragment using a speed-vacuum centrifuge for 10 min.
3. Add 15 µl of 10 µg/ml trypsin solution to the dried gel and then add 20 µl of 50 mM NH_4HCO_3 to cover the rehydrated gel. Incubate overnight at 37°C.
4. Extract the tryptic peptide twice from the gel by 45 min sonication each time, first in 200 ml of 5% acetic acid solution, and then in 200 ml 5% acetic acid/50% acetonitrile solution.
5. Combine the gel extracts, lyophilize, and resuspend in 0.1% TFA[b,c].

Notes

[a]Following cell lysis, a variety of methods such as 1D, SDS-PAGE-2DE or two-dimensional liquid chromatography (2DLC) can be used for protein separation prior to MS analyses.

[b]TFA-containing solutions should be made on the day of use. Extreme care should be taken when using TFA as it is an extremely hazardous chemical and the concentrated acid should only be used in a fume hood and dispensed wearing suitable protective clothing. Consult the safety datasheet for handling details and disposal.

[c]In the integrated 1D-SDS-µLC-MS/MS analytical mode, gel elutes of proteolytic digests will be directly subjected to further µLC separation before nanospray-MS/MS experiments. This process can also be conducted in a high-throughput mode by linking the autosampler to the LC apparatus that loads the digests of 1D bands automatically, e.g. the digest elute from 24 gel bands can be analyzed within 12 h. For MALDI-TOF MS analysis, additional desalting steps will be performed using C_{18} ZipTips (Millipore) with the final elution solution being 50% acetonitrile/0.1% TFA.

2.2 High-throughput MS for AACT-based quantitative proteomic analysis

MS analysis of proteolytic digests for AACT-based analyses is described in *Protocol 3* (see also 8, 12). As shown in *Fig. 1* (also available in the color section), the use of AACT provides 'in-spectra' markers for the quantitation of differentially expressed proteins or modifications. In MS spectra, a mono-isotopic distribution pattern of proteolytic peptides can be resolved for each individual isotope species as M^+, $(M+1)^+$, $(M+2)^+$, etc., where M refers to the mass of the most abundant isotopes. At natural abundance, biomolecules are composed of over 99% ^{12}C, ^{14}N, and ^1H, with the heavier isotopes of ^{13}C, ^{15}N, and ^2H (deuterium) being close to or less than 1%. Stable isotope enrichment or labeling is a process whereby the amount of the least abundant isotope(s) is enriched to a higher level. The incorporation of an amino acid enriched by heavy isotopes (such as Leu-d_3 or Lys-d_4) in any given peptide increases its mass. This mass shift leads to a pair of mono-isotopic peaks, the lower mass component corresponding to the naturally occurring isotope and the higher mass component containing the heavier isotope. The intensity ratios between the lower and upper mass components of these peak pairs will correlate to the relative abundance of the natural and heavy isotopes.

An AACT-based quantitative approach enables changes in a whole proteome to be identified and quantitated without the need for 2DE; *Protocol 3* describes the analysis of proteolytic digests from 1D-SDS-PAGE separations for this purpose.

Protocol 3

1D-SDS-µLC-MS/MS

Equipment and Reagents
- ABI MDS SCIEX QSTAR XL quadrupole TOF MS with both nanospray API (atmospheric pressure ionization) and orthogonal MALDI sources, coupled to an Ultimate micro-capillary LC system (LC Packings/Dionex)
- Mobile phase A: 0.1% formic acid, 5% acetonitrile
- Mobile phase B: 0.1% formic acid, 95% acetonitrile

Method
1. Analyze the in-gel digest of each 1D-SDS-PAGE gel slice in a high-throughput mode by µLC-nanospray-MS/MS using a QSTAR XL mass spectrometer coupled with an Ultimate micro-capillary LC system. Configure an autosampler using the partial loop injection mode with a 10 µl sampling loop; connect a pre-concentration C_{18} cartridge with the analytical column through a 10-port switch valve. The end of the analytical column is connected to a 10 µm ID PicoTip nanospray emitter (New Objective) by a stainless steel union mounted on the nanospray source. The spray voltage (usually set between 1800 and 2100 V) is applied to the emitter through the stainless steel union and tuned to get the best signal intensity using standard peptides. The two most intense ions with charge states between 2 and 4 in each survey scan are selected for the MS/MS experiment provided they pass the switching criteria for the MS/MS scan. The rolling collision energy feature is employed to fragment the peptide ions according to their charge states and m/z values.

2. Using the partial loop injection method, load a 3 µl sample into the 10 µl loop and pump this onto the pre-concentration column at a flow rate of 30 µl/min via a sample-loading pump.

3. Three minutes after the start of the sample loading, switch the 10-port valve to place the pre-concentration cartridge in line with the nano-flow solvent delivery system, thus enabling the trapped peptides to be eluted onto the analytical column. Keep the solvent at 5% mobile phase B for 5 min, then ramp a gradient linearly from 5 to 50% mobile phase B over 50 min, then jump the concentration to 75% mobile phase B and keep at this for 10 min. Finally, change the solvent back to 5% mobile phase B and equilibrate the column for 10 min. Throughout the elution, the flow rate should be 200 nl/min.

To further improve the sensitivity and throughput, 2DLC is another high-throughput technique and is not biased in favor of proteins with particular physical properties such as mass, pI, or hydrophobicity (19, 20). AACT-assisted gel-free 2DLC-MS/MS provides much more comprehensive protein profiles than is possible using gel-based approaches (19, 20). In each 2DLC run, a large number of AACT-containing peptides are identified and quantitated. Automated spectral and data analysis using PROICAT software (QSTAR XL) constrained by AACT can perform a data-dependent search to locate relevant isotope pairs and their derivative MS/MS spectra (see *Fig. 1*, also available in the color section). The 'light' peptide isotope peak in the MS spectrum, derived from the protein expressed in the normal ('light') medium, will form a pair with the peptide from the same protein expressed in the heavy medium. Isotope pairs of equal intensity show that the parental protein expression is the same in both cell pools, whereas isotope pairs of different intensities correspond to protein expression triggered by a particular biological process(s) or environmental stimuli. Proteolytic peptide masses are typically in the range of 500 to 4000 a.m.u., where both ESI and MALDI-TOF mass spectrometers have sufficient resolution and sensitivity to distinguish AACT peptides and their isotopic distribution patterns. As shown in *Fig. 2*, the amino acid-coded mass tag increases signal specificity in database searches. Both parameters, the *m/z* value of a fragment ion and its content of a particular amino acid(s), can be used to characterize the corresponding ions. Proteolytic peptides derived from the theoretical digestion of various proteins translated from the genomic sequence or expressed sequence tag databases can be filtered with the content of the labeled amino acid residue(s), resulting in an AACT-constrained proteolytic peptide library. The database search is then constrained by two parameters, allowing far more selective and confident protein sequence searches than those by peptide *m/z* values alone. The AACT approach can reveal the partial amino acid composition of particular AACT peptides (12).

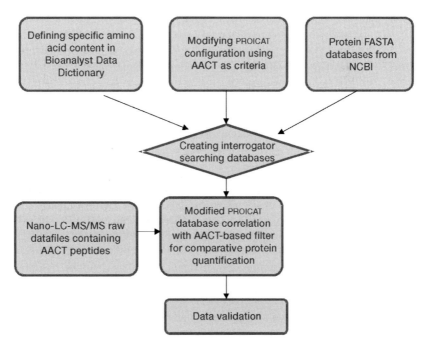

Figure 2. AACT-assisted modifications of the PROICAT program in QSTAR XL for simultaneous protein quantitation and identification.

All of the peptides with complete AACT patterns showing induced protein expressions on MS spectra, the observed m/z, and the peptide composition of the four labeled amino acids are submitted to the human National Center for Biotechnology Information (NCBI) non-redundant database to determine the protein identity with 500 p.p.m. mass tolerance using the MS-SEQ program (12). Both MASCOT and PROID loaded on the QSTAR instrument can be used to interpret the LC-MS/MS data by searching against the NCBI protein database. Peptides matches with a score greater than 39 in MASCOT or greater than 43 in PROID are considered to be a significant match or homologous ($P < 0.05$) (21).

Current proteomics is still mainly 'genome sequence-based', which can inevitably introduce ambiguities and false positives arising from mis-annotated genes in identifying human proteins. The quality of MS/MS spectra is critical, and rigid criteria for the fragment pattern matching are usually required to determine each peptide sequence unambiguously. In practice, the MS/MS signals originating from low-abundance proteins (e.g. signal or membrane proteins, or proteins modified post-translationally) tend to be weak (1, 22–26). Because of such 'poor-quality' MS/MS spectra, only subsets of proteins in a proteome (usually ~10–50% of annotated gene products) can be identified unambiguously using classical MS approaches. Without sufficient signal specificity, incomplete-fragment MS/MS spectra (e.g. weak or missing fragment signals in various ion series) may lead to ambiguous search results, and large numbers of meaningful spectra may be excluded from protein identification. In this regard, AACT can be used as a molecular signature to label the human proteome for high signal specificity in MS analysis. With greater accuracy, sensitivity, and throughput, this residue-specific labeling strategy provides information about the content of particular amino acids. AACT enhances signal specificity for large-scale analyses of post-translationally modified (11), low-abundance, and membrane proteins (12). The high-specificity tagging enables *de novo* peptide sequencing to distinguish protein mutants and the correction of mis-annotations resulting from genomic sequence errors or ambiguities (8, 9). This enhancement in signal specificity bypasses instrument limitations and allows the researcher to obtain more comprehensive protein profiles (27). AACT also resolves the ambiguities in 'poor-quality' MS spectra; it identifies proteins unambiguously through 'data-dependent' instead of 'genome sequence-based' approaches, thus avoiding false-positive protein identification from gene mis-annotations.

AACT has several advantages compared with other chemical or enzymatic tag methods that can be used to provide MS-recognizable quantitative markers (28). Unlike ICAT (6), AACT requires no further chemical modifications and the isotope-labeled amino acids are incorporated into cellular proteins during growth with no effect on protein function *in vivo*. More importantly, the presence of isotope-labeled amino acids in peptides has no effect on their MS/MS fragmentation and signal ionization, but the tags facilitate fragment assignment. Furthermore, due to the high sensitivity of MS, the cost of the isotope-labeled amino acids is minimal.

2.3 AACT/epitope dual-tagging strategy for pathway scale profiling of protein–protein interactions regulating gene expression (29–31)

A dual-tagging strategy (both epitope and isotope tags) has recently been developed for highly sensitive detection of signal transduction components as illustrated in *Fig. 3*. Compared with the commonly used tandem affinity purification tag strategy, which requires multi-step affinity purification of protein prior to MS

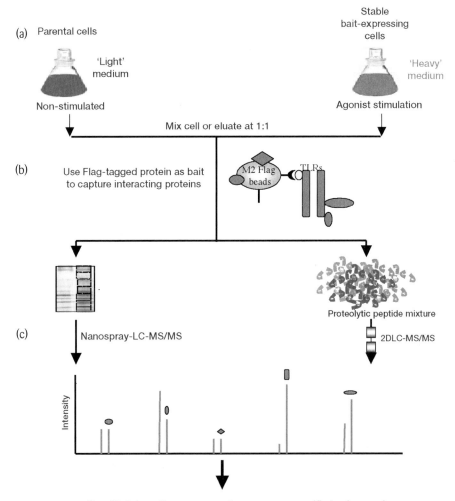

Figure 3. A dual-tagging proteomic platform for the characterization of protein complexes (see page xx for color version).
(*a*) Epitope-tagging, *in vivo* cell culturing and *in situ* stimulation; (*b*) affinity purification; and (*c*) MS-based quantitative identification of specific protein interacting partners. Background binding proteins are identified based on a 1 : 1 ratio between labeled and unlabeled states (e.g. peaks ● and ◆). Peaks that show a divergence from this ratio will represent interacting proteins (peaks ❶, ■, and ➖). TLRs, toll-like receptors.

identification, this dual-tagging proteomic method minimizes the need for multi-step purification. The AACT/epitope dual-tagging strategy combines a single epitope tag for isolation of the complexes and an AACT isotope tag for enhancing the MS signal specificity to distinguish genuine signal proteins in the background of nonspecific binding proteins. The epitope-tagged protein is used as a 'bait' to pull down proteins that interact with it. This approach (see *Protocol 4*) minimizes the need for multi-step purifications and preserves real-time interactomes at the naturally occurring level for systems analysis.

The sample consumption using the AACT approach, usually 100–300 µg for each run, is far less than that needed for the ICAT method, which usually requires milligram quantities. In addition, the AACT strategy can be coupled to 2DLC-based methods for higher throughput and less sample consumption.

In practice, because an overexpressed protein may affect phenotype (32, 33), the expression level of epitope-tagged bait protein (a protein to pull down its interacting partners) in a stable cell line first needs to be compared with that of its endogenous protein counterpart. A stable cell line expressing the bait protein at an endogenous level is then selected. Parental cells expressing untagged 'bait' protein and a derived stable cell line containing an epitope affinity-tagged bait protein are grown in 'light' and 'heavy' medium, respectively. Both light and heavy cell pools can be subjected to any stimulus for a paired comparison (stable cell line versus parental cells). Equal numbers of cells from each cell pool are then mixed for pull-down of the protein complex formed. The cells are lysed and subjected to affinity purification, and bound and eluted proteins are analyzed by MS-based quantitative proteomics to provide a complete assessment of each bait-containing complex with comprehensive coverage of possible signal proteins in all stable and transient interactions. The high sensitivity of this method (*Protocol 4*) depends on both the single-step enrichment of multi-protein signal complexes and the use of isotope tags for 'in-spectra' quantitative measurement. The analysis allows the identification of both stable and transient interactions with high sensitivity in cells in real time.

Protocol 4

Dual-tagging quantitative proteomic strategy for the analysis of protein–protein interaction complexes

Equipment and Reagents
- Stable cell lines expressing selected bait proteins
- Dulbecco's minimal essential medium containing Leu-d_3 instead of leucine
- AMJ-C8 cells
- Lipid A (500 ng/ml)
- Lysis buffer (50 mM Tris/HCl, pH 7, 150 mM NaCl, 1% NP-40) supplemented with protease inhibitor cocktail (1 tablet/50 ml; Sigma)
- M2 anti-Flag beads (Sigma)
- 3× Flag peptide (Sigma)

Method

1. Select a stable clone that expresses the tagged bait close to endogenous levels for the AACT-assisted signaling pathway study. Maintain the selected clone in Dulbecco's minimal essential medium with leucine replaced with Leu-d_3 until the Leu-d_3 has been 100% incorporated into each protein.

2. Mix 1×10^9 stable cells with 1×10^9 parental AMJ-C8 cells and treat with lipid A (500 ng/ml) for 5 min. Then lyse the cells in 10 ml of lysis buffer.

3. Incubate approximately 20 mg of protein from the mixed cell lysate with 100 µl of M2 anti-Flag beads at 4°C for 2–3 h.

4. Wash the beads twice in 5 ml of lysis buffer, followed by a wash in 5 ml of lysis buffer containing 300 mM NaCl.

5. Elute the bound proteins by adding 100 µg/ml of 3× Flag peptide.

6. Separate the eluted proteins by 1D-SDS-PAGE.

7. After staining for protein, cut out the visible bands for analysis by LC-MS/MS.

8. Carry out MS analysis as described previously (34). Briefly, the in-gel digest of each gel slice is analyzed by µLC-nanospray-MS/MS using a QSTAR XL mass spectrometer (Applied Biosystems) coupled with an Ultimate micro-capillary LC system (LC Packings/Dionex). Samples (5 µl) are loaded onto the column for separation. Mobile phase A is 0.1% formic acid and 5% acetonitrile, whereas mobile phase B is 0.1% formic acid and 95% acetonitrile. The gradient is kept at 5% mobile phase B for 5 min and ramped linearly from 5 to 50% mobile phase B in 50 min, and then jumped to 75% mobile phase B and maintained for 10 min. The gradient is then jumped back to the start point and the column equilibrated for 10 min. Next, the solvent is changed back to 5% mobile phase B and the column equilibrated for 10 min. The flow rate should be 200 nl/min. The spray voltage is tuned to get stable background signals with the best signal-to-noise intensity. The two most intense ions with charge states between 2 and 4 in each survey scan are selected for the MS/MS experiment provided they pass the switching criteria of the MS/MS scan. The rolling collision energy feature is employed to fragment the peptide ions according to their charge states and *m/z* values.

9. Use the PROID program loaded on the QSTAR instrument and an in-house licensed MASCOT server to interpret the LC-MS/MS data by searching against the relevant protein databases downloaded from the NCBI public ftp site (http://www.ncbi.nlm.nih.gov/). The Leu-d_3 modification is added in the configuration file as a static or variable modification in the database searching. The parameters for database searching are: (i) 100 p.p.m. mass error tolerance for both MS and MS/MS; (ii) fixed carboxamidomethylation of cysteine, variable modifications including phosphorylations on tyrosine/serine/threonine, oxidation of methionine and deuterated leucine (Leu-d_3); (iii) tryptic enzyme specificity and one mis-cleavage allowed. Protein with two or more peptides with a matching score over 98% confidence in PROID or a score of 40 in MASCOT is considered as positive identification.

2.4 Mapping 'real-time' phosphorylation sites of signal proteins involved in signal transduction

Reversible phosphorylation of signal proteins acts as the switch in both activating and repressing particular interactions in a time-dependent manner. To date, few reports have described the global phosphorylation status of critical proteins involved in signaling pathways.

The AACT/epitope dual-tagging technology is particularly useful for quantitatively monitoring real-time signal transduction-related phosphorylation. As shown in *Fig. 2*, AACT-assisted detection of post-translational modifications can be extended (11) to profile real-time serine/tyrosine/threonine phosphorylation occurring on specific complex components in a comprehensive and high-throughput manner. The enrichment approaches prior to MS analysis use reagents such as monoclonal anti-phosphoserine or anti-tyrosine agarose or immobilized metal affinity chromatography. During MS, because phosphoserine/phosphothreonine tend to lose H_3PO_4 or HPO_3 under collision-induced dissociation fragmentation, it is best to track the neutral loss of 98 or 80 Da for characterization of phosphoserine/phosphothreonine-containing peptides. Although we have shown the direct detection of serine phosphorylation, in some cases low-abundance modification sites will need to be enriched to eliminate the background of a larger population of unmodified proteins. Phosphotyrosine is more resistant to neutral loss. Therefore, precursor ion scanning, which monitors the immonium ion of phosphotyrosine (*m/z* 216.04 Da), must be carried out in parallel to follow phosphotyrosine changes.

3. CONCLUSIONS

The AACT strategy is superior to the popular ICAT-based approaches where cysteine-containing peptides are the sole reference signals for protein quantitation. AACT is amino acid-specific: most of the 20 common amino acid residues can be chosen as the labeled precursors, although the essential amino acids are ideal. Consequently, in comparison with the 1.8% natural distribution of cysteine in cellular proteins, the availability of a variety of amino acid signatures in cellular proteins gives much larger sequence coverage than ICAT and overcomes one of ICAT's disadvantages, namely that it frequently quantifies and identifies proteins based on a single peptide sequence. AACT allows the observation of multiple AACT-containing peptide signals for more accurate protein/modification quantitation and unambiguous identification. As a direct result, different isomers in the same protein family can be quantitated and identified because the signal specificity provided by these residue-specific tags helps to reduce spectral complexity in MS spectra of complex proteomes. These unique features facilitate comprehensive analysis of all possible regulated proteins for a whole proteome.

4. TROUBLESHOOTING

- During μLC separation runs, although most AACT peptides co-elute from the LC column with their unlabeled counterparts, some labeled peptides can appear slightly (usually 1–5 s) ahead of the unlabeled peptides. Therefore, the chromatographic profiles of both unlabeled and labeled peptides should be averaged rather than using single time spectra for quantitative analysis.

- When using AACT with mass tags less than 3 Da, for those peptides containing only one precursor residue, the first mono-isotopic ion peak (M$^+$) of the labeled peptide may overlap with the fourth isotope peak, (M+3)$^+$ of the unlabeled peptide. In this case, calculate the theoretical intensity of the (M+3)$^+$ isotope peak of each unlabeled peptide from the atomic compositions and then subtract this from the overall apparent peak intensity to give the intensity of the labeled peak.
- Generation of stable cell lines for expressing corresponding bait protein is a key step. However, it will be difficult to obtain appropriate cell lines that stably express every desired bait under physiological conditions. In problem cases, immunoprecipitation can be carried out coupled with AACT to study the pull-down complexes. The success of this alternative approach will be dependent on the specificity of the available antibodies.

5. REFERENCES

1. Wilkins MR, Gasteiger E, Gooley AA, *et al.* (1999) *J. Mol. Biol.* **289**, 645–657.
2. Mariappan SV, Silks LA III, Chen X, *et al.* (1998) *J. Biomol. Struct. Dyn.* **15**, 723–744.
3. Oda Y, Huang K, Cross FR, Cowburn D & Chait BT (1999) *Proc. Natl. Acad. Sci. U.S.A.* **96**, 6591–6596.
4. Veenstra TD, Martinovic S, Anderson GA, Pasa-Tolic L & Smith RD (2000) *J. Am. Soc. Mass Spectrom.* **11**, 78–82.
5. Chen X, Smith LM & Bradbury EM (2000) *Anal. Chem.* **72**, 1134–1143.
6. Gygi SP, Rist B, Gerber SA, Turecek F, Gelb MH & Aebersold R (1999) *Nat. Biotechnol.* **17**, 994–999.
7. Cagney G & Emili A (2002) *Nat. Biotechnol.* **20**, 163–170.
8. Gu S, Pan S, Bradbury EM & Chen X (2002) *Anal. Chem.* **74**, 5774–5785.
9. Gu S, Pan S, Bradbury EM & Chen X (2003) *J. Am. Soc. Mass Spectrom.* **14**, 1–7.
10. Zhu H, Pan S, Gu S, Bradbury EM & Chen X (2002) *Rapid Commun. Mass Spectrom.* **16**, 2115–2123.
11. Zhu H, Hunter TC, Pan S, Yau PM, Bradbury EM & Chen X (2002) *Anal. Chem.* **74**, 1687–1694.
12. Pan S, Gu S, Bradbury EM & Chen X (2003) *Anal. Chem.* **75**, 1316–1324.
13. Hunter TC, Yang L, Zhu H, Majidi V, Bradbury EM & Chen X (2001) *Anal. Chem.* **73**, 4891–4902.
14. Bae W & Chen X (2004) *Mol. Cell. Proteomics*, **3**, 596–607.
15. Zhu H, Hunter TC, Pan S, Yau PM, Bradbury EM & Chen X (2002) *Anal. Chem.* **74**, 1687–1694.
16. Chen X (2005) In *Methods in Molecular Biology/Methods in Molecular Medicine Book Series.* Edited by J.M. Walker. Humana Press.
17. Gu S, Chen J, Dobos KM, Bradbury EM, Belisle JT & Chen X (2003) *Mol. Cell. Proteomics*, **2**, 1284–1296.
18. Mortz E, Krogh TN, Vorum H & Gorg A (2001) *Proteomics*, **1**, 1359–1363.
19. Link AJ, Eng J, Schieltz DM, *et al.* (1999) *Nat. Biotechnol.* **17**, 676–682.
20. Mawuenyega KG, Kaji H, Yamuchi Y, *et al.* (2003) *J. Proteome Res.* **2**, 23–35.
21. Perkins DN, Pappin DJ, Creasy DM & Cottrell JS (1999) *Electrophoresis*, **20**, 3551–3567.
22. Pandey A & Mann M (2000) *Nature*, **405**, 837–846.
23. Qin J, Herring CJ & Zhang X (1998) *Rapid Commun. Mass Spectrom.* **12**, 209–216.
24. Zhang X, Herring CJ, Romano PR, *et al.* (1998) *Anal. Chem.* **70**, 2050–2059.
25. Zhou H, Watts JD & Aebersold R (2001) *Nat. Biotechnol.* **19**, 375–378.
26. Oda Y, Nagasu T & Chait BT (2001) *Nat. Biotechnol.* **19**, 379–382.
27. Lubeck O, Sewell C, Gu S, Chen X & Cai D (2002) *Proc IEEE*, **90**, 1868–1874.

28. Aebersold R & Goodlett, DR (2001) *Chem. Rev.* **101**, 269–295.
29. Wang T, Gu S, Ronni T, Du YC & Chen X (2005) *J. Proteome Res.* **4**, 941–949.
30. Wang T, Chuang TH, Ronni T, *et al.* (2006) *J. Immunol.* **176**, 1355–1362.
31. Du YC, Gu S, Zhou J, *et al.* (2006) *Mol. Cell. Proteomics*, **5**, 1033–1044.
32. Levine SS, Weiss A, Erdjument-Bromage H, Shao Z, Tempst P & Kingston RE (2002) *Mol. Cell. Biol.* **22**, 6070–6078.
33. Zhou Q, Lieberman PM, Boyer TG & Berk AJ (1992) *Genes Dev.* **6**, 1964–1974.
34. Gu S, Chen J, Dobos KM, Bradbury EM, Belisle JT & Chen X (2003) *Mol. Cell. Proteomics*, **2**, 1284–1296.

CHAPTER 3
Gel-based approaches

Stuart J. Cordwell, Ben Crossett, and Melanie Y. White

1. INTRODUCTION

Since the beginning of the post-genome era in 1995, the tools encompassed under the term 'proteomics' have been in high demand for the elucidation of how gene expression responds to changes in biological conditions or during disease processes. Proteomics offers researchers an opportunity to view the 'living genome' by monitoring the expression of genes in response to changes in the cellular or genetic environment (1). Furthermore, this field has evolved to provide information that cannot be supplied by transcript analysis alone, for example, analysis of the plethora of function-modifying post-translational modifications (2, 3), and the ability to decipher how proteins interact to form functional complexes (4–6).

The interest in proteomics has driven the improvement and industrialization of methods and technologies for protein separation. The traditional tool for the separation of proteins has been two-dimensional electrophoresis (2DE), relying on isoelectric focusing (IEF) in the first-dimension gel and sodium dodecyl sulfate-polyacrylamide gel electrophoresis (SDS-PAGE) in the second (7). This technique, combined with sensitive staining and image detection and computational analysis, has allowed protein abundance and expression differences to be detected across identical samples after changes in biological conditions. Unfortunately, however, and despite many substantial improvements in resolution and sensitivity, 2DE is plagued by several well-known limitations:

1. The inability to solubilize and separate highly hydrophobic proteins, especially those with many transmembrane-spanning regions.
2. The effects of dynamic range issues between high- and low-abundance proteins and the sensitivity of available staining reagents (this is particularly problematic in samples where one or two individual protein species account for a significant fraction of total protein, for example albumin in human plasma; 8).
3. The poor reproducibility of separations specific for highly alkaline proteins.
4. The poor resolution of very high- and low-molecular-mass proteins (9, 10).

Proteomics: *Methods Express* (C.D. O'Connor and B.D. Hames, eds)
© Scion Publishing, 2008

These difficulties have led many researchers to examine non-gel-based technologies for high-throughput proteomics (11–13), as well as several innovative pre-fractionation methods to improve protein recovery, in association with 2DE gels. For example, the resolving power of 2DE has been enhanced by improved biological sample preparation techniques, technologies for IEF-based pre-fractionation, subcellular fractionation, improved micro-range (1.0–1.5 pH units) immobilized pH gradients (IPGs) (14–16), and better fluorescent dyes (17). Pre-fractionation to purify a subcellular compartment or organelle (18, 19), or the use of the physical, chemical, and functional properties of a protein or class of proteins to enrich for a particular subset prior to 2DE or two-dimensional liquid chromatography (2DLC), is a particularly useful approach as it serves two purposes:

1. The group of proteins associated with a given subcellular fraction or organelle can be studied with relative specificity.
2. The overall complexity of the 2DE pattern is reduced when compared with whole-cell preparations, so that lower-abundance proteins can be visualized.

The major non-gel-based approach used to date has been multi-dimensional chromatography, coupled with tandem mass spectrometry (MS/MS) of peptide samples (20), in conjunction with peptide labels including isotope-coded affinity tags (21) and stable isotope labeling of amino acids in cell culture (22). Despite these advances in alternative technologies, 2DE remains the method of choice for performing high-resolution separation of complex protein mixtures (7). 2DE is a relatively cost-effective approach for researchers entering the proteomics field and is also capable of simultaneously resolving 3000–10 000 protein species. This method is also suitable for visualizing and purifying protein 'isoforms' and variants, such as those containing post-translational modifications that subtly alter the molecular mass (M_r) and isoelectric point (pI) of a predicted protein. Due to the complementary nature of these technologies, several groups have shown that utilizing the power of both 2DE-MS and 2DLC-MS/MS results in substantially improved proteome coverage (23).

This chapter describes 2DE gel-based approaches for the separation of complex protein mixtures.

2. METHODS AND APPROACHES

2.1 Sample preparation

The most critical phase in any proteomics project, especially those relying on 2DE, is the method used to extract proteins from whole cells (24). *Protocol 1* describes a one-step solubilization of protein mixtures from bacterial and cell culture cells, and from myocardial tissue.

Protein solubilization requires two major factors: an ability to maximize those proteins amenable to electrophoretic separation, whilst reducing the presence of ionic and other IEF-interfering substances within the sample buffer. The critical steps involve efficient cellular lysis by physical means including osmotic stress, freeze-thawing, a French press or tip-probe sonication, and disruption with zirconium or glass beads prior to solubilization with 2DE-compatible sample buffers. Clearly, the nature of the sample under study dictates which method is utilized, with mammalian cells typically being relatively easy to extract using chemical lysis methods, whilst many bacteria, mammalian tissues, and particularly plant samples require much harsher methods of extraction.

Once the cellular contents have been released, the removal of interfering substances becomes critical, as well as the inactivation of enzymes, particularly proteases. Salts (especially following washing of cells in buffers such as PBS), nucleic acids, lipids, and carbohydrates may all interfere with the separation process, leading to poorly focused or 'streaky' 2DE gels. Other samples may contain more-specific interfering compounds, for example phenolics associated with plant tissue. Traditionally, a cocktail of protease inhibitors and an excess of endonuclease may also be added. Removal of interfering substances is usually achieved by protein precipitation by ice-cold methanol, trichloroacetic acid (TCA)/acetone, or methanol/chloroform, followed by centrifugation (7).

The sample solution consists of several components:

- A denaturant, typically 7–9 M urea, often supplemented with 2 M thiourea for optimal recovery of membrane proteins (25)
- A reductant, most commonly dithiothreitol (DTT; 60–100 mM) or tributylphosphine (TBP; 2 mM)
- A series of detergents: 2% (v/v) 3-((3-cholamidopropyl)dimethylammonio)-1-propanesulfonate (CHAPS) is used most often, although many protocols also employ stronger solubilizing reagents such as 1–2% sulfobetaine 3-10 (SB3-10), or 1% amidosulfobetaine-14 (ASB-14; 26, 27)
- 40 mM Tris/HCl (pH 7.8)
- Carrier ampholytes (CAs; 0.2%, v/v)
- Bromophenol blue dye

The components that are most debated by 2DE researchers are generally the choice of reductant and the specific detergents to be employed. DTT is an ionic reducing agent and hence needs to be used at higher concentrations to avoid drift from the IPG strip during focusing, especially when using alkaline pH gradients, whilst TBP is nonionic, yet poorly stable in aqueous solutions and must therefore be used fresh when making sample buffers (7). The choice of detergent is typically based on the desire to recover more hydrophobic proteins: SB3-10 and ASB-14 recover more membrane-associated proteins, especially following membrane isolation and purification.

Protocol 1

Protein extraction for whole-cell lysate

Equipment and Reagents
- Tip-probe sonicator (Branson) or hand-held homogenizer (Omni)
- Urea, thiourea, TBP, DTT, CHAPS, SB3-10, CAs, and Tris/HCl (all from Bio-Rad individually or pre-made as 'Whole protein isolation buffer')
- IPG strips
- Protease inhibitor cocktail (Sigma)
- Okadaic acid (Sigma)
- Endonuclease (Benzonase, 150 U/ml; Sigma)
- Low-salt washing buffer (50 mM Tris/HCl, pH 7.8, or phosphate-buffered saline (PBS))
- 2DE sample buffer (5–7 M urea, 2 M thiourea, 2 mM TBP, 2% (w/v) CHAPS, 2% (w/v) SB3-10, 40 mM Tris/HCl, 0.2% (v/v) carrier ampholytes, and 0.002% (w/v) bromophenol blue dye. All components can be batch made and TBP added when required. The TBP can be replaced by 100 mM DTT)

Method
1. Pellet the cells from culture (bacterial or cell culture) or prepare fresh tissue in liquid nitrogen.
2. *For cells*, wash three times in low-salt washing buffer or PBS. Following the final wash, invert and tap the tube against a dry tissue to remove excess salt-containing buffer. *For tissue*, remove the tissue from storage and bring to room temperature in 2DE sample buffer, supplemented with protease inhibitor cocktail and phosphatase inhibitor (0.2%, v/v, okadaic acid).
3. Add 1 ml of fresh 2DE sample buffer.
4. *For cells*, sonicate the cells using the tip-probe sonicator (the manufacturer's instructions should provide a guide to the necessary length of time and sonication output). In order to avoid heating and possible carbamylation of the proteins in the urea, sonicate in replicate cycles of 30 s with 1 min on ice between cycles. *For tissue*, homogenize the tissue samples in 3 vols of 2DE sample buffer for 10 s replicate cycles until the tissue is completely disrupted.
5. Add 150 U of endonuclease and leave at room temperature for 20 min.
6. Centrifuge the lysate at 12 000 g for 15 min at 4°C to remove insoluble material. Place the supernatant in a fresh tube.
7. The amount to be loaded for analysis depends on the sample type. For IPGs, typically 250 µg to 1 mg of protein can be loaded by passive rehydration of the IPG strip (see section 2.3).

2.2 Protein pre-fractionation

Proteomics is a tool for identifying proteins related to events that can be assayed at the phenotypic level, for example diagnosis of disease or understanding disease processes. The aim of such experiments will always be to determine those unique proteins that are associated with performing a particular biological process amongst a complex variety of housekeeping proteins. Therefore, it is essential that the amount of the functional proteome that can be viewed at any given point in time is maximized, or to fractionate samples such that an increasingly specific subset of that functional proteome most relevant to the biological process under investigation is purified. Several methods for pre-fractionation of protein samples currently exist (28) and are based on one or more of the following characteristics:

- Subcellular location of the relevant protein
- Physical property (protein mass and pI)
- Chemical property (presence of a particular post-translational modification)
- Functional property (ligand- or protein-binding capability)

These approaches are considered in detail in the following sections.

2.2.1 Differential solubility

Sequential solubilization, or extraction of proteins from a complex cell or tissue, can aid in the visualization of a greater percentage of the proteome (14, 27). *Protocol 2* describes a method that has been utilized to separate outer-membrane proteins from Gram-negative bacteria (27) and to remove abundant myofilament-associated proteins from rabbit myocardium. *Fig. 1* shows a typical analysis

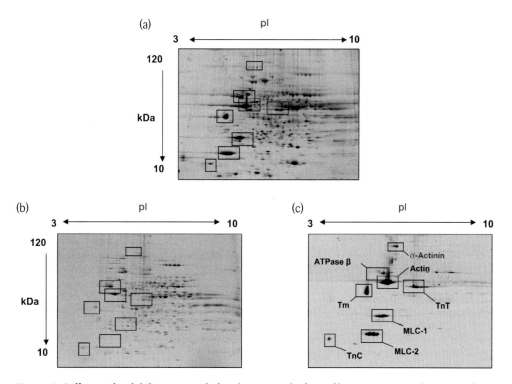

Figure 1. Differential solubility approach for the removal of myofilament-associated proteins from rabbit myocardium.
(a) Whole-tissue proteins; (b) Tris-soluble proteins; (c) Tris-insoluble, myofilament-associated proteins. Boxed regions represent protein spots identified as abundant myofilament-associated proteins on both whole-tissue and Tris-insoluble 2DE gels. Proteins were identified by MALDI-TOF MS and MS/MS following tryptic digests.

using this differential solubility approach. This technique uses the different solubilities of proteins, allowing them to be extracted by a series of progressively harsher extraction reagents. In the first step, highly soluble proteins are extracted with 40 mM Tris/HCl (pH 7.8). The insoluble pellet from this step is then washed extensively and resolubilized in 2DE sample buffer containing 2% (w/v) CHAPS as the detergent. The remaining insoluble pellet derived after centrifugation of this mixture can then be subjected to one of two methods: (i) solubilization in SDS-containing sample buffer and separation by SDS-PAGE alone, or (ii) solubilization in 2DE sample buffer supplemented with 1% (w/v) ASB-14. Whilst this overall approach is undoubtedly effective at enriching for lower-abundance and membrane-associated proteins, it is still highly probably that highly hydrophobic proteins (those with positive grand average of hydropathy, or GRAVY, values and more than three transmembrane-spanning regions) will be under-represented (29).

Protocol 2

Sequential extraction of proteins

Equipment and Reagents
- Tip-probe sonicator (Branson)
- 2DE sample buffer (see *Protocol 1*)
- 2DE sample buffer containing 1% (w/v) ASB-14 (Bio-Rad)
- 40 mM Tris/HCl (pH 7.8) containing protease inhibitor cocktail (Sigma) and 0.2% (v/v) okadaic acid (Sigma)

Method
1. Homogenize the tissue in 3 vols of ice-cold 40 mM Tris/HCl (pH 7.8) containing protease inhibitors and 0.2% okadaic acid.

2. Centrifuge at 12 000 *g* for 15 min at 4°C to remove insoluble material. Collect the supernatant and keep the pellet.

3. Place the supernatant in a 50 ml centrifuge tube and add ice-cold methanol to a final volume of 40 ml. Incubate at −80°C for a minimum of 2 h.

4. Centrifuge at 16 000 *g* for 30 min at 4°C. Carefully remove and discard supernatant and add 1 ml of 2DE sample buffer to the precipitated proteins. This contains highly soluble, predominantly cytosolic proteins.

5. Wash the pellet from step 2 twice in 40 mM Tris/HCl (pH 7.8) containing protease inhibitors and 0.2% okadaic acid to remove any contaminating Tris-soluble proteins and centrifuge at 12 000 *g* for 15 min.

6. Add 1 ml of 2DE sample buffer supplemented with 1% (w/v) ASB-14, vortex strongly, and sonicate with a tip-probe sonicator for 30 s.

7. Centrifuge at 12 000 *g* for 15 min at 4°C.

8. Collect the supernatant and remove to a fresh 1.5 ml centrifuge tube. This fraction contains myofilament-associated proteins.

Optional

9. Wash the insoluble pellet twice in 2DE sample buffer supplemented with 1% (w/v) ASB-14.
10. Centrifuge at 12 000 *g* for 15 min at 4°C.
11. Add SDS-PAGE sample buffer to the remaining insoluble pellet and perform one-dimensional SDS-PAGE.

2.2.2 Subcellular location

Organellar enrichment and purification prior to sample solubilization and protein separation has previously been performed using a variety of centrifugal and chemical processes, including sucrose density gradients, differential solubility, phase partitioning, and detergent extraction. This approach relies heavily on the ability to generate subcellular fractions that are close to true purity and relatively free of contaminating abundant proteins from other fractions. Subcellular organelles that have been purified and subjected to proteomics include the nucleus (30), mitochondria (31–33), and prokaryotic and eukaryotic membranes (34–37). In recent times, these methods have been supplemented by performing IEF to enrich further for a particular organelle. Zischka et al. (38) utilized free-flow electrophoresis (FFE) to further purify mitochondria from *Saccharomyces cerevisiae* following density-gradient ultracentrifugation. From this approach, only 2% of the identified proteins were not of mitochondrial origin in comparison with the 16% of nonmitochondrial proteins identified following centrifugation and 2DE alone. In addition, 43 unique mitochondrial proteins were only identified on the centrifugation/FFE-derived 2DE gels. This method clearly has substantial potential for providing highly pure organellar fractions prior to 2DE. For more detailed analysis of proteomes following subcellular fractionation, see Chapter 2.

2.2.3 Physical properties

The principles of protein separation based on pI and molecular mass are central to the use of 2DE gels for resolving individual proteins from amongst a complex mixture. Furthermore, IEF is now also making a significant contribution towards a third dimension of pre-fractionation prior to conventional separation, 2DE and 2DLC approaches. Pre-fractionating devices based on IEF allow the separation of proteins into pI fractions suitable for use with SDS-PAGE one-dimensional gels or micro-range (1 pH unit) IPG-2DE gels (14, 39) or as a first dimension prior to 2DLC-MS/MS, thus aiding in the identification of lower-abundance proteins. Several pre-fractionating devices are now commercially available, and a wealth of data exists to prove the value of this approach. Commercially available devices based on IEF include the Gradiflow (Life Therapeutics), Rotofor (Bio-Rad), FFE (BD), and multi-compartment electrolyzers (Proteome Systems).

Rotofor is a liquid-phase IEF-based pre-fractionation device that separates proteins in free solution using the buffering capability of CAs to create a pH gradient (40). Proteins are separated by IEF and collected into up to 20 sample chambers separated by permeable screens. Interfering CAs are removed from the resulting fractions prior to 2DE or 2DLC (41). Excellent separation can be achieved in the acidic to neutral pH ranges, whilst fractions from the basic range are generally overlapping due to the poor stability of CAs at alkaline pH, typically referred to as 'cathodic drift'. The Rotofor is now also capable of performing liquid-phase IEF without CAs by using polyacrylamide membranes with embedded immobilines (42), and is also available in a micro-scale version for handling smaller sample volumes.

Gradiflow is a unique preparative electrophoresis apparatus that uses thin polyacrylamide membranes with variable pore sizes providing both size- and charge-based separation (43). This device has been used predominantly for removing abundant proteins including serum albumin (44), for pre-fractionating complex mixtures prior to micro-range 2DE (45, 46), and for separating 'difficult' proteins incompatible with 2DE such as basic proteins (47).

Multi-compartment electrolyzers (48–50) may be more suitable for separating proteins into pH fractions compatible with micro-range 2DE. Multi-compartment electrolyzers utilize isoelectric membranes with embedded immobiline technology and have several distinct advantages, including their downstream compatibility with IPG-2DE without further sample clean-up or removal of CAs, reduced sample precipitation, and the flexibility to create membranes specific for the removal of a given protein or protein fraction (29). Similar technology utilizing thin immobiline-containing polyacrylamide membranes has been developed by Zuo and Speicher (51–53). These authors have also pre-fractionated proteins into 0.5 pH unit fractions in conjunction with single pH unit IPGs, although there is obviously the potential to use even narrower IPG pH ranges, especially in those areas where substantial numbers of expressed proteins are usually found (e.g. the pH range between 4.5 and 5.5). This principle was also used by Görg et al. (54), who utilized Sephadex IEF prior to IPG-2DE. The advantage of this system is its simplicity and low cost. Sephadex fractions are removed with a scalpel and applied to the compatible IPG strip.

The pre-fractionation technique of FFE has significant potential for improving separation prior to 2DE and 2DLC, for both proteomic and peptidomic approaches. FFE is a solution-based preparative IEF technique that fractionates proteins and peptides according to charge. The high resolution enables 96 fractions to be collected, each separated by only 0.02–0.10 pH units, depending on the gradient (55). As described above, FFE has also been used to separate organelles and protein complexes (56). Whilst pre-fractionation is undoubtedly useful for improving proteome coverage, the creation of many fractions makes undertaking a gel-based approach cumbersome, such that fractions are generally digested with a protease and subjected to MS-only-based approaches that utilize LC for peptide separation.

2.2.4 Chemical and functional properties

Subproteomics approaches are necessary to enable examination of a particular fraction associated with a biological question in near-to-total purity, without major contamination from proteins not associated with that particular phenomenon. Such fractions may be based on subcellular location, but may also be based on post-translational modifications, ligand binding, or other functional characteristics. These approaches rely on affinity chromatography (29, 57) and have many applications including: (i) the removal of abundant proteins, especially serum albumin, from plasma (58, 59); (ii) the binding of proteins with a selectable property, for example phosphoproteins and metal-binding proteins using immobilized metal-affinity chromatography (IMAC) (60), or various ligand-binding proteins such as those that bind ATP (61) or calcium (62); and (iii) the study of protein complexes. Furthermore, once the proteins themselves have been bound and eluted, the same chromatography resins can often be employed to bind the peptide responsible prior to MS.

2.3 Isoelectric focusing

IEF of proteins has allowed 2DE to become the core technology for protein separation and purification of individual components from complex mixtures (63). Two methods exist for this process:

1. The separation of proteins in capillary-diameter, self-cast, rod gels where the acrylamide matrix contains CAs that are 'pre-focused' to create the pH gradient (for review, see 64).
2. IPG technology that utilizes well-characterized acrylamido buffers to create a stable pH gradient in a thin, dried, acrylamide strip (for review, see 7).

The CA method suffers from several drawbacks, including an inability to load preparative amounts of sample, reproducibility problems associated with batch-to-batch variations of CA preparations, and difficulties in pouring and handling of IEF gels, as well as cathodic drift caused by the destabilization of the pH gradient at alkaline pH.

IPG technology has proved to be a revolution for 2DE and was one of the critical developments, along with the advent of sensitive, high-throughput matrix-assisted laser desorption ionization time of flight (MALDI-TOF) and electrospray ionization MS, leading to the proliferation of proteomics studies (7). IPG-IEF in the first dimension of 2DE provides substantially improved reproducibility between runs, as well as interlaboratory standardization due to the availability of pre-cast IPG strips (65, 66) and pre-made reagents. IPG strips have an increased loading capacity suitable for adding up to 1 mg of protein sample per gel, and samples can readily be applied during in-gel rehydration. These features combine to make IPG strips extremely easy to use. They are available from a number of vendors in a wide variety of pH gradients ranging from very wide (pH 4–12 and 3–10) and medium range (pH 3–6, 4–7, 5–8, 7–10, 6–9, 6–11, etc.) through to micro-range (single pH unit) gradients and in a variety of strip lengths suitable for either high-resolution (17, 18, 24, and 30 cm) or rapid analysis (7 and 11 cm) proteomics applications.

In our laboratory, we find it most useful to use wide-range IPG strips (e.g. pH 3–10) for initial screening of newly prepared samples and then combined sets of mid-range pH gradients (pH 3–6, 4–7, 5–8, and 7–10) for higher-resolution separations and comparative analyses. Micro-range pH gradients (e.g. pH 4–5) may be used to enrich for low-abundance proteins allowed by the increased loading capacity to separating area available, or perhaps, more importantly, for the enhanced resolution of proteins that have undergone subtle post-translational

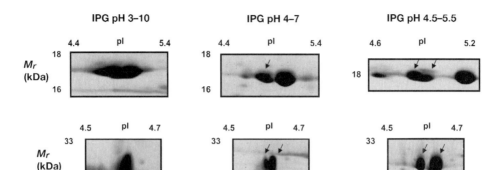

Figure 2. Separation of proteins using wide- (pH 3–10), medium- (pH 4–7) and micro-range (pH 4.5–5.5) IPG 2DE gels for the visualization of protein variants caused by post-translational modifications.
Upper panels: myosin light chain 2; *lower panels*: tropomyosin; both derived from whole-tissue lysates of rabbit myocardium.

modifications (see *Fig. 2*). IPG strips are supplied in a dehydrated format when purchased (major suppliers of IPG strips include Bio-Rad and GE Healthcare) and must be rehydrated in sample buffer either with or without the presence of the protein sample. For many sample types, 'passive' rehydration of the protein sample into each individual IPG strip is the best method available, although more even gel strip re-swelling may be obtained using a rehydration cassette without the presence of the sample. Protein lysates (diluted to 250 µg per gel for analytical loads and up to 1 mg for preparative loads) are applied to the gel surface of the dry strip and allowed to re-swell the strip for a minimum of 6 h. Plastic strip trays are now commercially available; however, 2 ml disposable plastic pipettes are also useful.

The alternative is to rehydrate the strip in an equal volume of sample buffer and add protein to a cup at either end of the IPG strip. For basic pH strips (pH 6–11, 9–12, and 7–10 gradients), cup loading at the anodic end appears to be essential for the generation of high-quality patterns due to the uneven osmotic exchange caused by the precipitation of acidic proteins at the anodic end. IEF of proteins in IPG strips mainly involves the stepwise increase of voltage from an initial low voltage (to remove current-inducing salts) to high voltage for efficient and rapid protein separation. The total kVh used to focus proteins can be between 30 and 100 kVh depending on sample composition, choice of IPG strip, and the sample load.

2.4 Difference in-gel electrophoresis

Another method has recently been described for the separation of proteins using 2DE. Difference in-gel electrophoresis (DIGE) (see *Protocol 3*) utilizes the cyanine dyes Cy2, Cy3, and Cy5 to label samples subjected to different biological conditions (see *Fig. 3*, also available in the color section; 67, 68). The samples are then mixed and run on a single 2DE gel, followed by imaging at different wavelengths and image analysis based on the relative ratios of Cy3- and Cy5-labeled spots. Cy2 is

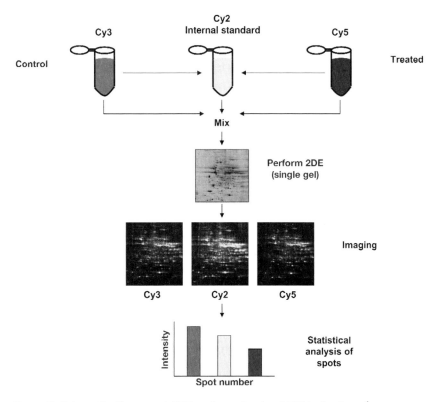

Figure 3. Schematic diagram of 2DE performed using DIGE technology (see page xxi for color version).
Two samples can be compared by labeling with Cy3 and Cy5 dyes, whilst a pooled internal standard is labeled with Cy2. Fluorescent images are then taken at the respective dye wavelengths and the images overlapped to quantify changes in protein abundance. The pooled standard is used to normalize the resulting image data.

generally used as an internal control, where the two samples to be compared are mixed in equal volumes and then labeled. This step allows the samples to be 'normalized'. Whilst DIGE is suitable for the comparison of two samples in a single experiment, it is not as suitable for large-scale studies involving tissue or serum plasma from individuals. DIGE experiments have the advantage of reducing the number of gel replicates required for a statistically viable data set, as well as the necessary sensitivity in comparison with visible dyes such as silver diamine and Coomassie blue. On the other hand, the DIGE approach is also limited by several technical factors including the following:

1. The requirement for no more than 1–3% of lysine residues to be bound by the respective labels, in order to maintain protein solubility.
2. The molecular mass of the Cy-labeled protein is only slightly altered in comparison with unlabelled protein, resulting in difficulties in downstream MS analysis and the need for a second staining prior to spot excision, although this has been overcome by modifications to the software in matching algorithms (69).

3. The high comparative costs of the DIGE reagents, especially when considering the need for biological replicates.

Despite this, the DIGE approach has shown considerable promise for comparative proteomics based on 2DE gel technology (70).

Protocol 3

Difference in-gel electrophoresis

Equipment and Reagents
- 2D Protein Clean-up kit (GE Healthcare)
- 2D sample buffer (GE Healthcare; must be free of primary amines and with pH 8-9)
- 2D Quant kit (GE Healthcare)
- CyDyes (CyDye DIGE fluors Cy2, Cy3, and Cy5; GE Healthcare)
- Dimethylformamide (DMF; Sigma)
- 10 mM Lysine (Sigma)

Method
1. Clean up the sample using the 2D Protein Clean-up kit as described in the manufacturer's instructions.

2. Check the pH of the sample using pH paper. It should be between pH 8 and 9.

3. Determine the protein concentration using the 2D Quant kit as described in the manufacturer's instructions.

4. Next, prepare the CyDyes. CyDyes are light-sensitive, so perform the following steps in the dark at all times. Allow the CyDyes to warm to room temperature and add 5 μl of DMF to each dye (Cy2, Cy3, and Cy5) creating a 1 mM stock of each. Vortex for 30 s and then centrifuge at 12 000 g for 30 s. The stock is stable for 2 months at −20°C. Prepare a working CyDye solution (400 μM) by mixing the stock solution with DMF at a ratio of 1 : 1.5.

5. Add 50 μg of each sample into an appropriately labeled microcentrifuge tube. Prepare the pooled standard by adding equal microgram amounts of the two comparable samples into one tube.

6. Label the protein samples by mixing 1 μl of CyDye solution with 50 μg of protein, mix, centrifuge briefly, and leave on ice (in the dark) for 30 min. Label the pooled standard with Cy2, and the test and control samples with Cy3 and Cy5 (see *Fig. 3*). For multiple samples, randomly allocate CyDyes to test and control samples to avoid dye bias.

7. Add 1 μl of 10 mM lysine to stop the reaction, mix, and spin briefly. Leave on ice for 10 min.

8. Mix an equal volume of each sample with 2D sample buffer and make up to the appropriate volume for the size of IPG strip being used.

9. Carry out IEF as described in section 2.3.

10. Equilibrate the IPG strips as described in *Protocol 4* and analyze these by second-dimension SDS-PAGE as described in *Protocol 5*.

11. Detect images using a suitable fluorescent scanner. Once scanned, the gel can be stained with a second dye such as Coomassie blue.

2.5 Reduction, alkylation, and detergent exchange (equilibration)

Following IEF, the proteins separated in IPG strips must be reduced, alkylated, and the detergent exchanged with SDS buffer prior to second-dimension SDS-PAGE (see *Protocol 4*). This step is often referred to as 'equilibration'. SDS coats the proteins giving them the same net negative charge and allows separation to occur on the basis of molecular mass. The equilibration procedure is mostly commonly performed in two steps: first a reduction using DTT and then an alkylation step using iodoacetamide (IAA). A one-step protocol (incubation period of 20 min), where DTT is replaced by TBP and IAA is replaced by acrylamide monomer in solution, is also sometimes performed. The acrylamide monomer allows subsequent MS searches to be performed with the cysteine/acrylamide option checked.

Other essential components of the equilibration solution include urea and glycerol, which reduce electroendosmotic effects. Surprisingly, some reports suggest that up to 20% of the separated protein is lost in the first few minutes of equilibration (7). However, this is believed to occur evenly and reproducibly within and between gel runs.

Protocol 4

Reduction, alkylation, and detergent exchange

Equipment and Reagents
- IPG strips from *Protocol 3*
- 1× Tris/HCl gel buffer (pH 8.8) (192 mM glycine, 0.1% SDS, 24.8 mM Tris/HCl, pH 8.3)
- Solution I (6 M urea; 2%, w/v, SDS; Tris/HCl gel buffer; 20%, v/v, glycerol; 60 mM DTT; and a trace of bromophenol blue)
- Solution II (6 M urea; 2%, w/v, SDS; Tris/HCl gel buffer; 20%, v/v, glycerol; 5 mM IAA; and a trace of bromophenol blue)
- One-step equilibration buffer (6 M urea; 2%, w/v, SDS; Tris/HCl gel buffer; 20%, v/v, glycerol; 5 mM TBP; 2.5%, v/v, acrylamide solution; and a trace of bromophenol blue)[a]
- SDS-PAGE second-dimension gel
- 0.5% (w/v) Agarose in Tris/glycine running buffer (192 mM glycine; 0.1%, w/v, SDS; 24.8 mM Tris/HCl, pH 8.3; and a trace of bromophenol blue)

Method
1. Incubate the IPG strips (gel side upwards) in 5–10 ml of solution I for 20 min with gentle shaking[b].
2. Remove solution I and replace with 5–10 ml of solution II for 10 min with gentle shaking.
3. Remove solution II and quickly transfer the IPG strip to the top of an SDS-PAGE second-dimension gel (cast previously).
4. Embed the strip with 0.5% (w/v) agarose in Tris/glycine running buffer.

Notes
[a] An alternative to using solutions I and II is to use the one-step equilibration buffer as described in this protocol.
[b] Alternatively, for the one-step equilibration, incubate the IPG strip in one-step equilibration buffer for 20–25 min.

2.6 SDS-PAGE

The pore size of the second-dimension slab gel used for SDS-PAGE depends on the application and range of anticipated protein molecular masses. For high-molecular-mass proteins, pore sizes as low as 4–7.5% T (total concentration of acrylamide monomer plus cross-linker) may be utilized. However, the tensile strength of such gels is poor and gel handling may become difficult. To visualize the greatest number of proteins in the molecular mass range 10–120 kDa, gradient gels of 8–18% can be used, although for ease of pouring a single 12.5% gradient is often employed. *Protocol 5* describes the preparation of 8–18% gradient slab gels. After polymerization, the equilibrated IPG strips are placed on top of the SDS-PAGE slab gel, embedded with agarose, and the second dimension performed using 3 mA per gel for 2 h and 12 mA per gel overnight or until the bromophenol blue dye front reaches the end of the gel. The gel tank running buffer is 198.4 mM Tris/HCl (pH 8.3), 1.536 M glycine, and 0.8% (v/v) SDS.

Protocol 5

Second-dimension SDS-PAGE

Equipment and Reagents
- Gradient gel pourer (Bio-Rad)
- Gel casting chamber (Bio-Rad)
- Protean II Cell (two gels), Multi-Cell (six gels), or Dodeca Cell (12 gels) (all from Bio-Rad)
- 50% Glycerol
- Ultrapure water
- 5× Tris/HCl gel buffer (1.875 M Tris/HCl, pH 8.8)
- 40% (v/v) Acrylamide solution (40%, w/v, acrylamide; 1%, w/v, bis-acrylamide)
- Tetramethylethylenediamine (TEMED)
- 10% (w/v) Ammonium persulfate
- Water-saturated isobutanol

Method
The following reagent volumes allow the casting of six 8–18% slab gels.

1. Prepare 200 ml of 8% acrylamide mixture by mixing 40 ml of 5× Tris/HCl gel buffer (pH 8.8), 40 ml of 40% acrylamide solution, and 120 ml of ultrapure water.

2. Prepare 200 ml of 18% acrylamide mixture by mixing 40 ml of 5× Tris/HCl gel buffer, 90 ml of 40% acrylamide solution, and 70 ml of 50% glycerol.

3. Just prior to pouring the gel, degas the 8 and 18% acrylamide mixtures and to each solution add 35 µl of TEMED and 360 µl of 10% (w/v) ammonium persulfate.

4. Add the 8 and 18% acrylamide mixtures to the separate chambers of the gradient gel pourer and stir the solution gently in the 8% chamber.

5. Connect the gradient gel pourer to the gel casting chamber containing the gel cassettes and open the valve from the 8% chamber.

6. Allow the 8% acrylamide solution to fill the lower 0.5 cm of the gel cassettes within the gel casting chamber and then open the valve separating the 8 and 18% solutions, allowing them to mix.
7. Allow the acrylamide to fill to a level approximately 0.5–1 cm from the top of the gel cassettes.
8. When the gels have been poured, cover the top of the unpolymerized gels with water-saturated isobutanol and allow at least 6 h for complete polymerization.

2.7 Staining 2DE gels

A plethora of options currently exists for staining 2DE gels (17). Several will be considered here, and the advantages and disadvantages of each will be discussed.

2.7.1 Visible dyes (Coomassie blue and silver staining)

When choosing a stain for 2DE (or SDS-PAGE) analysis, several important considerations must be taken into account. These are:

- Sensitivity, i.e. the ability of the chosen stain to detect low-abundance proteins
- Reproducibility, including the linear order of magnitude over which the stain can detect different protein abundances
- Compatibility with down-stream analytical techniques, especially MS
- Relative cost and ease of use

Coomassie blue stains are the most commonly used dyes for 2DE. The dye comes in two forms, the triphenylmethane R-250 form and the more pronounced colloidal G-250 form, both of which bind to primary amines on lysine, arginine, and histidine residues. The colloidal form is up to three times more sensitive (71). Coomassie dyes have a good quantitative relationship to the relative amount of protein present, but are generally less sensitive (200–500 ng of protein) than either silver or fluorescent dyes. Colloidal G-250 has a very high affinity for SYPRO Ruby; hence, double-staining is possible.

A variety of silver staining approaches exists (72), but they can be placed into two broad categories:

1. Silver nitrate in weakly acidic solution with development at alkaline pH (generally using formaldehyde).
2. Silver diamine in alkaline solution (usually ammonia and sodium hydroxide), where image development is achieved by dilute acidic solutions of formaldehyde.

Protocol 6 describes silver nitrate staining using formaldehyde. This method is not as sensitive as the diamine silver staining method, but gives a better MS performance as no glutaraldehyde is used in the sensitization step. Whilst silver staining is undoubtedly more sensitive than either Coomassie dye, there are several disadvantages that need to be considered. Firstly, the silver staining process is step-intensive and laborious, often involving an overnight incubation. The reproducibility is also poor due to difficulties in controlling the development step across gels, as well as a poor linear quantitative range representative of protein quantity. Furthermore, some proteins, especially highly acidic or glycosylated proteins, may be 'negatively' stained, making image analysis problematic. Finally, stains using glutaraldehyde as a sensitization agent are poorly compatible with MS, as this chemical is an effective cross-linker. Before MS of proteins stained previously with silver, the protein must be washed extensively to remove any bound stain (73).

Protocol 6

Silver nitrate staining

Equipment and Reagents
- Fixer solution (40%, v/v, ethanol; 10%, v/v, acetic acid)
- Wash solution (50%, v/v, ethanol)
- Sensitization solution (0.02%, w/v, sodium thiosulfate)
- Silver stain (0.1%, w/v, silver nitrate; store at 4°C and protect from light)
- Developer (0.04%, v/v, formaldehyde (use 0.2 ml of a 37% stock solution); 2%, w/v, sodium carbonate)
- Stop solution (1%, v/v, acetic acid; make fresh each time)

Method:
1. Fix the gel in fixer solution for a minimum of 1 h (17 cm gels) or up to overnight.
2. Wash the fixed gel in wash solution for approx. 5 min and then twice in water for 5 min each.
3. Wash in the sensitization solution for 1 min (approx. 150 ml per gel).
4. Wash the gel twice in water for 1 min each.
5. Place the gel in silver stain for 30 min at 4°C.
6. Wash the gel twice in water for 1 min each.
7. Remove the last wash and add developer. Change the developer as required but before it starts to turn yellow (normally about 1 min).
8. Once the gel has developed (5–10 min), pour off the developer and add the stop solution. The gel can then be stored in stop solution indefinitely.

2.7.2 Fluorescent dyes (SYPRO Ruby and Deep Purple)

The advent of easy-to-use fluorescent dyes has been welcomed by researchers using 2DE as their protein separation technology. SYPRO Ruby is a ruthenium complex that binds to proteins through Coomassie-like interactions with primary amines (74). The staining process essentially involves a short fixation step in methanol/acetic acid and overnight (minimum of 2 h) incubation in the dye, prior to imaging. Deep Purple is another fluorescent dye. It is based on a wholly natural fungal by-product called epicocconone (75), which binds to primary amines. Both dyes are 'end-point' stains, referring to their capability to stain to their limit of detection (subnanogram quantities of protein) without possible overstaining, and have an excellent linear quantitation range. An obvious disadvantage of fluorescent dyes is the need for specialized imaging equipment, although double-staining with Coomassie blue is possible (see *Protocol 7*) (76).

Protocol 7

Fluorescent and Coomassie blue double staining

Equipment and Reagents
- Nalgene (polymethylpentene) staining tray
- Fixer solution (10%, v/v, methanol; 7%, v/v, acetic acid)
- Destain solution (same composition as fixer solution in *Protocol 6*)
- SYPRO Ruby (Sigma or BioRad)
- Coomassie blue stock (Coomassie Brilliant Blue G-250 (0.5 g dissolved in 10 ml of water); 50 g of ammonium sulfate; 6 ml of 85%, v/v, phosphoric acid; made up to 500 ml with water)
- Coomassie blue working solution (4 : 1, stock solution : methanol)
- Coomassie destain solution (500 ml) (1%, v/v, acetic acid)

Method
1. Place the gel in the Nalgene (polymethylpentene) staining tray, add fixer solution, and leave with gentle shaking for a minimum of 1 h.
2. Remove the fixer and add SYPRO Ruby (50 ml for a 7 cm IPG 2DE gel, 100 ml for an 11 cm gel, or 350 ml for a 17 cm gel). Leave the gel immersed in stain overnight.
3. Remove the stain (the stain can be reused once with only a slight decrease in performance).
4. Immerse the gel in the destain solution for a minimum of 1 h. Gels can be left for several days in the destain solution.
5. Wash gel for approximately 10 min in water and then capture the fluorescent image.
6. Transfer the gel to a glass staining tray and immerse in colloidal Coomassie blue working solution. Leave the gel overnight with gentle shaking at room temperature. Gels can be left for several days in this solution.
7. Remove and discard the stain. Place the gel in Coomassie destain solution for a minimum of 2 h to enhance detection.

2.7.3 Protein 'expression' versus protein 'abundance'

When a gel-based proteomics experiment is performed, what do the resulting data actually represent? Many researchers use the phrases 'expression' and 'up/downregulation' to refer to changes seen between samples on 2DE gels stained with visible or fluorescent dyes. Such terminology is not strictly correct. Dyes measure protein 'abundance' at a given point in time, and abundance is a function not only of expression at the transcriptional/translational level, but is also a measure of protein lifespan, or half-life. Thus, a protein that is the product of a gene that is actively being transcribed (and therefore is well expressed), but has a very short half-life (due to specific protease degradation for example) may not be seen on stained 2DE gels because it is not sufficiently abundant. One method that has been used to overcome this problem has been suggested by the group of Michael Hecker and colleagues in Greifswald, Germany, and the corresponding software has been commercialized by DeCodon (77). This method uses dual-channel color imaging of 2DE gels. Two gels are analyzed; the first is generated by pulse radiolabeling using [^{35}S]methionine (thus measuring expression),

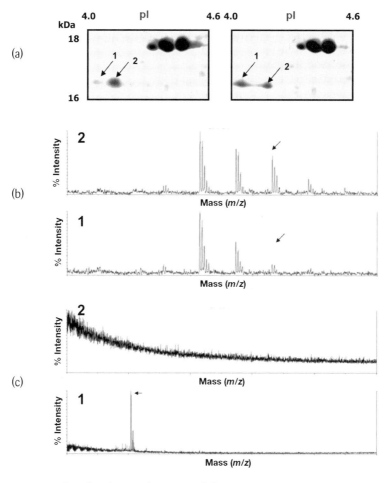

Figure 4. Phosphorylation of a myocardial protein in response to ischemia/reperfusion injury.
(*a*) Arrowheads indicate the position of the phosphorylated (1) and native (2) forms of the protein. (*b*) The protein was excised and digested with trypsin, and the resulting peptide profiles compared by MS. (*c*) Specific phosphopeptides were then enriched using affinity chromatography and the bound peptides again analyzed by MS. Arrowheads indicate the presence/absence of peaks.

whilst the second gel is stained with silver stain (measuring abundance). The spots on each gel are then given a color (e.g. red for the radiolabeled gel and green for the silver-stained gel) and the images overlaid. Protein spots appearing in red are those that are actively expressed but rapidly turned over, those in green are not being expressed but are long-lived, and those in yellow (where the two colors overlap) are expressed and turned over at equal rates. This approach has been used to examine heat-shock and other stresses in Bacillus subtilis (78, 79).

2.7.4 Modification-specific dyes

In more recent times, a series of fluorescent stains with differing specificities for chemical post-translational modifications, including phosphorylation and glycosylation, have become available. These are especially useful when combined with enrichment strategies (see *Fig. 1*).

Phosphorylation results in an acidic pH shift and an increase in mass of 80 Da (see *Fig. 5*, also available in the color section). Traditionally, ^{32}P-labeling of phosphopro-

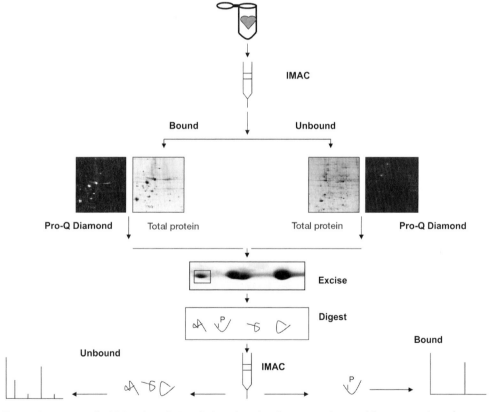

Figure 5. Large-scale 2DE gel analysis of phosphorylated proteins from rabbit myocardium (see page xxi for color version).
Affinity chromatography (IMAC) was used to bind phosphorylated proteins. Both bound and unbound proteins were then separated by 2DE and the resulting gels multiplex-stained with Pro-Q Diamond and SYPRO Ruby. Proteins of interest were then excised, digested with trypsin, and subjected to IMAC. Bound and unbound peptides were then analyzed using MS.

teins has been a well-established and robust methodology for visualization of phosphorylated proteins, following which phosphopeptides can be sequenced by Edman degradation or MS. Common problems associated with this technique include the limited experimental settings in which ^{32}P can be applied, and the labile nature and poor sensitivity of phosphopeptides under Edman sequencing. A second technique that has commonly been used to purify tyrosine-phosphorylated proteins is immunoaffinity chromatography using anti-phosphotyrosine antibodies (80), which are more reliable than either anti-phosphoserine or anti-phosphothreonine antibodies. The phosphorylated proteins may be separated by 2DE (for review, see 81), LC, or by direct coupling to MS.

Enrichment by IMAC prior to 2DE (or following enzymatic digestion prior to MS) provides a chromatographic method with which to 'pull down' phosphorylated proteins and peptides from complex protein mixtures (82). IMAC exploits the affinity of phosphorylation residues for metal ions such as Fe^{3+}. Using IMAC for the separation of phosphorylated peptides by MS and MS/MS, without prior gel-based separation, is becoming widespread (83–85). A second phosphor-specific affinity technique utilizes titanium dioxide to enrich for phosphopeptides prior to MS/MS via electrospray ionization or MALDI sources (86, 87). All of these enrichment processes are compatible with downstream gel-based separation and phospho-

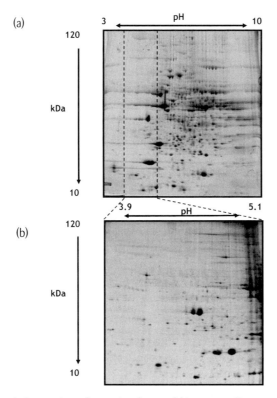

Figure 6. Separation of proteins from rabbit myocardium.
Proteins were separated using a wide-range (pH 3–10) IPG 2DE gel (*a*) and a microrange (pH 3.9–5.1) IPG 2DE gel (*b*). Dotted lines indicate the region of overlap.

specific fluorescent staining using Pro-Q Diamond (Molecular Probes) (see *Fig. 6*). This phospho-specific stain allows a 'multiplexing' approach to be taken for gel staining. The 2DE gels are initially stained with Pro-Q Diamond and the fluorescent images detected. The gel is then washed and restained with total protein dyes, such as SYPRO Ruby, and the gel is re-imaged. The two images are then overlapped to determine which proteins are phosphorylated (88, 89).

Glycoproteins contain two possible types of sugar attachment, either *N*-linked via the amido nitrogen of asparagine residues or *O*-linked via the hydroxyl group of serine and threonine residues. The addition of sugars to proteins results in an acidic pI shift, often accompanied by a slight but discernible increase in molecular mass, as sugar structures tend to be complex and relatively large when present in extended chains. Unlike phosphorylation and the majority of other post-translational modifications, a near-limitless possibility of different glycostructures exists. The most common approach for the gel-based separation of glycoproteins involves their affinity capture using a variety of lectins (90–92), which can be used to capture the proteins of interest as well as the glycosylated peptides following protease digestion prior to MS (93). Several approaches can be used to detect glycoproteins on 2DE gels (94) including enzymatic or chemical deglycosylation (95), for example using sialidase to remove sialic acids, peptide:*N*-glycosidase F to cleave the bond between *N*-acetylglucosamine and asparagines in *N*-linked sugars, and *N*-acetyl-β-D-glucosaminidase to remove *O*-linked sugars. Glycoproteins in 2DE gels can be visualized using Pro-Q Emerald fluorescent stain (Molecular Probes) and multiplexed as described above (96, 97).

3. TROUBLESHOOTING

3.1 Interfering molecules during IEF

Several 'contaminants' caused by poor or inadequate sample preparation can lead to a reduced quality of separation in the IEF dimension of 2DE. The major problem leading to substandard 2DE separation is the presence of high concentrations of salt. Salts can be present during sample preparation for many reasons, often due to the buffers used to wash cells prior to solubilization, or due to media components. High levels of salt (generally >50–100 mM for cup loading, with a significantly lower tolerance for in-gel rehydration) cause elevated conductivity within the gel strip, prolonging separation times and potentially stopping the separation altogether if salt loads are high. In traditional IEF systems, IPG strips could burn due to salt overload, but this problem has been overcome by the use of limitation of the electrical current in new IEF separation systems.

Other interfering molecules that are incompatible with protein IEF include polysaccharides, nucleic acids, and lipids. Lipids are a particular problem where whole-cell lysis is undertaken using a highly efficient solubilizing detergent such as ASB-14, and can often lead to vertical streaking on the second-dimension gel. We have found that such detergents are often better employed on specific subcellular fractions for this reason. Generally, lipids are removed effectively using centrifugation, but organic solvents have sometimes been employed to do this. Nucleic acids and polysaccharides generally result in severe streaking in the acidic horizontal axis of 2DE gels due to interactions with CAs (7). Nucleic acids can be removed by using an excess of endonuclease or Benzonase, whilst polysaccharides are removed by protein precipitation and centrifugation.

Salts (and other interfering molecules) are best removed using dialysis or precipitation of proteins using TCA, cold acetone, TCA/acetone, or cold methanol. However, it is important to note that all of these methods are likely to result in significant protein sample losses and/or difficulties in resolubilizing the precipitated proteins. An alternative method for salt removal involves the periodic removal and replacement of the paper wick filters used to connect the IPG strips to the respective electrodes, as this is the major site where salts collect during IEF. Several commercially available clean-up kits are also now available.

3.2 'Difficult' proteins in 2DE

Several problems remain with 2DE technology for high-resolution proteomics applications. The final section of this chapter will describe these problems and potential solutions to overcome them.

3.2.1 Basic proteins

A major problem that has traditionally been associated with 2DE-based protein separation is the inability to array highly alkaline proteins (pI > 9) reproducibly. These proteins may be lost or appear as vertical 'streaks' due to cathodic drift during IEF or the lack of suitable buffering molecules at highly alkaline pH (98). Poor separation of basic proteins is due to one or more of the following:

- Active water transport towards the anode (reverse electroendosmosis) caused by the strong positive charge of basic acrylamido buffers.
- Hydrolysis of acrylamide to acrylic acid at alkaline pH.
- Migration of reducing agents, mainly DTT, leading to reduced gel quality ('streaky' 2DE patterns).

Prior to the advent of IPG technology, nonequilibrium pH gradient electrophoresis was used to provide separation of basic proteins (99). IPG-IEF is possible in the basic pH range, even up to a pI of 12.0 (47, 98, 100). However, some precautions are needed to ensure quality separation, including:

- Anodic cup loading.
- Treatment of protein lysates with hydroxyethyl disulfide, allowing oxidation of thiol groups in disulfide-containing proteins (101).
- The use of dimethylacrylamide (98) or *N*-acryloylaminoethoxyethanol (102) instead of acrylamide, as these matrices have been shown to resist alkaline hydrolysis at basic pH.
- The use of nonionic reducing agents such as TBP to replace DTT (103), thus minimizing the transport of reducing agent out of the IPG during IEF.

- The addition of isopropanol or glycerol to sample buffers to reduce the electroendosmotic effects (103).

3.2.2 Low-abundance proteins

2DE technology currently fails to resolve many lower-abundance proteins for two major reasons:

- The 'dynamic range' of protein abundances in most cells and tissues, which is too broad to accommodate focusing of abundant proteins where high sample loads are added to discern lower-abundance proteins.
- The limit of detection of currently available stains.

There has been substantial research into abundant protein depletion using immunoaffinity approaches, especially for removing abundant proteins from human plasma (104–106), particularly albumin. Removal of such proteins may allow lower-abundance proteins, such as disease biomarkers, to become visible using 2DE or chromatography-based separation (58, 59). However, there is also evidence that the removal of abundant proteins results in the loss of bound low-abundance proteins (106, 107).

One gel-based approach that has come to prominence is the use of micro-range or 'zoom' IPG strips (1 pH unit). These are often employed in conjunction with IEF pre-fractionation. Micro-range IPGs increase the separating area to pH range ratio of the first dimension, and also allow a higher concentration of protein sample to be applied to the IPG (see *Fig. 2*). Protein samples that have not been pre-fractionated are difficult to separate using micro-range IPG strips in some pH ranges, due to sample build-up and precipitation at the ends of the strips (1 mg of total protein may only contain 500 mg of proteins with pIs in the pH range 4.5–5.5, or less than 100 mg in the very alkaline range). Several studies have examined the utility of micro-range IPG 2DE both alone and in conjunction with other pre-fractionation techniques such as subcellular fractionation or differential solubility and IEF-based pre-fractionation as described above (14–16, 51).

The disadvantage of this approach is that 'composite' 2DE gels, consisting of many subcellular or IEF fractions, combined with multiple micro-range IPG gels, must be generated. This is a substantial time and cost commitment, and realistically few laboratories can afford, or are inclined, to follow this approach to answer real biological questions. Therefore, it seems that the most useful application for micro-range 2DE gels is to provide unparalleled separation of protein isoforms for MS characterization of post-translational modifications (*Fig. 2*; 108).

3.2.3 High- and low-molecular-mass proteins

Another significant problem associated with 2DE is the under-representation of high- and low-molecular-mass proteins. This is because of the acrylamide pore size for separation within the IPG strip, and also the % T gradient used in the SDS-PAGE second dimension. The IPG strip pore size is 4% T in commercially available strips, which excludes higher-molecular-mass proteins >120–150 kDa, which are either unable to pass into the strip during passive rehydration or precipitate at the sample application point when cup loading. Recent work has attempted to improve the recovery of high-molecular-mass proteins using lower % T IPG strips in the IEF dimension, down to as low as 3% T (109, 110). Whilst lower % T IPG strips are more difficult to cast and handle, they improve the total amount of protein that enters the strip, as well as increasing the proportion of high-molecular-mass proteins on the ensuing 2DE gels. One research group has utilized agarose gels in the IEF dimension to improve the separation of high-molecular-mass proteins (111). Lower-molecular-mass proteins (<10 kDa) are typically resolved using the well-characterized Tris/tricine SDS-PAGE system (112).

3.3 Alternatives to 2DE gel-based approaches

3.3.1 SDS-PAGE/LC-MS/MS

Whilst 2DE gel technology has undoubtedly been the most common approach for the separation of protein mixtures for proteomics, an alternative approach utilizing simple SDS-PAGE has also been proposed. This approach relies on the unsurpassed solubilizing efficiency of SDS (which is incompatible with 2DE IEF) and one-dimensional separation. Despite the poor purification, gel bands (either those that are stained or simply evenly spaced slices) are excised and the proteins within each band are digested proteolytically. The released peptides from each band are then separated by reverse-phase chromatography and identified by MS/MS. This technique is also promising for examining integral membrane proteins, immunoprecipitated protein complexes, and subcellular fractions (113).

4. CONCLUSIONS

With the advent of new technology such as multi-dimensional LC-MS/MS and isotope tags for comparative analysis, and the well-documented flaws with gel technology for the separation of extremely alkaline, hydrophobic, and low-abundance proteins, the inevitable question must be asked: what is the future of gel-based approaches in high-throughput proteomics? Researchers must weigh many factors when choosing a technology. The 2DE gel approach offers a relatively low set-up cost and remains highly accessible to all researchers ranging from junior researchers through to established laboratories. The visual data format of these gels remains a distinct advantage, especially for groups without ready access to state-of-the-art MS. Therefore, laboratories establishing proteomics for the first time tend to start by investing in gel technology and progressing to higher-throughput approaches following these initial experiments. However, these reasons alone will not maintain gel technology forever. Despite this, there remain areas where 2DE gel technology remains necessary. No other approach is able to assay for proteolytic or physical cleavage or proteins, an effect that is often seen as a result of disease pathology (108, 114), or due to co- and post-translational modification. Visualization of other modifications of proteins, such as glycosylation and phosphorylation, is possible due to charge variation in IEF and has been assisted by specific, novel fluorescent dyes. In the future, methodological improvements in gel-based approaches, combined with the accessibility and visual power of the technology, should ensure the use of 2DE for many years to come.

Acknowledgements

This work was supported by the Australian Research Council and the University of Sydney, Australia.

5. REFERENCES

1. de Hoog CL & Mann M (2004) *Annu. Rev. Genomics Human Genet.* **5**, 267–293.
2. Reinders J & Sickmann A (2005) *Proteomics*, **5**, 4052–2061.
3. Veenstra TD (2003) *Adv. Protein Chem.* **65**, 161–194.
4. Dziembowski A & Seraphin B (2004) *FEBS Lett.* **556**, 1–6.
5. Gavin AC, Bosche M, Krause R, *et al.* (2002) *Nature*, **415**, 141–147.
6. Ho Y, Gruhler A, Heilbut A, *et al.* (2002) *Nature*, **415**, 180–183.
★★★ 7. Görg A, Weiss W & Dunn MJ (2004) *Proteomics*, **4**, 3665–3685. – *Comprehensive review of 2DE technology of proteomics analysis.*
8. Anderson NL & Anderson NG (2002) *Mol. Cell. Proteomics*, **1**, 845–867.
9. Garbis S, Lubec G & Fountoulakis M (2005) *J. Chromatogr. A*, **1077**, 1–18.
10. Rabilloud T (2002) *Proteomics*, **2**, 3–10.
11. Lin D, Tabb DL & Yates JR III (2003) *Biochim. Biophys. Acta*, **1646**, 1–10.
12. Ong SE & Mann M (2005) *Nat. Chem. Biol.* **1**, 252–262.
13. Zhang H, Yan W & Aebersold R (2004) *Curr. Opin. Chem. Biol.* **8**, 66–75.
14. Cordwell SJ, Nouwens AS, Verrills NM, Basseal DJ & Walsh BJ (2000) *Electrophoresis*, **21**, 1094–1103.
15. Wildgruber R, Harder A, Obermaier C, *et al.* (2000) *Electrophoresis*, **21**, 2610–2616.
16. Westbrook JA, Yan JX, Wait R, Welson SY & Dunn MJ (2001) *Electrophoresis*, **22**, 2865–2871.
17. Patton WF (2002) *J. Chromatogr. B Analyt. Technol. Biomed. Life Sci.* **771**, 3–31.
18. Yates JR III, Gilchrist A, Howell KE & Bergeron JJ (2005) *Nat. Rev Mol. Cell. Biol.* **6**, 702–714.
19. Brunet S, Thibault P, Gagnon E, Kearney P, Bergeron JJ & Desjardins M (2003) *Trends Cell Biol.* **13**, 629–638.
20. Washburn MP, Wolters D & Yates JR III (2001) *Nat. Biotechnol.* **19**, 242–247.
★ 21. Gygi SP, Rist B, Gerber SA, Turecek F, Gelb MH & Aebersold R (1999) *Nat. Biotechnol.* **17**, 994–999. – *Original paper describing the ICAT method for LC-MS/MS.*
22. Ong SE, Blagoev B, Kratchmarova I, *et al.* (2002) *Mol. Cell. Proteomics*, **1**, 376–386.
★★ 23. Koller A, Washburn MP, Lange, BM, *et al.* (2002) *Proc. Natl. Acad. Sci. U.S.A.* **99**, 11969–11974. – *Excellent comparison between 2DLC-MS/MS and 2DE datasets for plant proteins.*
24. Herbert B (1999) *Electrophoresis*, **20**, 660–663.
25. Rabilloud T (1998) *Electrophoresis*, **19**, 758–760.
26. Molloy MP, Herbert BR, Slade MB, *et al.* (2000) *Eur. J. Biochem.* **267**, 2871–2881.
27. Molloy MP, Herbert BR, Walsh BJ, *et al.* (1998) *Electrophoresis*, **19**, 837–844.
★★ 28. Righetti PG, Castagna A, Antonioli P & Boschetti E (2005) *Electrophoresis*, **26**, 297–319. – *Comprehensive review of pre-fractionation strategies used prior to 2DE.*
29. Cordwell SJ (2006) *Curr. Opin. Microbiol.* **9**, 320–329.
30. Schirmer EC & Gerace L (2002) *Genome Biol.* **3**, 1008.1–1008.4.
31. McDonald TG & Van Eyk JE (2003) *Basic Res. Cardiol.* **98**, 219–227.
32. Fountoulakis M & Schlaeger EJ (2003) *Electrophoresis*, **24**, 260–275.
33. Taylor SW, Fahy E, Zhang B, *et al.* (2003) *Nat. Biotechnol.* **21**, 281–286.

34. Molloy MP (2000) *Anal. Biochem.* **280**, 1–10.
35. Nouwens AS, Cordwell SJ, Larsen MR, *et al.* (2000) *Electrophoresis*, **21**, 3797–3809.
36. Ferro M, Salvi D, Brugiere S, *et al.* (2003) *Mol. Cell. Proteomics*, **2**, 325–345.
37. Navarre C, Degand H, Bennett KL, Crawford JS, Mørtz E & Boutry M (2002) *Proteomics*, **2**, 1706–1714.
38. Zischka H, Weber G, Weber PJA, *et al.* (2003) *Proteomics*, **3**, 906–916.
39. Wildgruber R, Harder A, Obermaier C, *et al.* (2000) *Electrophoresis*, **21**, 2610–2616.
40. Davidsson P & Nilsson CL (1999) *Biochim. Biophys. Acta*, **1473**, 391–399.
41. Hochstrasser AC, James RW, Pometta D & Hochstrasser DF (1991) *Appl. Theor. Electrophor.* **1**, 333–337.
42. Shang TQ, Ginter JM, Johnston MV, Larsen BS & McEwen CN (2003) *Electrophoresis*, **24**, 2359–2368.
43. Corthals GL, Molloy MP, Herbert BR, Williams KL & Gooley AA (1997) *Electrophoresis*, **18**, 317–323.
44. Rothemund DL, Locke VL, Liew A, Thomas TM, Wasinger V & Rylatt DB (2003) *Proteomics*, **3**, 279–287.
45. Locke VL, Gibson TS, Thomas TM, Corthals GL & Rylatt DB (2002) *Proteomics*, **2**, 1254–1260.
46. Pang L, Fryksdale BG, Chow N, Wong DL, Gaertner AL & Miller BS (2003) *Electrophoresis*, **24**, 3484–3492.
47. Bae SH, Harris AG, Hains PG, *et al.* (2003) *Proteomics*, **3**, 569–579.
48. Righetti PG, Wenisch E, Jungbauer A, Katinger H & Faupel M (1990) *J. Chromatogr.* **500**, 681–696.
49. Righetti PG, Bossi A, Wenisch E & Orsini G (1997) *J. Chromatogr. B Biomed. Sci. Appl.* **699**, 105–115.
50. Herbert B & Righetti PG (2000) *Electrophoresis*, **21**, 3639–3648.
51. Zuo X & Speicher DW (2000) *Anal. Biochem.* **284**, 266–278.
52. Zuo X & Speicher DW (2002) *Proteomics*, **2**, 58–68.
53. Tang HY & Speicher DW (2005) *Expert Rev. Proteomics*, **2**, 295–306.
54. Görg A, Boguth G, Kopf A, Reil G, Parlar H & Weiss W (2002) *Proteomics*, **2**, 1652–1657.
55. Moritz RL, Ji H, Schutz F, *et al.* (2004) *Anal. Chem.* **76**, 4811–4824.
★ 56. Moritz RL & Simpson RJ (2005) *Nat. Methods* **2**, 863–873. – Excellent guide to protocols used for FFE separations.
57. Lee WC & Lee KH (2004) *Anal. Biochem.* **324**, 1–10.
58. Pieper R, Su Q, Gatlin CL, Huang ST, Anderson NL & Steiner S (2003) *Proteomics*, **3**, 422–432.
59. Steel LF, Trotter MG, Nakajima PB, Mattu TS, Gonye G & Block T (2003) *Mol. Cell. Proteomics*, **2**, 262–270.
60. Jungblut P, Baumeister H & Klose J (1996) *Electrophoresis*, **14**, 638–643.
61. Besant PG, Lasker MV, Bui CD, Tan E, Attwood PV & Turck CW (2004) *J. Proteome Res.* **3**, 120–125.
62. Lopez MF, Kristal BS, Chernokalskaya E, *et al.* (2000) *Electrophoresis*, **21**, 3427–3440.
63. O'Farrell PH (1975) *J. Biol. Chem.* **250**, 4007–4021.
64. Lopez MF (1999) *Methods Mol. Biol.* **112**, 111–127.
65. Choe LH & Lee KH (2003) *Electrophoresis*, **24**, 3500–3507.
66. Nishihara JC & Champion KM (2002) *Electrophoresis*, **23**, 2203–2215.
67. Tonge R, Shaw J, Middleton B, *et al.* (2001) *Proteomics*, **1**, 377–396.
68. Yan JX, Devenish AT, Wait R, Stone T, Lewis S & Fowler S (2002) *Proteomics*, **2**, 1682–1698.
69. Karp NA, Kreil DP & Lilley KS (2004) *Proteomics*, **4**, 1421–1432.
★ 70. Lilley KS & Friedman DB (2004) *Expert Rev. Proteomics*, **1**, 401–409. – Review of 2D-DIGE technology for comparative proteomics.
71. Candiano G, Bruschi M, Musante L, *et al.* (2004) *Electrophoresis*, **25**, 1327–1335.
72. Rabilloud T, Vuillard L, Gilly C & Lawrence JJ (1994) *Cell. Mol. Biol.* **40**, 57–75.
73. Shevchenko A, Wilm M, Vorm O & Mann M (1996) *Anal. Chem.* **68**, 850–858.
74. Lopez MF, Berggren K, Chernokalskaya E, Lazarev A, Robinson M & Patton WF (2000) *Electrophoresis*, **21**, 3673–3683.

75. Mackintosh JA, Choi HY, Bae SH, *et al.* (2003) *Proteomics*, **3**, 2273-2288.
76. Cordwell SJ (2002) *Methods Enzymol.* **358**, 207-227.
★ 77. Bernhardt J, Buttner K, Scharf C & Hecker M (1999) *Electrophoresis*, **20**, 2225-2240. – *Paper describing a dual-channel color imaging method to identify protein abundance versus protein expression in response to changes in environmental conditions.*
78. Voigt B, Schweder T & Sibbald MJ, *et al.* (2006) *Proteomics*, **6**, 268-281.
79. Hoper D, Bernhardt J & Hecker M (2006) *Proteomics*, **6**, 1550-1562.
80. Mann M, Ong SE, Gronborg M, Steen H, Jensen ON & Pandey A (2002) *Trends Biotechnol.* **20**, 261-268.
81. Kaufmann H, Bailey JE & Fussenegger M (2001) *Proteomics*, **1**, 194-199.
82. Ficarro SB, McCleland ML, Stukenberg PT, *et al.* (2002) *Nat. Biotechnol.* **20**, 301-305.
83. Li S & Dass C (1999) *Anal. Biochem.* **270**, 9-14.
84. Moser K & White FM (2006) *J. Proteome Res.* **5**, 98-104.
85. Dubrovska A & Souchelnytskyi S (2005) *Proteomics*, **5**, 4678-4683.
★ 86. Pinkse MW, Uitto PM, Hilhorst MJ, Ooms B & Heck AJ (2004) *Anal. Chem.* **76**, 3935-3943. – *Analytical method for the capture and identification of phosphorylated peptides by titanium dioxide.*
87. Larsen MR, Thingholm T, Jensen ON, Roepstorff P & Jorgensen TJ (2005) *Mol. Cell. Proteomics*, **4**, 873-886.
88. Steinberg TH, Agnew BJ, Gee KR, *et al.* (2003) *Proteomics*, **3**, 1128-1144.
★★★ 89. Schulenberg B, Aggeler R, Beechem JM, Capaldi RA & Patton WF (2003) *J. Biol. Chem.* **278**, 27251-27255. – *Description of a fluorescent dye specific for phosphorylated proteins.*
90. Geng M, Zhang X, Bina M & Regnier F (2001) *J. Chromatogr. B Biomed. Sci. Appl.* **752**, 293-306.
91. Young NM, Brisson JR, Kelly J, *et al.* (2002) *J. Biol. Chem.* **277**, 42530-42539.
92. Ghosh D, Krokhin O, Antonovici M, *et al.* (2004) *J. Proteome Res.* **3**, 841-850.
93. Larsen MR, Højrup P & Roepstorff P (2005) *Mol. Cell. Proteomics*, **4**, 107-119.
★★ 94. Kuster B, Krogh TN, Mørtz E & Harvey DJ (2001) *Proteomics*, **1**, 350-361. – *Excellent review of strategies for glycoprotein analysis following 2DE separation.*
95. Fryksdale BG, Jedrzejewski PT, Wong DL, Gaertner AL & Miller BS (2002) *Electrophoresis*, **23**, 2184-2193.
96. Hart C, Schulenberg B, Steinberg TH, Leung WY & Patton WF (2003) *Electrophoresis*, **24**, 588-598.
97. Schulenberg B, Beechem JM & Patton WF (2003) *J. Chromatogr. B Biomed. Sci. Appl.* **793**, 127-139.
98. Görg A, Obermaier C, Boguth G, Csordas A, Diaz JJ & Madjar JJ (1997) *Electrophoresis*, **18**, 328-337.
99. O'Farrell PZ, Goodman HM & O'Farrell PH (1977) *Cell*, **12**, 1133-1141.
100. Cordwell SJ, Basseal DJ, Bjellqvist B, Shaw DC & Humphery-Smith I (1997) *Electrophoresis*, **18**, 1393-1398.
101. Olsson I, Larrson K, Palmgren R & Bjellqvist B (2002) *Proteomics*, **2**, 1630-1632.
102. Chiari M, Micheletti C, Nesi M, Fazio M & Righetti PG (1994) *Electrophoresis*, **15**, 177-186.
103. Hoving S, Gerrits B, Voshol H, Müller D, Roberts RC & van Oostrum J (2002) *Proteomics*, **2**, 127-134.
★ 104. Bjorhall K, Miliotis T & Davidsson P (2005) *Proteomics*, **5**, 307-317. – *Comparison of depletion strategies for the removal of abundant proteins from serum and plasma.*
105. Cho SY, Lee EY, Lee JS, *et al.* (2005) *Proteomics*, **5**, 3386-3396.
106. Yocum AK, Yu K, Oe T & Blair IA (2005) *J. Proteome Res.* **4**, 1722-1731.
★ 107. Granger J, Siddiqui J, Copeland S & Remick D (2005) *Proteomics*, **5**, 4713-4718. – *Paper showing that cytokines and other low-abundance proteins bind to serum albumin and are removed alongside this protein by immunodepletion.*
108. White MY, Cordwell SJ, McCarron HC, Tchen AS, Hambly BD & Jeremy RW (2003) *J. Mol. Cell Cardiol.* **35**, 833-840.
109. Candiano G, Musante L, Bruschi M, *et al.* (2002) *Electrophoresis*, **23**, 292-297.
110. Bruschi M, Musante L, Candiano G, *et al.* (2003) *Proteomics*, **3**, 821-825.

111. Oh-Ishi M & Maeda T (2002) *J. Chromatogr. B Analyt. Technol. Biomed. Life Sci.* **771**, 49–66.
112. Schagger H & von Jagow G (1987) *Anal. Biochem.* **166**, 368–379.
113. Simpson RJ, Connolly LM, Eddes JS, Pereira JJ, Moritz RL & Reid GE (2000) *Electrophoresis*, **21**, 1707–1732.
114. Djordjevic SP, Cordwell SJ, Djordjevic MA, Wilton J & Minion FC (2004) *Infect. Immun.* **72**, 2791–2802.

CHAPTER 4

Peptide sorting by reverse-phase diagonal chromatography

Kris Gevaert and Joël Vandekerckhove

1. INTRODUCTION

A common route to proteome analysis is to separate proteins by two-dimensional (2D) polyacrylamide gel electrophoresis. Following comparison of 2D patterns from different proteome states, spots appearing at different positions or intensities suggest altered protein modification, degradation, or synthesis (1). Protein spots of interest are excised, in-gel digested, and the generated peptide mixture is analyzed by matrix-assisted laser desorption ionization (MALDI) mass spectrometry (MS) peptide mass fingerprinting or liquid chromatography (LC)-coupled electrospray ionization (ESI)-based tandem MS (MS/MS). Whilst this approach has proved to be successful in finding a proteomic answer for explaining several phenotypes and complex biological processes (e.g. 2), it tends to visualize only highly expressed and soluble proteins (3). This implies that low-copy-number proteins and hydrophobic, integral membrane proteins are not readily amenable to 2D-gel-based analyses.

To counterweigh these intrinsic 2D-gel problems, so-called gel-free, non-gel, or, more precisely, peptide-centric proteome analytical techniques have arisen. Here, instead of considering the protein as the central information, peptides are considered the primary information carriers. This shift was made possible by technical advances in multi-dimensional LC, as well as MS. It was further fueled by genomics, which has provided tools for rapid genome sequencing, thereby loading databases with proteomic blueprints.

Almost all peptide-centric approaches start by digesting an isolated proteome using either specific or nonspecific proteases (4). Although the resulting peptides are generally more soluble than their precursor proteins, the overall complexity of the analyte mixture is considerably increased, as a tryptic digest generates on average 20 different analyzable peptides for each protein. As present-day analytical instruments are incapable of probing such a complicated peptide mixture, two different approaches were introduced to reduce the analyte complexity to a more manageable level.

Proteomics: *Methods Express* (C.D. O'Connor and B.D. Hames, eds)
© Scion Publishing, 2008

The first approach combines different, so-called orthogonal, chromatographic separations that split up a peptide mixture to a higher degree than reverse-phase (RP) chromatography alone. Conveniently, strong cation exchange and RP peptide separations are combined (5). Although the number of proteins identified is remarkably high, so is the flux of peptides to the mass spectrometer, resulting in an 'undersampling' phenomenon whereby peptide ions are often randomly selected for MS/MS analysis. The net result is that a given complex set of peptides needs several rounds of analysis before acceptable proteome coverage is reached (6) and, equally important, there is low sample-to-sample reproducibility, which hinders routine proteome analysis for the discovery or monitoring of biomarkers.

The second type of gel-free approach attempts to reduce the complexity of the peptide mixture by isolating characteristic sets of peptides and only using these for MS/MS analysis. Such characteristic peptides typically contain a sufficiently rare amino acid (in order to reduce the complexity significantly) but with this being well distributed (in order to cover all proteins). What one could already now call archetypal examples of this approach are the isotope-coded affinity tag (ICAT) (7) molecule for isolation of cysteinyl peptides and the use of immobilized metal ion affinity chromatography (IMAC; e.g. using Fe^{3+}) for the isolation of phosphorylated peptides (8).

One disadvantage of these approaches is that they are predestined to isolate only one type of peptide, which might not always be the most appropriate choice for analyzing a given proteome. For example, about 15% of all *Escherichia coli* proteins cannot be identified or quantified by ICAT, as they do not contain any cysteines (9). This implies that either additional technologies need to be implemented or drastic modifications of existing peptide isolation strategies and/or instrumentation are needed (10). In view of this, we have recently created an assembly of different gel-free procedures, all centered around only one separation technique: diagonal RP chromatography (9, 11). As described in this chapter, the general concept is specifically to alter the column retention behavior of selected peptides between two identical chromatographic runs such that a particular set of characteristic peptides is sorted. By simply changing the sorting reaction, different sets of peptides are isolated. This eventually allows global analysis of protein expression and changes thereof, including protein processing and protein phosphorylation.

2. METHODS AND APPROACHES

2.1 Principles of combined fractional diagonal chromatography

The general principle of our gel-free proteomics approach is based on the technique of diagonal paper electrophoresis described by Brown & Hartley (12). In this original publication, a paper strip in which peptides were separated in an electrical field was exposed to vapors of performic acid, which converts cysteine to cysteic acid, thus introducing an additional negative charge on cysteinyl peptides. When this strip was turned 90° and the peptides were re-separated using the same buffer and electrical field, non-cysteinyl peptides migrated on a diagonal line whereas cysteinyl peptides migrated outside this diagonal and were thus specifically visualized; hence the name, diagonal electrophoresis. Later on, similar approaches were introduced for the isolation of methionyl peptides (13), peptides carrying free amino groups (14), histidinyl peptides (15), and tyrosinyl peptides (16).

Primarily because of the eletrophoretic/chromatographic platform (at the time, mainly paper strips and large-diameter chromatographic columns), such techniques were not directly amenable to isolating peptide sets from whole-proteome digests for further MS/MS analysis. Exploiting both the reproducibility and the high-throughput of contemporary micro-scale RP high-performance liquid chromatography (HPLC), we adapted diagonal chromatography for peptide-centric proteomics. As will be discussed in detail in section 2.2, our technology starts by digesting an isolated protein mixture with a specific protease (trypsin). The complex peptide mixture – often containing tens of thousands of different peptides – is then fractionated first by RP-HPLC: this is known as the primary combined fractional diagonal chromatography (COFRADIC) run. In each primary fraction, selected peptides (see below) are either chemically or enzymatically modified such that their column-retention characteristics are considerably altered. When such an altered primary fraction is re-separated on the same column and under identical chromatographic conditions (solvent gradient, column temperature, etc.), altered peptides undergo a relatively predictable chromatographic shift, as they are either more weakly (hydrophilic shift) or more strongly (hydrophobic shift) retained on the chromatographic resin. Thus, during this secondary COFRADIC run, peptides of interest are separated from the bulk of unaltered peptides and are collected based on their changed column-retention time.

According to this procedure, each altered primary fraction must thus be separated a second time. Therefore, the total HPLC time is directly proportional to the number of primary fractions and will grow accordingly. As the extents of the chromatographic shifts are predictable, several altered primary fractions may be combined per secondary chromatographic separation, thereby decreasing the HPLC analysis time and increasing the sample throughput (see below). As our technology closely resembles the techniques of diagonal electrophoresis or diagonal chromatography (selected, altered peptides elute at different positions during the two separation steps) and as different primary fractions may be combined, we termed this technology combined fractional diagonal chromatography, or COFRADIC (17).

In its essence, COFRADIC is a combination of three consecutive steps: an initial separation of peptides, a chemical or enzymatic modification of selected peptides, and finally a separation of (combined) primary peptide fractions. As such (and at least in theory), every class of peptide carrying an amino acid or, more generally, a functional group that can be specifically modified, can be sorted by COFRADIC. Currently, COFRADIC sorting procedures enable routine isolation of methionyl, cysteinyl, N-terminal, and phosphorylated peptides. These procedures are principally different both in the way that proteins are prepared prior to the actual COFRADIC sorting step and in the chemicals or enzymes that are applied between the two chromatographic separations. *Fig. 1* provides an overview of the sorting schemes for COFRADIC. For instance, both methionyl and cysteinyl peptides are

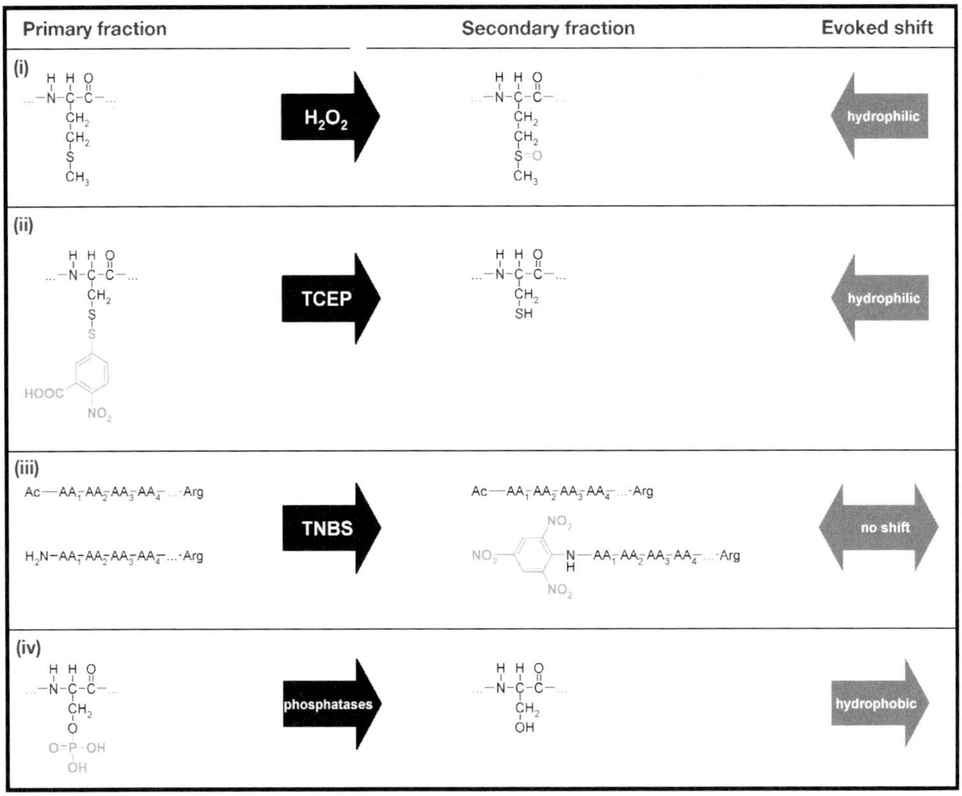

Figure 1. Overview of the central sorting schemes used in COFRADIC.
Currently, four different types of peptide can be sorted routinely by COFRADIC. (i) Central to the sorting of methionyl peptides is oxidation with H_2O_2, leading to methionine sulfoxide. Peptides carrying such amino acids display a shift towards earlier retention times on an RP column (hydrophilic shift) and are hence sorted. (ii) Cysteine groups are first modified by Ellman's reagent, which is removed between the two COFRADIC separations by reduction with TCEP. Here again, peptides display a hydrophilic shift. (iii) For the isolation of N-terminal peptides, all free protein thiol and amine groups are blocked by alkylation and acetylation, respectively. Following trypsin digestion and primary separation, two types of peptide are present: N-terminal acetylated peptides and internal peptides carrying a free α-amine. The latter react with trinitrobenzene sulfonic acid (TNBS) resulting in highly hydrophobic trinitrophenyl peptides that are shifted out of the primary collection interval. In this scenario, the peptides of interest (the N-terminal ones) are not sorted; instead, the bulk of (internal) unwanted peptides is sorted. (iv) Sorting of phosphorylated peptides is based on an enzymatic dephosphorylation reaction using a cocktail of nonspecific phosphatases and the resulting 'ex-phosphorylated' peptides undergo a hydrophobic shift.

sorted following a hydrophilic shift evoked by an oxidation (hydrogen peroxide) or a reduction (phosphines) reaction, respectively. On the other hand, phosphorylated peptides are sorted in their 'ex-phosphorylated' form following a hydrophobic shift introduced after dephosphorylation, and N-terminal peptides are sorted after modifying all peptides with a free α-amine.

In addition to monitoring the nature of the proteins present in a proteome, one of the major tasks of proteomics is discovering proteins whose concentration or modification status (or both) differs between two proteomic states. In peptide-centric proteome studies, most conveniently stable heavy isotopes are introduced in one (or several) sample(s) either metabolically (18, 19) or post-metabolically (e.g. 20, 21). These heavier isotopes are chosen such that (almost) all peptides obtained from one proteome are labeled ('heavy peptides') and when mixed with an unlabeled peptide mixture (i.e. 'light peptides' reflecting the natural distribution of isotopes), the light and heavy isotope envelopes are discernible and can be weighed individually. The ratio of these weights relates to the individual abundances of the peptides in the two proteome digests and, following a statistical assessment of the observed ratio values, a confidence interval can be calculated that identifies peptide ions whose concentration is significantly different in one of the proteome digests (22). When multiple peptides of one protein are weighed, a more direct transformation of the calculated peptide ratio values to protein abundances can be made, leading to the assignment of proteins whose concentration differs significantly in a proteome.

Several isotope-labeling strategies have been described (for recent reviews, see 23, 24). However, we sought to introduce a general approach that labels peptides irrespective of the nature of the protein mixture, whether derived from culture-grown cells (allowing metabolic labeling) or body fluids and biopsies (almost always restricted to post-metabolic labeling). To do this, we used the fact that trypsin exchanges the oxygen atoms of the C-terminal carboxyl group of tryptic peptides during proteolysis (25–27). In our approach (see section 2.2), two ^{18}O isotopes are incorporated at the C terminus of each of the peptides following the proteome digestion step (28). We have found this procedure to be quantitative, and have not seen any difference in ^{18}O incorporation into arginine-ending compared with lysine-ending peptides. Furthermore, following inactivation of trypsin, the isotope label remains stably fixed to the peptides. As this labeling strategy is compatible with every COFRADIC sorting procedure, it is our current method of choice for differential gel-free proteomics.

2.2 General applications of COFRADIC for gel-free proteomics

Although the exact conditions to be used are largely dependent on the nature of the samples and the biochemical events one wants to analyze, here we present some general guidelines on which COFRADIC technology may be preferentially suited to a given proteome analysis. In particular, we emphasize the differences between a more general COFRADIC approach (sorting of methionyl or cysteinyl peptides) and a more specific approach (sorting of N-terminal peptides).

When the aim is to identify the proteins in a given proteome ('telephone book proteomics'), we recommend isolating methionyl peptides, both because of the simplicity of the procedure (see below) and because in many cases several sorted peptides may be linked to proteins, thereby avoiding very high numbers of 'one-hit wonders' (29), which are often considered untrustworthy. On the other hand, one needs to realize that the expected drop in sample complexity by sorting methionyl peptides is a factor of about five, as, on average, about one-fifth of all analyzable tryptic peptides carry at least one methionine (9). When confronted with highly complex proteomes – in general, proteomes from higher eukaryotes – this complexity reduction might be too low, leading to significant undersampling effects. However, as every protein contains only one N-terminus, sorting of N-terminal peptides leads to the highest possible reduction of complexity and thus to a higher proteome coverage. Furthermore, given the importance of protein processing both in health and disease (e.g. 30), sorted protein N-termini hint at processing events and to the actual processing sites, as novel N-terminal peptides will be identified in a proteome digest. Thus, both for reasons of complexity reduction and immediate access to protein processing, we recommend using sorted N-terminal peptides. In fact, when combined with isotope labeling, we recently showed that this particular COFRADIC procedure allows a global analysis of *in vivo* protease substrates (31) and thus needs to be considered as an important technique for visualizing protein processing events more generally.

2.3 Methodology

This section describes the methods used for COFRADIC in some detail. However, before doing so, we would like to stress that the analytical equipment (HPLC instrumentation and mass spectrometer) used in our laboratory and described in the protocols is not the only equipment that can be used; instruments from other companies with comparable analytical performance work equally well. Furthermore, we would also like to point out that, next to ESI-based analyses, MALDI-based analyses may also be used to assess the identity of sorted peptides. In fact, a method involving the exhaustive use of MALDI time-of-flight (TOF)/TOF-MS for identifying sorted methionyl peptides has been developed in our laboratory and was applied to the identification of over 2000 proteins in human multipotent adult progenitor cells (32). MALDI-MS analyses have several advantages compared with LC-ESI-MS/MS; samples are archived onto a support and, when necessary, may be reassessed several times and, at least in our hands, the sample-to-sample reproducibility is much higher compared with ESI-based analyses (on average, more than 80% of peptide ions with a signal : noise ratio of greater than 60 are present in different but analogous samples).

Protocol 1 describes how to set up an RP diagonal chromatographic system (RP-HPLC) for peptide sorting.

Protocol 1

Setting up an RP diagonal chromatographic system for peptide sorting

Equipment and Reagents
- Analytical RP-HPLC column (2.1 mm internal diameter × 150 mm length 300SB-C18 column, Zorbax; Agilent)
- Agilent 1100 Series HPLC system
- HPLC-grade water (e.g. Baker HPLC-analyzed; Mallinckrodt Baker)
- HPLC-grade acetonitrile (e.g. Baker HPLC-analyzed; Mallinckrodt Baker)
- Trifluoroacetic acid (TFA; Rathburn)[a]
- HPLC solvent A (10 mM ammonium acetate, pH 5.5, or 0.1% TFA in water : acetonitrile, 98 : 2, v/v)[b]
- HPLC solvent B (10 mM ammonium acetate, pH 5.5, or 0.1% TFA in water : acetonitrile, 30 : 70, v/v)[b]

Method
1. Inject the peptide sample onto the HPLC column.
2. Apply a 10 min isocratic run with 100% solvent A at a constant flow rate of 80 µl/min.
3. Apply a linear, binary gradient over 100 min from 0% solvent B (i.e. 100% solvent A) to 100% solvent B, at a constant flow rate of 80 µl/min.
4. Apply a 10 min isocratic wash with 100% solvent B, followed by a linear, binary gradient over 5 min to 0% solvent B (100% solvent A).
5. Equilibrate the column for another 20 min with 100% solvent A (at 80 µl/min) before injection of another sample.
6. Depending on the type of peptide isolated (see *Protocols 3–6*) and thus the preceding protein preparation steps, peptides typically elute at between 20 and 100 min of gradient time, corresponding to acetonitrile concentrations of 7 and 63%, respectively. Collect the primary fractions as indicated in *Protocols 3–7*, starting at the pre-defined collection starting point.
7. Proceed with the collected primary fractions as indicated in *Protocols 3–6*.

Notes
[a]TFA-containing solutions should be made on the day of use. Extreme care should be taken when using TFA, as it is an extremely hazardous chemical, and the concentrated acid should only be used in a fume hood and dispensed wearing suitable protective clothing. Consult the safety datasheet for handling details and disposal.

[b]The choice of ammonium acetate or TFA as the pairing ion in the HPLC buffers generally depends on two factors. If a differential proteome analysis with peptides labeled with ^{18}O isotopes (see *Protocol 2*) will be performed, then TFA cannot be used as it can lead to acid-catalyzed exchange of oxygen atoms in carboxyl groups (28). Furthermore, as the extent of either hydrophilic or hydrophobic shifts depends on the composition of the HPLC buffer (17), for cataloging proteomes (thus not for differential, ^{18}O-based proteomics) we suggest the use of TFA when sorting methionyl, cysteinyl, or N-terminal peptides, and ammonium acetate when sorting phosphopeptides (see below).

In the overall set-up of COFRADIC, the reproducibility of peptide separations is the 'Achilles' heel' of the technique. Adequate HPLC instrumentation is available nowadays, creating highly reproducible solvent gradients and thus equally reproducible peptide separations. In the system that we use on a daily basis, for instance, we use Agilent's electronic flow controller for maintaining a constant solvent flow through the column independent of the overall back pressure and we tend to thermo-control as many parts of the system as possible (e.g. the column compartment, as well as the tubing delivering the solvent to the column and the fraction collector). By taking care of these issues, we generally observe a standard deviation of only a few seconds on the retention time of peptides in a complex peptide mixture over a gradient of nearly 2 h.

As indicated in *Protocol 1*, the interval during which peptides elute from the column and should thus be collected depends on a number of factors, such as the preparation procedure of the proteins prior to digestion (e.g. when isolating N-terminal peptides, the average hydrophobicity of the peptide mixture will be higher compared with isolation of methionyl peptides), the type of pairing ion in the HPLC buffers (e.g. in ammonium acetate systems, peptides tend to elute at lower concentrations of organic solvent than in TFA systems), and the type and dimensions of the RP column used. Therefore, we recommend pre-defining the starting point of peptide collection prior to every primary COFRADIC run. The most direct way of doing this is by fractionating a small part of the available peptide material onto the column and determining when the most hydrophilic peptides elute. Therefore, the collection start and end points given here and below are those that are best suited for the analytical set-up used in our laboratory and should thus be modified when using any other system.

Protocol 2 describes the labeling of peptides with two ^{18}O isotopes using trypsin. As this is applicable to all COFRADIC protocols, it is described here first and referred to in subsequent protocols and where relevant.

Protocol 2

Peptide labeling with ^{18}O atoms

Equipment and Reagents
- 0.1 M KH_2PO_4 (pH 4.5)
- ^{18}O-rich water (93.7% $H_2^{18}O$, w/w, pure; ARC Laboratories)
- 10 mM Tris(2-carboxyethyl)phosphine (TCEP; Pierce)
- 100 mM Iodoacetamide (Fluka) in 2 M guanidinium hydrochloride

Method
1. Following digestion (see *Protocols 3–6*) in 50 mM ammonium bicarbonate (pH 7.6), dry the peptide mixture completely under vacuum.

2. Redissolve the peptides in 25 µl of 0.1 M KH_2PO_4 (pH 4.5) and dry again.

3. Add 100 µl of ^{18}O-rich water ('heavy' peptides) or 100 µl of natural water ('light' ^{16}O-labeled peptides) and incubate overnight at 37°C.

4. Transfer 10 µl of 10 mM TCEP solution to an Eppendorf tube and dry.

5. Transfer 10 µl of 100 mM iodoacetamide solution in 2 M guanidinium hydrochloride to a second Eppendorf tube and dry.
6. Transfer the peptide mixture to the TCEP tube, mix thoroughly, and incubate at 37°C for 1 h.
7. Transfer the reduced peptide mixture to the iodoacetamide tube and incubate again for 1 h at 37°C, this time in the dark.
8. Store at –20°C until further analysis.

Protocol 3 describes the methodology for sorting methionyl peptides and a typical example of this type of analysis is shown in *Fig. 2*.

Protocol 3

Sorting of methionyl peptides

Equipment and Reagents
- Suitable RP column
- 50 ml Falcon tube
- Phosphate-buffered saline (PBS)
- 3-((3-Cholamidopropyl)dimethylammonio)-1-propanesulfonate (CHAPS; Sigma-Aldrich)
- 50 mM Tris/HCl (pH 8.0)
- Lysis buffer (0.7%, w/v, CHAPS and 2 M guanidinium hydrochloride (Fluka) in PBS containing protease inhibitors (added using Complete Protease Inhibitor Cocktail Tablets (Roche Applied Science) following the manufacturer's instructions)
- PD-10 desalting column (Amersham Biosciences)
- Sequencing-grade modified trypsin (Promega)
- TFA (Rathburn)[a]
- 30% (w/w) H_2O_2 (Sigma-Aldrich)

Method
1. Transfer the cells[b] to a 50 ml Falcon tube and wash three times with 10 ml of PBS. Between washes, centrifuge for 2 min at 2000 *g* and carefully resuspend the pellet in PBS.
2. Following the last washing step, lyse the cells in 2.5 ml of lysis buffer. Incubate for 5 min at room temperature and transfer to an ice bath for another 10 min.
3. Centrifuge the lysed cells for 10 min at 10 000 *g* to remove insoluble cellular debris.
4. Desalt the supernatant on a PD-10 column in 3.5 ml of 50 mM Tris/HCl (pH 8.0)[c].
5. Concentrate the protein mixture to 1 ml by vacuum drying in a centrifugal concentrator.
6. Boil the protein mixture for 10 min at 95°C and immediately transfer for 10 min to an ice bath.
7. Add 20 µg of trypsin (the enzyme : substrate ratio is about 1 : 50) and incubate overnight at 37°C.
8. Inactivate the trypsin digestion by lowering the pH to 2 by adding 10 µl of TFA[a,d].
9. Centrifuge the peptide mixture for 10 min at 10 000 *g* to remove any insoluble material. Transfer the supernatant to an HPLC sample vial.
10. Load the peptide mixture onto the RP column and fractionate into 60 fractions of 1 min each as described in *Protocol 1*[e]. Label the primary fractions 1–60.

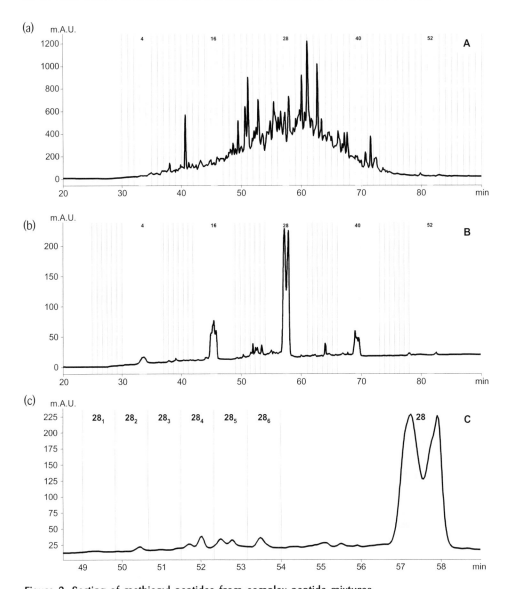

Figure 2. Sorting of methionyl peptides from complex peptide mixtures.
A tryptic digest of a proteome preparation from human multipotent adult progenitor cell cells was separated by RP-HPLC. From 30 min onwards, 60 fractions of 1 min each were collected (i.e. the primary COFRADIC run); the UV absorbance chromatogram (at 214 nm) is shown in (*a*). In a secondary COFRADIC run, five fractions separated by 12 min each (fractions 4, 16, 28, 40, and 52) indicated in the chromatogram shown in (*b*) were combined, dried, and oxidized with H_2O_2. This oxidation converted methionine to methionine sulfoxide and hence rendered methionyl peptides more hydrophilic. As the evoked shift is semi-predictable, during each secondary run, secondary fractions containing the methionyl sulfoxide peptides were collected in a time frame between 3 and 8 min prior to the collection start of the corresponding primary fraction. The chromatogram shown in (*c*) shows the collection of methionyl sulfoxide peptides (labeled 28_1 to 28_6) isolated from the oxidized primary fraction 28. Clearly, the majority of peptides in this fraction did not shift, as about 80% of these peptides did not contain any methionine residues (see shaded box, indicating the collection interval of the primary fraction). However, in the time interval during which the secondary fractions were collected, small traces of UV absorbance were observed from the sorted methionyl sulfoxide peptides. mAU, milli-absorbance unit.

11. Combine primary fractions that are separated by 12 min (e.g. fractions 1, 13, 25, 37, and 49) and dry them under vacuum.
12. Redissolve one pool of peptides in 100 µl of 1% TFA[a] and add 2 µl of 30% (w/w) H_2O_2 solution[f].
13. Incubate for 30 min at 30°C and immediately inject the peptide mixture onto the RP column for the second, identical COFRADIC separation.
14. For each primary fraction, collect methionine sulfoxide peptides at an interval of 3–8 min preceding the original, primary collection interval. Typically, for each primary fraction, sorted peptides are collected in six secondary fractions. For example, methionine sulfoxide peptides from primary fraction 1 are collected in six secondary fractions named 1_1 to 1_6. A typical example of COFRADIC sorting of methionyl peptides is shown in *Fig. 2*.
15. Repeat steps 12–14 for each pool of primary fractions.
16. For each secondary separation, pool identically indexed secondary fractions (e.g. 1_1, 13_1, 25_1, 37_1, and 49_1) and dry completely under vacuum. Store these pooled fractions at −20°C until further LC-MS/MS analysis (see *Protocol 7*).

Notes

[a]TFA-containing solutions should be made on the day of use. Extreme care should be taken when using TFA as it is an extremely hazardous chemical and the concentrated acid should only be used in a fume hood and dispensed wearing suitable protective clothing. Consult the safety datasheet for handling details and disposal.

[b]The required minimal amount of starting material using the HPLC set-up described in *Protocol 1* is about 1 mg. When working with cultured animal cells, depending upon the cell type, between 1×10^7 and 5×10^7 cells are needed.

[c]When peptides are to be labeled by ^{18}O (see *Protocol 2*), it is recommended that the Tris/HCl solution is replaced with freshly prepared 50 mM ammonium bicarbonate (pH 7.6).

[d]When peptides are to be labeled with ^{18}O isotopes, the digestion reaction must not be stopped by acidification. Instead, as indicated in *Protocol 2*, the peptide mixture must be dried under vacuum.

[e]If peptides are labeled with ^{18}O isotopes, ammonium acetate (pH 5.5) should be used in the HPLC buffers instead of TFA.

[f]For ^{18}O-labeled peptides, carry out the oxidation reaction in 100 µl of 10 mM ammonium acetate (pH 5.5) instead of TFA.

The sorting of cysteinyl peptides is described in *Protocol 4*.

Protocol 4

Sorting of cysteinyl peptides

Equipment and Reagents
- PD-10 desalting column (Amersham Biosciences)
- 50 mM TCEP (prepare just before use; Pierce)
- 10 mM 5,5′-dithiobis(2-nitrobenzoic acid (Ellman's reagent, prepare just before use; Fluka)
- 10 and 50 mM Tris/HCl (pH 8.7)
- Sequencing-grade modified trypsin (Promega)

- TFA (Rathburn)[a]
- 2 M Guanidinium hydrochloride in 50 mM Tris/HCl (pH 8.7)
- 0.3 M Guanidinium hydrochloride in 50 mM Tris/HCl (pH 8.7)
- 30% (w/w) H_2O_2 (Sigma)

Method

1. Prepare the cells and proteins as described in steps 1–3 of *Protocol 3*.

2. Desalt the protein mixture on a PD-10 column in 3.5 ml of 2 M guanidinium hydrochloride in 50 mM Tris/HCl (pH 8.7).

3. Dry the protein mixture under vacuum and reconstitute the pellet in 2.5 ml of a freshly prepared 50 mM TCEP solution. Let the reduction reaction proceed for 1 h at 37°C.

4. Desalt the reduced protein mixture on a PD-10 column in 3.5 ml of 2 M guanidinium hydrochloride in 50 mM Tris/HCl (pH 8.7).

5. Dry the protein mixture under vacuum and reconstitute the pellet in 2.5 ml of a freshly prepared 10 mM solution of Ellman's reagent. Let the modification reaction proceed for 1 h at 37°C.

6. Desalt the mixture of modified proteins on a PD-10 column in 3.5 ml of 0.3 M of guanidinium hydrochloride in 50 mM Tris/HCl (pH 8.0).

7. Reduce this volume to 1 ml by vacuum drying and digest the protein with trypsin as described in steps 7–9 of *Protocol 3*.

8. Add 20 µl of 30% H_2O_2 and incubate for 30 min at 30°C[b].

9. Load the sample on the RP column (see *Protocol 1*) for the primary COFRADIC separation and fractionate in 64 consecutive fractions of 1 min and label these 1–64.

10. Pool primary fractions that are separated by 16 min (e.g. fractions 1, 17, 33, and 49). Dry the pooled fractions in a vacuum concentrator.

11. Redissolve each pool of primary fractions in 70 µl of 10 mM Tris/HCl (pH 8.7) and add 30 µl of a freshly prepared 50 mM TCEP solution. Incubate for 1 h at 37°C and acidify by adding 2 µl of TFA[a,c].

12. Using one pool of reduced primary fractions at a time, load the peptides on the same column and use the same solvent gradient to fractionate the peptides (secondary COFRADIC run). Collect cysteinyl peptides in six equal-volume fractions in a time interval between 3 and 10 min prior to the original collection interval of each primary fraction. Label the secondary fractions as, for example, 1_1 to 1_6.

13. For each secondary COFRADIC separation, pool identically indexed secondary fractions (e.g. 1_1, 17_1, 33_1, and 49_1). Dry to complete dryness and store at −20°C until LC-MS/MS analysis (see *Protocol 7*).

Notes

[a]TFA-containing solutions should be made on the day of use. Extreme care should be taken when using TFA as it is an extremely hazardous chemical and the concentrated acid should only be used in a fume hood and dispensed wearing suitable protective clothing. Consult the safety datasheet for handling details and disposal.

[b]This makes the peptide mixture more uniform as it oxidizes methionines to their sulfoxides (see *Protocol 3*) and also prevents accidental shifts of methionyl peptides.

[c]When peptides are labeled with ^{18}O isotopes, TFA cannot be used as a pairing ion in HPLC buffers or to acidify the peptide mixture. Instead, add acetic acid to lower the pH to about 5 before injecting peptides onto the HPLC column.

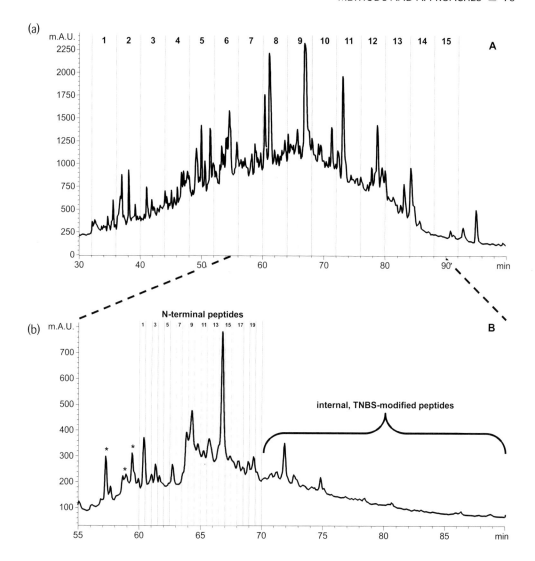

Figure 3. Sorting of N-terminal peptides from complex peptide mixtures.
A section of lung tissue from a patient suffering from chronic obstructive pulmonary disease was frozen in liquid nitrogen, pulverized in a mortar, and following acetone precipitation the proteome was extracted with 4 M guanidinium hydrochloride. This protein mixture was then treated as described in *Protocol 5*. The RP-HPLC chromatogram (UV absorbance at 214 nm) of the separation of the tryptic digest of this protein mixture (the primary COFRADIC run) is shown in (*a*). This peptide mixture was fractionated into 15 consecutive primary fractions of 4 min each. The RP-HPLC chromatogram of the secondary, identical COFRADIC run of primary fraction 9 after treatment with TNBS is shown in (*b*). Unaltered N-terminal peptides were collected in 20 secondary fractions over a 10 min time window starting 4 min prior to the original, primary elution interval of fraction 9 (shaded gray). TNBS-modified peptides (i.e. internal peptides that carried a free α-amine group) obtained a hydrophobic trinitrophenyl group and were thus shifted to later elution times. As indicated in (*b*), we chose to expand the secondary collection interval compared with the primary interval, as the capacity of the RP column was generally too low to resolve all peptides present in highly complex peptide mixtures such as a proteome digest, resulting in peptides apparently 'shifting' between the primary and secondary runs independent of the COFRADIC sorting chemistry (see also 33). Background peaks due to impurities in TNBS are indicated by asterisks. mAU, milli-absorbance unit.

N-terminal peptides can be analyzed using the procedure described in *Protocol 5*, and an example of COFRADIC sorting of N-terminal peptides is given in *Fig. 3*.

Protocol 5

Sorting of N-terminal peptides

Equipment and Reagents
- PD-10 desalting column (Amersham Biosciences)
- 2 M Guanidinium hydrochloride in 50 mM sodium phosphate (pH 7.5)
- 10 mM TCEP (Pierce)/100 mM iodoacetamide (Fluka) solution (prepare just before use)
- 1.4 M Guanidinium hydrochloride in 50 mM sodium phosphate (pH 8.0)
- 50 mM Tris/HCl (pH 8.0)
- 5 mM Sulfo-*N*-hydroxysuccinimide acetate (prepare just before use; Perbio)
- Hydroxylamine (Fluka)
- Sequencing-grade modified trypsin (Promega)
- 30% H_2O_2 (Sigma)
- 50 mM Sodium borate (pH 9.5)
- 15 mM 2,4,6-Trinitrobenzenesulfonic acid (TNBS, supplied as a 1 M solution in water, Fluka)
- TFA (Rathburn)[a]

Method
1. Prepare the cells and proteins as described in steps 1–3 of *Protocol 3*.
2. Desalt the protein mixture on a PD-10 column in 3.5 ml of 2 M guanidinium hydrochloride in 50 mM sodium phosphate (pH 7.5).
3. Vacuum dry the protein mixture and reconstitute the pellet in 2.5 ml of a freshly prepared 10 mM TCEP/100 mM iodoacetamide solution. Let the reduction/alkylation reaction proceed for 1 h at 37°C.
4. Desalt the protein mixture on a PD-10 column in 3.5 ml of 1.4 M guanidinium hydrochloride in 50 mM sodium phosphate (pH 8.0).
5. Vacuum dry the protein mixture and reconstitute the pellet in 2.5 ml of freshly prepared 5 mM sulfo-*N*-hydroxysuccinimide acetate. Incubate the protein mixture for 90 min at 30°C.
6. Revert the partial acetylation of tyrosine, serine, and threonine by adding 2 µl of hydroxylamine to the protein mix and incubate for 15 min at 30°C.
7. Desalt the mixture of modified proteins on a PD-10 column in 3.5 ml of 50 mM Tris/HCl (pH 8.0).
8. Reduce the volume to 1 ml by vacuum drying and digest the protein with trypsin as described in steps 6–9 of *Protocol 3*.
9. Add 20 µl of 30% H_2O_2 and incubate for 30 min at 30°C.
10. Load the sample on the RP column (see *Protocol 1*) for the primary COFRADIC separation and fractionate in 15 consecutive fractions of 4 min each.
11. Dry these primary fractions completely and redissolve each primary fraction in 50 µl of 50 mM sodium borate (pH 9.5).

12. Add 10 μl of a 15 mM TNBS solution and incubate for 1 h at 37°C.
13. Repeat the previous step three times to assure near-quantitative TNBS modification of free α-amino groups.
14. Acidify the modified primary fractions by adding 2 μl of TFA[a,b] and centrifuge the peptide mixtures for 10 min at 10 000 g to remove insoluble material. Transfer the supernatant to an HPLC sample vial.
15. Starting with the most hydrophobic primary fraction, load the TNBS-treated fraction onto the RP column and fractionate using the same solvent gradient as during the primary run. Collect the N-terminal peptides in 20 equal-volume secondary fractions in a 10 min time interval starting 4 min prior to the primary collection interval[c]. An example of COFRADIC sorting of N-terminal peptides is depicted in *Fig. 3*.
16. Dry the collected N-terminal peptides and store at −20°C until LC-MS/MS analysis (see *Protocol 7*).

Notes

[a]TFA-containing solutions should be made on the day of use. Extreme care should be taken when using TFA as it is an extremely hazardous chemical and the concentrated acid should only be used in a fume hood and dispensed wearing suitable protective clothing. Consult the safety datasheet for handling details and disposal.

[b]When applied in a differential ^{18}O set-up, as noted above, TFA must be swapped with acetic acid to lower the pH to about 5 before injection onto the column, which is here run in an ammonium acetate system.

[c]In theory, the acetylated N-terminal peptides should elute within exactly the same time interval as during the primary run. In practice, however, this secondary collection interval is stretched as LC is never absolutely reproducible (33) and, in particular, abundant N-terminal peptides tend to smear out over larger elution intervals. Therefore, peptides are collected both before (4 min) and after (2 min) the primary collection interval. Nevertheless, as both the number and the amount of peptides collected in these particular intervals tend to be lower than those collected within the expected elution window, such secondary fractions may be pooled (e.g. in twos) to reduce the number of LC-MS/MS analyses.

Protocol 6 provides the step-by-step methodology for sorting phosphorylated peptides. Note the deliberately longer time window for collecting the secondary fractions enriched for ex-phosphorylated peptides (step 17), which is necessary to maximize the recovery of these peptides.

Protocol 6

Sorting of phosphorylated peptides

Equipment and Reagents
- Lysis buffer for phosphoproteomics: 0.7% (w/v) CHAPS, 4 M guanidinium hydrochloride, 10 mM TCEP and 100 mM iodoacetamide in PBS supplemented with protease inhibitors (see *Protocol 3*) and phosphatase inhibitors (10 mM NaF, 200 µM sodium orthovanadate, 20 mM β-glycerophosphate, 5 µM phenylvalerate, and 2 mM levamisole hydrochloride)
- Desalting solution (2 M of a fresh urea solution in 100 mM Tris/HCl, pH 8.7, containing 10 mM NaF, 200 µM sodium orthovanadate, 20 mM β-glycerophosphate, 5 µM phenylvalerate, and 2 mM levamisole hydrochloride)
- 1 M HCl
- PHOS-Select iron affinity gel (Sigma)
- 10 mM and 0.25 M Acetic acid
- 0.4 M NH$_4$OH
- Calf intestinal alkaline phosphatase (CIP) and reaction buffer (New England Biolabs)
- Lambda protein phosphatase (Upstate)
- *E. coli* alkaline phosphatase (Sigma-Aldrich)
- 0.1 M KH$_2$PO$_4$
- Sequencing-grade modified trypsin (Promega)

Method

1. Prepare the cells as described in steps 1–3 of *Protocol 3* using the lysis buffer for phosphoproteomics.
2. Desalt the protein solution in 3.5 ml of desalting solution.
3. Add 20 µg of trypsin to the protein solution and incubate overnight at 37°C.
4. Centrifuge the generated peptide solution for 10 min at 10 000 g to remove insoluble material and transfer the supernatant to a new tube.
5. Adjust the pH of the peptide solution to 3 using 1 M HCl and add 50 µl of washed PHOS-Select iron affinity gel slurry (50 µl of gel slurry is expected to bind 50 nmol of phosphopeptides).
6. Incubate overnight at 4°C in a rotating device.
7. Centrifuge for 10 min at 10 000 g and wash with 2 ml of 0.25 M acetic acid. Repeat this step three times.
8. Elute the phosphopeptides by incubating the gel slurry for 1 h at room temperature with 2 ml of 0.4 M NH$_4$OH.
9. Split the sample into two equal parts and dry the peptide mixture under vacuum.
10. Redissolve one sample (the control sample) in 50 µl of CIP reaction buffer and add 10 units of CIP, 200 units of lambda protein phosphatase, and 1 unit of *E. coli* alkaline phosphatase[a]. Let the dephosphorylation reaction proceed for 1 h at 37°C.
11. Redissolve the second sample (this will be the sample containing the phosphopeptides) in 50 µl of CIP reaction buffer and incubate for 1 h at 37°C.
12. For both samples, lower the pH to 5 by adding 80 µl of 0.1 M KH$_2$PO$_4$, add 4 µg of trypsin, and dry completely.

13. Label the peptides from the control sample with two ^{18}O isotopes and those from the phosphopeptide sample with two ^{16}O isotopes as described in steps 2–7 of *Protocol 2*.

14. Mix the two peptide mixtures together and separate first by RP-HPLC as described in *Protocol 1*[b]. Collect 64 consecutive primary fractions of 1 min each and label these 1–64.

15. Pool primary fractions that are separated by 16 min (e.g. fractions 1, 17, 33, and 49) and dry these in a vacuum concentrator.

16. Redissolve each pool of primary fractions in 50 µl of CIP reaction buffer to which 10 units of CIP, 200 units of lambda protein phosphatase and 1 unit of *E. coli* alkaline phosphatase have been added. Allow dephosphorylation to proceed for 1 h at 37°C and terminate the reaction by adding 50 µl of 10 mM acetic acid.

17. Using one pool of dephosphorylated primary fractions at a time, load the peptides on the same column and use the same solvent gradient to fractionate the peptides (secondary COFRADIC run). Collect ex-phosphorylated (or dephosphorylated) peptides in six equal-volume fractions at a time interval between 1 and 15 min after the original collection interval of each primary fraction[c]. Label secondary fractions as, for example, 1_1 to 1_6.

18. For each secondary COFRADIC separation, pool identically indexed secondary fractions (e.g. 1_1, 17_1, 33_1, and 49_1). Dry completely and store at −20°C until LC-MS/MS analysis (see *Protocol 7*).

Notes

[a] The definition of enzymatic units is according to the information supplied by the corresponding manufacturers.

[b] For the sorting of dephosphorylated peptides, we recommend using 10 mM ammonium acetate (pH 5.5) as the pairing ion, as this gives a significantly larger shift compared with TFA (34). Furthermore, ammonium acetate is completely compatible with ^{18}O labeling as used in this protocol.

[c] The time window for collecting the secondary fractions enriched for ex-phosphorylated peptides might seem quite large. The reason for this is that, although the shift of a singly phosphorylated peptide is quite predictable, IMAC strategies such as the one used here preferentially enrich multi-phosphorylated peptides whose hydrophobic shifts on dephosphorylation are very difficult to predict (34). Thus, a wider collection interval is recommended to avoid losing too many ex-phosphorylated peptides.

Once the sorted peptides have been prepared by any of *Protocols 3–6*, they can be analyzed by LC-MS/MS according to *Protocol 7*. Note that the data storage space on the data acquisition computer is a limiting factor here, and it is this that is a major consideration in deciding when to start collecting spectra (see *Protocol 7* note a).

Protocol 7
LC-MS/MS analysis of sorted peptides

Equipment and Reagents
- Sorted peptides from *Protocols 3–6*
- CapLC system (Waters Corp.)
- Q-TOF Premier mass spectrometer (Micromass)
- PicoTip needle (New Objective Inc.)
- Formic acid (puriss. p.a. ~98% (T); Fluka)
- HPLC-grade water (e.g. Baker HPLC-analyzed; Mallinckrodt Baker)
- HPLC-grade acetonitrile (e.g. Baker HPLC analyzed; Mallinckrodt Baker)
- Solvent A (0.05% formic acid in water : acetonitrile, 98 : 2)
- Solvent B (0.05% formic acid in water : acetonitrile, 30 : 70)
- RP trapping column (0.3 mm internal diameter × 5 mm length PepMap column; LC Packings)
- Nano-scale RP analytical column (75 µm internal diameter × 150 mm length PepMap column; LC Packings)

Method
1. Connect the CapLC system set-up with the trapping and analytical column (35) in line with the Q-TOF Premier mass spectrometer using a metal-coated fused silica needle (PicoTip).
2. Redissolve dried secondary fractions in 20 µl of solvent A.
3. Using the CapLC system, load half of this solution onto the trapping column at a flow rate of 20 µl/min of solvent A for 5 min.
4. Load the sample onto the analytical nano-scale column by back-flushing the trapping column.
5. Use the HPLC system to apply a linear gradient to 100% solvent B over 50 min at a constant flow rate of 200 nl/min.
6. Start data-dependent acquisition 20 min after the solvent gradient was started, such that doubly and triply charged ions are preferentially selected for further fragmentation[a].
7. At the end of the gradient, the HPLC is equilibrated with 100% solvent A for 15 min after which a new sample can be analyzed.

Note
[a]In this particular LC-MS set-up, start to collect spectra 20 min after starting the gradient, as very few peptides ions are detected earlier in the gradient. In this way, we make sure that we do not waste too much data storage space on the data-acquisition computer. For example, an LC-MS/MS run such as the one described above on a Q-TOF Premier mass spectrometer requires on average 200 Mb of storage space. Calculating an average of 16 LC-MS/MS analyzes a day, 3.2 Gb of Q-TOF data is generated daily that needs interpretation and archiving. As Fourier transform *m/z* analyzers tend to generate even more data (especially when run unattended), any attempts to reduce the data size should be considered favorable for further processing.

3. TROUBLESHOOTING

- When sorting methionyl peptides (see *Protocol 3*), it is extremely important to respect the oxidation time (30 min) and temperature (30°C) as given in step 12 of *Protocol 3*, otherwise unwanted oxidation reactions on methionine and other amino acids take place. Upon longer incubation, methionine sulfoxide starts to convert to methionine sulfone. As peptides carrying the latter amino acids are not as hydrophilic as their sulfoxide forms, the difference in column retention is much lower, leading to an inadequate sorting of methionyl peptides. Furthermore, the sorted peptides will not be uniform, as both the sulfoxide and the sulfone form of methionyl peptides will be isolated, leading to unnecessary sample dilution.
- It is important to emphasize that a peptide oxidation mixture cannot be stored in freezers awaiting separation by RP-HPLC. Storage, even at −80°C, leads to almost complete conversion of methionine sulfoxides to methionine sulfones in addition to oxidation of cysteine and tryptophan. After the oxidation step, the peptide mixture must be separated directly by RP-HPLC; thus, peptide samples cannot be treated in parallel. Nevertheless, contemporary HPLC instruments have liquid-handling robotics (e.g. sample injection systems) that may be programmed and are thermostatically controlled so that the oxidation reaction is performed automatically within the instrument and this allows high sample throughput.
- As described in *Protocol 4*, cysteinyl peptides are isolated in their reduced form, i.e. carrying a free thiol group. In order to overcome oxidation of cysteines leading to cysteine bridges, one may consider an alkylation reaction blocking free cysteines prior to LC-MS/MS analysis. This is carried out most conveniently using a mixture of TCEP and iodoacetamide, as neither component is retained by the trapping column and will not hinder further peptide separation and MS/MS analysis.
- In theory, the N-terminal COFRADIC protocol (*Protocol 5*) should only sort the N-terminal peptides of proteins. However, in practice, a number of other types of peptide are unavoidably co-sorted, for example peptides carrying (or acquiring) a blocked, nonacetylated N-terminal amino acid such as proline, a pyrrolidone carboxylic acid, or cyclic S-carbamoylmethylcysteine. Such peptides are simply co-sorted by the chemical nature of their N terminus. To some extent, they 'pollute' the mixture of sorted peptides. However, for differential proteomics purposes, their presence may be beneficial, as several peptides per protein are monitored, potentially increasing the accuracy of determining the abundance ratio of the proteins. The number of sorted peptides still carrying free α-amines is quite low (about 5% of all sorted peptides) and mainly points to highly abundant structural proteins and enzymes. Any attempts to further reduce their prevalence will lead to improved detection of true N-terminal peptides and thus to broader and more sensitive proteome coverage. However, one must also realize that chemical modification reactions are never 100% quantitative, and particularly abundant and readily ionizable peptides might appear in their free α-amine form following sorting.

Table 1. Recommended parameters for searching databases with MS/MS spectra of peptides sorted by the different COFRADIC strategies

Fixed modifications	Variable modifications
Methionyl peptides	
Oxidation (Met)	Acetylation (N terminus)
	Deamidation (Asn, Gln)
	Pyroglutamic acid (N-terminal Gln)
Cysteinyl peptides	
None	Acetylation (N terminus)
	Deamidation (Asn, Gln)
	Pyroglutamic acid (N-terminal Gln)
N-terminal peptides	
Acetylation (Lys)	Acetylation (N terminus)
Carbamidomethyl (Cys)	Deamidation (Asn, Gln)
	Oxidation (Met)
	Pyrocarbamidomethyl cysteine (Cys)
	Pyroglutamic acid (N-terminal Gln)
Phosphorylated peptides	
Carbamidomethyl (Cys)	Acetylation (N terminus)
	Deamidation (Asn, Gln)
	Oxidation (Met)
	Pyrocarbamidomethyl cysteine (Cys)
	Pyroglutamic acid (N-terminal Gln)

- As evident from section 2.2 and *Fig. 1*, depending on the type of COFRADIC strategy, sorted peptides may be 'decorated' with different types of modification in addition to those one can foresee, as they are the result of any protein/peptide preparation procedure (e.g. formation of pyroglutamic acid). In order to link MS/MS spectra of COFRADIC-sorted peptide ions efficiently to peptide/protein sequences in databases, search engines such as Mascot (36) need to consider the (potential) presence of these modifications on the analyzed peptide ions. An overview of both the fixed modifications (i.e. due to the protein preparation method and/or the sorting chemicals used) and potential variable modifications (i.e. modifications that are likely to be present in some of the sorted peptides) is presented in *Table 1*.
- In the context of the N-terminal COFRADIC procedure (*Protocol 5*) visualizing protein processing (31), the sequence of a sorted peptide that indicates such an irreversible modification is not always exactly predicted by search engines, as they do not consider *in vivo* 'ragging' of N termini. Hence, identification of these particular peptides will be missed. To overcome such flaws, we constructed DBTOOLKIT (freely available via http://www.proteomics.be), an algorithm that essentially uses protein databases as input, imitates the *in vivo* and *in vitro* processing events, and creates FASTA-formatted peptide databases (37). This algorithm also deals with the unavoidable increase in complexity (many *in silico*-predicted protein digestion products link to multiple protein entries in databases). We noted an increase of at least 30% in identified MS/MS spectra when using Mascot to identify sorted N termini (38).

4. REFERENCES

★ 1. Fey SJ & Larsen PM (2001) *Curr. Opin. Chem. Biol.* **5**, 26-33. – *This paper, bearing the appropriate title '2D or not 2D. Two-dimensional gel electrophoresis', discusses several advances on visualizing low-abundance and highly hydrophobic proteins by 2D-PAGE, as well as dealing with the intrinsic disadvantages of this technique.*
2. Celis JE, Gromov P, Ostergaard M, et al. (1996) *FEBS Lett.* **398**, 129-134.
3. Wilkins MR, Gasteiger E, Sanchez JC, Bairoch A & Hochstrasser DF (1998) *Electrophoresis*, **19**, 1501-1505.
4. Wu CC, MacCoss MJ, Howell KE & Yates JR III (2003) *Nat. Biotechnol.* **21**, 532-538.
★★★ 5. Washburn MP, Wolters D & Yates JR III (2001) *Nat. Biotechnol.* **19**, 242-247. – *Description of the combination of strong cation exchange and RP chromatography for routine gel-free proteome analysis using multi-dimensional protein identification technology to identify 1484* Saccharomyces cerevisiae *proteins.*
6. Liu H, Sadygov RG & Yates JR III (2004) *Anal. Chem.* **76**, 4193-4201.
★★★ 7. Gygi SP, Rist B, Gerber SA, Turecek F, Gelb MH & Aebersold R (1999) *Nat. Biotechnol.* **17**, 994-999. – *The original application of ICATs for the isolation of cysteinyl peptides.*
★★ 8. Ficarro SB, McCleland ML, Stukenberg PT, et al. (2002) *Nat. Biotechnol.* **20**, 301-305. – *First description of the combination of IMAC (after esterification of carboxyl groups) and RP-HPLC for high-throughput, non-gel analysis of protein phosphorylation sites.*
★★ 9. Gevaert K, Damme PV, Martens L & Vandekerckhove J (2005) *Anal. Biochem.* **345**, 18-29. – *Review of the different COFRADIC sorting technologies, focusing on their possible applications.*
10. Oda Y, Nagasu T & Chait BT (2001) *Nat. Biotechnol.* **19**, 379-382.
11. Gevaert K & Vandekerckhove J (2004) *Drug Discov. Today : Targets*, **3**, S16-S22.
12. Brown JR & Hartley BS (1966) *Biochem. J.* **101**, 214-228.
13. Tang J & Hartley BS (1967) *Biochem. J.* **102**, 593-599.
14. Butler PJ, Harris JI, Hartley BS & Leberman R (1967) *Biochem. J.* **103**, 78P-79P.
15. Cruickshank WH, Radhakrishnan TM & Kaplan H (1971) *Can. J. Biochem.* **49**, 1225-1132.
16. Cruickshank WH, Malchy BL & Kaplan H (1974) *Can. J. Biochem.* **52**, 1013-1017.
★★★ 17. Gevaert K, Van Damme J, Goethals M, et al. (2002) *Mol. Cell. Proteomics*, **1**, 896-903. – *Description of the first application of COFRADIC, used to map the proteome of* E. coli *growing in the exponential phase.*
18. Ong SE, Blagoev B, Kratchmarova I, et al. (2002) *Mol. Cell. Proteomics*, **1**, 376-386.
19. Krijgsveld J, Ketting RF, Mahmoudi T, et al. (2003) *Nat. Biotechnol.* **21**, 927-931.
20. Munchbach M, Quadroni M, Miotto G & James P (2000) *Anal. Chem.* **72**, 4047-4057.
21. DeSouza L, Diehl G, Rodrigues MJ, et al. (2005) *J. Proteome Res.* **4**, 377-386.
22. MacCoss MJ, Wu CC, Liu H, Sadygov R & Yates JR III (2003) *Anal. Chem.* **75**, 6912-6921.
23. Moritz B & Meyer HE (2003) *Proteomics*, **3**, 2208-2220.
24. Julka S & Regnier F (2004) *J. Proteome Res.* **3**, 350-363.
25. Rose K, Simona MG, Offord RE, Prior CP, Otto B & Thatcher DR (1983) *Biochem. J.* **215**, 273-277.
26. Schnolzer M, Jedrzejewski P & Lehmann WD (1996) *Electrophoresis*, **17**, 945-953.
27. Gevaert K, De Mol H, Verschelde JL, Van Damme J, De Boeck S & Vandekerckhove J (1997) *J. Protein Chem.* **16**, 335-342.
★★★ 28. Staes A, Demol H, Van Damme J, et al. (2004) *J. Proteome Res.* **3**, 786-791. – *The original publication from our laboratory describing the combination of enzymatic incorporation of ^{18}O with COFRADIC analyses.*
29. Veenstra TD, Conrads TP & Issaq HJ (2004) *Electrophoresis*, **25**, 1278-1279.
30. Belvisi MG & Bottomley KM (2003) *Inflamm. Res.* **52**, 95-100.
★★ 31. Van Damme P, Martens L, Van Damme J, et al. (2005) *Nat. Methods*, **2**, 771-777. – *Description of the potential of combining N-terminal COFRADIC with stable isotope (^{18}O) labeling for specific characterization of* in vivo *processing events. This approach is unique in being generally and globally applicable to studying protein processing or protein degradomics.*
32. Gevaert K, Pinxteren J, Demol H, et al. (2006) *J. Proteome Res.* **5**, 1415-1428.

★ 33. Liu P, Feasley CL & Regnier FE (2004) *J. Chromatogr. A.* **1047**, 221–227. – *Description of the possible pitfalls of diagonal chromatography in enriching specific peptide sets.*
★★★ 34. Gevaert K, Staes A, Van Damme J, *et al.* (2005) *Proteomics*, **5**, 589–599. – *Description of the original COFRADIC protocol for isolating phosphopeptides and results of a phosphoCOFRADIC analysis on HepG2 cells treated with forskolin.*
35. Vissers JP, Chervet JP & Salzmann JP (1996) *J. Mass Spectrom.* **31**, 1021–1027.
36. Perkins DN, Pappin DJ, Creasy DM & Cottrell JS (1999) *Electrophoresis*, **20**, 3551–3567.
★★★ 37. Martens L, Vandekerckhove J & Gevaert K (2005) *Bioinformatics*, **21**, 3584–3585. – *Description of DBTOOLKIT and details of potential applications and benefits.*
★★★ 38. Gevaert K, Goethals M, Martens L, *et al.* (2003) *Nat. Biotechnol.* **21**, 566–569. – *The first publication describing the COFRADIC technology for isolation of N-terminal peptides.*

CHAPTER 5
Mass spectrometry strategies for protein identification

David R. Goodlett and Garry L. Corthals

1. INTRODUCTION

At the end of the 1980s, two novel ionization methods were developed that allowed the nondestructive measurement of proteins and peptides. The impact of this on biological sciences was immediate and far-reaching. Traditional protein chemistry and proteomics laboratories were transformed by the speed of analysis and sophistication of results provided. Biological mass spectrometry (MS) received its second big boost from the serendipitous completion and annotation of genome sequences, which enabled the correlation of MS data with genome sequence information. The result was rapid, reliable, and facile protein identification. Since then, data on more than 2000 genomes have become available, in part due to their use for functional genomics, which MS provided. The impact of modern biological MS on life sciences has been tremendous and was publicly acknowledged through the 2002 Nobel prizes for the development of laser desorption ionization and electrospray ionization (ESI) methods for the analysis of biological macromolecules (http://nobelprize.org/).

Biological MS with sophisticated bioinformatics is now the driving force behind proteome research. It has transformed the field of protein chemistry from serial *de novo* association of tandem mass spectra to amino acid sequences for protein identification to methods where, in parallel, thousands of proteins can be identified in a single proteomics experiment. Proteomics methods exist that enable the routine cataloging of hundreds to thousands of proteins present in biological samples (1). Due to this technological feat, one of the main development themes in proteomics research for biosciences is biomarker discovery, where one aims to distinguish which proteins represent healthy or disease states in models of disease, ultimately moving society away from reactive medicine and towards preventative/predictive medicine. Systems biology (2–4) is also an area of bioscience that relies heavily on proteomic information. Here, one aims to transform proteomics information into knowledge, and fuse this knowledge with other sets of 'global' biological information in model systems or organisms, ultimately to predict biological events. This information can exist in the form of DNA, RNA (e.g. siRNA, microarrays), protein, protein interactions, biomodules, protein and gene regulatory networks, cells, organs, individuals, populations, or ecologies. To facilitate the acquisition of large amounts of information in a short time frame, it is critical to understand the technologies used to probe and report biological events. A successful approach that has emerged to capture and define proteomes without the use of gel electrophoretic separation of proteins prior to MS analysis is known as high-throughput or 'shotgun' proteomics.

In this chapter, we provide an insight into so-called shotgun proteomics technology through the description of selected methodologies. The application of these methods allows the reliable and comprehensive cataloging of proteomes. *Fig. 1* shows one example of a typical shotgun proteomic workflow where, for

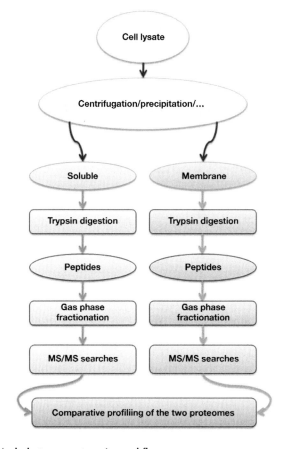

Figure 1. Typical shotgun proteomic workflow.
Cells are lysed and soluble proteins separated from insoluble (e.g. microsomal) proteins by centrifugation. The proteins are denatured and proteolytically digested using trypsin. Iterative gas-phase fractionation is carried out to increase the number of unique peptides selected for collision-induced dissociation during liquid chromatography-MS/MS with ion selection controlled by automated data-dependent computer routines. Fragmentation data from the tandem mass spectra are used for database searching to identify matching peptide sequences and hence the parental proteins.

example, bacterial cells are separated into a soluble and membrane fraction and each of these is processed independently, after which the data are linked together. The skill level and strategies employed for coupling biological samples to MS analysis are significant but not insurmountable. A three-part description delineates what could be considered a routine shotgun proteomic experiment. We hope that this will provide the reader with directions on how to prepare a sample, set the instrument parameters for a liquid chromatography (LC)-MS experiment, and finally analyze the data. First, protein digestion and sample preparation are discussed. Secondly, a shotgun proteomic experiment is described that uses gas-phase fractionation (5), a simple analytical approach that can be used to increase the number of identifications without the need for protein fractionation or any hardware modifications. Finally, the process of associating peptide tandem mass spectral data with sequences in databases is discussed.

2. SAMPLE PREPARATION

2.1 Digestion

In shotgun proteomics, analysis is performed directly from the solution phase without prior separation of proteins via two-dimensional electrophoretic methods. However, various simple methods for proteome fractionation are often utilized such as one-dimensional gel electrophoresis. We will consider here the most basic form of shotgun proteomics: the so-called 'lyse and go' direct analysis approach (see Chapter 1 for additional sample preparation protocols). Proteins in mixtures are denatured, subjected to proteolysis, and analyzed by LC-MS/MS. Proteins are identified by matching the results of tandem mass spectrometry (MS/MS) of peptides to sequences in databases, most often by correlative analysis algorithms. Thus, shotgun proteomics directly identifies peptides and indirectly identifies the proteins from which they were derived. Whilst any endoprotease can be used, in practice trypsin is used most commonly, as it generates many peptides of an ideal length and charge for MS analysis. It also cleaves C-terminal to lysine (K) and arginine (R) amino acids, resulting in an N-terminal amine that can accept a proton and a basic side-chain residue of the K/R that may also sequester a proton allowing facile detection in positive ion mode by most mass analyzers. For ESI, this typically translates into multiply charged peptides such as $[M + 2H]^{2+}$, $[M + 3H]^{3+}$, etc., and with matrix-assisted laser desorption ionization (MALDI) one usually typically observes singly charged peptides as $[M + 1H]^{1+}$. Nevertheless, the choice of enzyme is sample-dependent. *Protocol 1* is a typical proteolytic procedure for proteins that are optimally digested with trypsin.

Protocol 1

Proteolysis with trypsin

Equipment and Reagents
- Urea
- 1.5 M Tris/HCl (pH 8.8)
- 200 mM Tris(2-carboxyethyl)phosphine (TCEP)
- 200 mM Iodoacetamide
- 200 mM Dithiothreitol (DTT)
- 1.5 ml Microcentrifuge tubes
- 25 mM Ammonium bicarbonate
- Methanol (HPLC-grade)
- Sequencing-grade modified trypsin (Promega)
- Vacuum concentration centrifuge, e.g. SpeedVac centrifuge (Thermo Electron Co.)
- Freezer at –80°C for sample storage prior to MS analysis

Method
1. Adjust the protein solution to 6 M urea by, for example, adding 108 mg of urea to 300 µl of solution[a].
2. Add 20 µl of 1.5 M Tris/HCl (pH 8.8)[b].
3. Add 7.5 µl of 200 mM TCEP and incubate for 1 h at 37°C[c].
4. Add 60 µl of 200 mM iodoacetamide. Vortex and then incubate for 1 h at room temperature in the dark.
5. Add 60 µl of 200 mM DTT. Vortex and then incubate for 1 h at room temperature[d].
6. Dispense 150 µl aliquots into 1.5 ml microcentrifuge tubes and add 800 µl of 25 mM ammonium bicarbonate to each tube to dilute the urea.
7. Add 200 µl of methanol (HPLC-grade) to each tube and add trypsin at a protein : trypsin ratio of 50 : 1 (w/w). Incubate overnight at 37°C.
8. Pool the contents of the tubes containing the same sample[a] into a single sample vial and wash the tube walls with water. Using a vacuum concentration centrifuge, evaporate to near dryness.
9. Add 200 µl of water to the sample vials and evaporate to near dryness to remove the ammonium bicarbonate.
10. Repeat the water additions and evaporation step (step 9) twice more.
11. Evaporate to dryness and use immediately or store at –80°C.

Notes
[a]This protocol is for a 300 µl protein mixture; e.g. a cell lysate or a cell fraction, or a semi-purified mixture of proteins. This method can be scaled up or down, but scaling up requires division of the sample into multiple aliquots.
[b]This raises the pH, as the TCEP used in the next step is acidic.
[c]The amount of Tris/HCl used should be proportional to the amount of TCEP – use pH paper to check that the pH is not acidic.
[d]This eliminates excess iodoacetamide.

2.2 Sample clean-up prior to LC-MS

As mentioned above, peptide samples 'acquire' a proton during the ionization step. Other molecules present in the sample can also be ionized as well as peptides and consequently may interfere with the identification process, as this relies on effective ionization. Hence, one key to the successful identification of peptides is their analysis in relatively pure form. For both LC-MALDI and LC-ESI applications, clean-up occurs in a microcapillary column prior to peptide separation by incorporating a washing step in the protocol. As the MALDI and ESI methods are identical with respect to their separation capabilities, we will focus only on the ESI method.

Inorganic salts (sodium, potassium, etc.) and organic molecules such Tris ionize extremely well and hinder the desired ionization of peptides. The presence of these molecules in a sample causes two problems. First, they can lead to signal suppression of the peptides of interest, and secondly, they can form multiple adducts with the peptides, producing a peptide signal that is distributed into several m/z values, instead of a single m/z; e.g. where M = a peptide and $[M + 2H]^{2+}$ the '2+ ion' form of that peptide, then with contaminants present we may also observe, for example, $[M + 2NH_4^+]^{2+}$ and $[M + 2K^+]^{2+}$ etc. These latter two forms of the peptide, where H^+ is replaced with some other cation, will provide identification only if one is aware of their presence, as the standard search engines consider H^+ to be the *de facto* cation found in such acidic proteomic solutions. Failure to remove these nonstandard forms of the peptide adduct prior to microcapillary LC-MS/MS will also reduce the overall signal of the expected protonated form, possibly even preventing detection. Additionally, the presence of chemical compounds intentionally added to denature the proteins allowing the protease to cleave amide bonds effectively (e.g. urea, guanidine hydrochloride, or sodium dodecyl sulfate) and unintentionally added chemicals (e.g. pervasive plasticizers) will both have a deleterious effect, preventing the mass spectrometer from 'observing' the peptide ions of interest.

Positively charged, zwitterionic or even nonionic detergents are also known to ionize well during both MALDI and ESI, and can substantially suppress the analyte signal. In addition, they have a detrimental effect on the performance of reverse-phase columns or cartridges by preventing peptide interaction with the C18 solid phase, or by irreversibly binding to and modifying the surface properties of the resin. Their presence in a sample can cause erratic and unreliable separations. Thus, one should omit their use if possible, but for proteins and complexes that are difficult to denature, this is not always practicable. Note though that 3-((3-cholamidopropyl)dimethylammonio)-1-propanesulfonate (CHAPS) or octyl-β-glucopyranoside can be used at low concentrations during LC-MS, as they elute at the end of gradients. Triton X-100 and NP-40 ionize extremely well and should be avoided, as they are chemically heterogeneous, containing a distribution of molecular species that can span hundreds or thousands of daltons and completely dominate all mass spectra across an LC separation. *Protocol 2* describes a typical 'clean-up' procedure to remove salts, chaotropic agents, and detergents before carrying out LC-MS.

Protocol 2

Sample clean-up after digestion in preparation for LC-MS/MS analysis

Equipment and Reagents
- Protein digest (from *Protocol 1*)
- UltraMicro spin cartridge (Nest Group, Inc.)
- 5% Acetonitrile, 0.1% trifluoroacetic acid (TFA)[a]
- Solvent A (80% acetonitrile, 0.1% TFA)[a]
- Solvent B (5% acetonitrile, 0.1% TFA)[a]
- 0.1 % Formic acid
- 5% Acetonitrile, 0.05% heptafluorobutyric acid (HFBA), 0.4% acetic acid
- Vacuum concentration centrifuge, e.g. SpeedVac

Method

1. Dissolve the dry protein digest in 5% acetonitrile, 0.1% TFA[a] in water.
2. Saturate the spin cartridge column by adding 100 µl of solvent A and centrifuging for 4 min at 3000–3500 r.p.m. in a bench-top centrifuge. Repeat once.
3. Equilibrate the column with 100 µl of solvent B. Centrifuge for 4 min at 3000–3500 r.p.m. Repeat twice.
4. Blot dry the tip of the column and change the collecting tube.
5. Add up to 50 µg of protein digest per column[b]. Collect the flow-through as a precaution.
6. Wash the column with 100 µl of solvent B. Repeat twice.
7. Change the collecting tube and collect the clean, desalted sample with 50 or 100 µl of solvent A.
8. Using a vacuum concentration centrifuge, evaporate to dryness in a glass tube.
9. Resuspend in 0.1% formic acid to a final concentration of 1 µg/µl.
10. Dilute 1 : 100 with 5% acetonitrile, 0.05% HFBA, 0.4% acetic acid, or with another LC-MS-compatible solvent.

Notes

[a]TFA-containing solutions should be made on the day of use. Extreme care should be taken when using TFA, as it is an extremely hazardous chemical, and the concentrated acid should only be used in a fume hood and dispensed wearing suitable protective clothing. Consult the safety datasheet for handling details and disposal.

[b]The maximum protein capacity is 5–50 µg and the maximum volume is 100 µl.

3. MS ANALYSIS

3.1 Microcapillary LC-MS/MS

The breakthrough in interfacing microcapillary columns directly with ESI-MS/MS for peptide 'sequencing' by low-energy collision-induced dissociation (CID) was originally popularized by Hunt and colleagues in 1992 (6). The methods have become automated since then, allowing one to load samples via an autosampler (e.g. from 96-well plates) or from centrifuge tubes (0.6–1.5 ml) or vials. Column loading and elution into the mass spectrometer follow a two-step process where the sample is directly loaded onto the column in the first step and peptides are eluted directly into the mass spectrometer in the second step. This set-up is flexible and lends itself to the configuration of many different types of LC/ESI-MS applications. *Fig. 2(a)* (also available in the color section) shows a typical LC/ESI set-up. Typically, a sample is loaded on a C18 column and washed with a hydrophilic solution (i.e. so-called solvent A) to remove high concentrations of contaminants prior to initiation of the hydrophobic solvent (solvent B), which elutes the peptides away from the C18 resin and into the mass spectrometer.

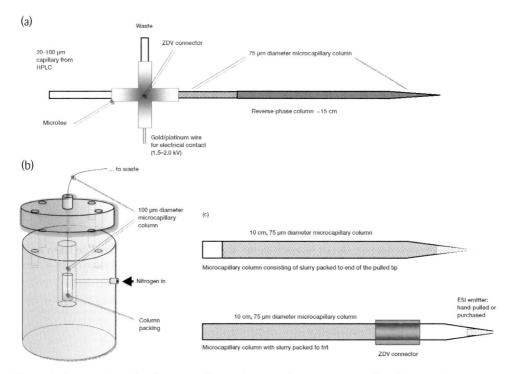

Figure 2. Construction of in-house capillary columns using a pressure cell and connection to an HPLC system using a zero dead volume cross-connection (see page xxii for color version).
(a) A zero dead volume union 'cross' showing connection to a capillary column (right), the waste flow (top), the flow from HPLC (left), and an electrode (bottom). (b) A pressure cell (Brechbuehler, Inc.) for packing capillary columns or loading samples onto a capillary column. (c) Examples of packed capillary columns.

Shotgun proteomic high-performance liquid chromatography (HPLC) separations are carried out in capillaries with internal diameters ranging from 50 to 100 µm, as shown in *Fig. 2(c)*. This is often referred to as microcapillary LC but also as nano-LC, reflecting flow rates in the nl/min regime of 150–300 nl/min. These capillaries are interfaced with the ESI source, or with an instrument that allows the spotting of fractions onto a MALDI target plate. Decreasing the flow rate to the nl/min range (from ml/min) results in an effective increase in sensitivity, as ESI is a concentration-dependent process. Excellent capillary columns, as detailed in *Fig. 2(c)*, are simple to construct in the laboratory from inexpensive commercial parts and pressure vessels (5, 7). Whilst they are inexpensive, without extraordinary care they are not as robust as the commercially available capillary columns, which also have more reproducible retention times. A suitable method for constructing in-house capillary columns is available at http://proteomics.btk.fi and their performance has been assessed with standards and complex samples (5, 8).

Flow rates in the nl/min regime may be generated by either a pump (typically very expensive) capable of operating down to this low level of flow or a pump capable only of much higher flow rates of µl/min where the flow is split pre-column from µl/min down to nl/min (see *Fig. 2a* for an example). Flow splitting is the less-expensive alternative that we have used for many years (5, 8) and can be accomplished with many different pump styles. Yates and co-workers have also written an excellent guide to preparation and operation of capillary LC columns coupled with ESI-MS (9).

3.2 Data-dependent MS/MS allowing automated ion selection

Current automated LC-MS/MS procedures result in the acquisition of massive amounts of MS/MS data that can be used to identify hundreds to thousands of proteins, depending on the complexity of the sample. For approximately the last 10 years, LC-MS/MS analyses have been conducted using software that controls the many operations of a tandem mass spectrometer during an LC analysis. The key steps are:

1. Acquisition of available precursor ions at a given moment in chromatographic time.
2. Decisions about which of the many co-eluting precursor ions to select for CID.
3. Fragmentation by CID of one precursor ion at a time.
4. Acquisition and storage of an MS/MS spectrum for a single peptide.

This is a complex, iterative process that happens many thousands of times during a typical 60–180 min LC separation. Such a generic process conducted during LC-MS/MS analysis of a complex peptide mixture might have the following steps, where ions are selected from a predefined *m/z* range; e.g. 400–1800 *m/z*:

1. A precursor ion mass spectrum is recorded.
2. From the precursor ion mass spectrum, the computer makes a decision about which ion to select for CID based on rules, some of which might include the

following: (i) select only ions with a charge state equal to $[M + 2H]^{2+}$; (ii) select the nth most intense ion, which is typically the base peak (i.e. the highest signal : noise ratio in a single mass spectrum); and (iii) select no ion that is in either a predefined static exclusion list or was recently observed (within the last 1 min) in the rolling list of ion values maintained by the mass spectrometer's data system during the experiment, i.e. the dynamic exclusion list.
3. Isolate the *m/z* value of interest from above and fragment based on a collision energy that is a function of the observed *m/z* values, e.g. higher *m/z* ions receive higher CID energy.
4. Record a tandem mass spectrum for that single *m/z* value.
5. Repeat steps 1–4.

In step 1, the computer acquires information on the precursor ions present at a single moment in chromatographic time, i.e. a precursor mass spectrum (see *Fig. 3*, also available in the color section), whilst in step 4, a product ion or tandem mass spectrum is recorded for a single precursor ion (see *Fig. 4*, also available in the color section). The benefit of this data-dependent approach is that, as peptides elute from an LC column into the mass spectrometer, they are selected for

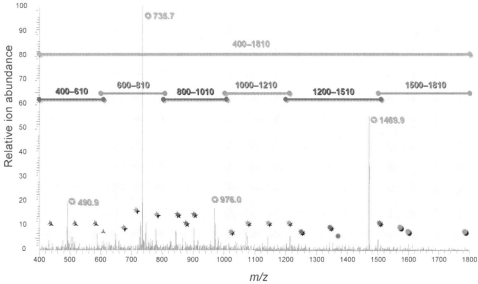

Figure 3. Gas-phase fractionation (see page xxii for color version).
Comparison of the increase in ions subjected to CID among six randomly chosen *m/z* ranges used for gas-phase fractionation and one normal wide *m/z* range.

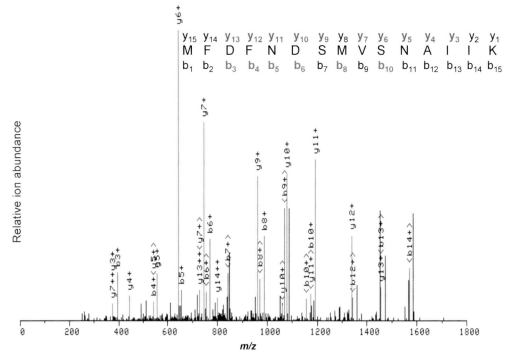

Figure 4. Example of a tandem mass spectrum of peptide ion fragments (see page xxiii for color version).
The mass spectrum shows fragment ions generated after CID of a specific parental peptide ion. The spectrum was generated using a linear ion-trap mass spectrometer (ThermoElectron) controlled by data-dependent rules during LC introduction for ESI-MS/MS. Bracketed ions (<>) indicate that they have lost either H_2O or NH_3.

CID in a logical, automated process that allows thousands of tandem mass spectra to be recorded without operator attendance or intervention. One aim of this automated process is to acquire tandem mass spectra for as many unique peptides as possible with as little redundancy as possible. Doing so effectively enables higher proteome coverage or, if using the same method to analyze peptides extracted from a gel, individual protein sequence coverage will increase by preventing the repeated selection of the same peptides. Until about 2003, this process had, without exception, exclusively been performed on LC-ESI-configured instruments. However, it is now also possible to perform this type of analysis on MALDI-MS/MS systems (Bruker Daltonics and Applied Biosystems). Although the actual procedural events differ, the concepts and goals remain the same: nonredundant and high proteome coverage. There are numerous variations on this theme of data-dependent tandem MS/MS (and names depending on the MS vendor), depending on the goal of the experiment and the instrument being used.

3.3 Iterative gas-phase fractionation increasing proteome coverage

Gas-phase fractionation (GPF) (10) involves iterative LC-MS/MS analysis of a sample prepared as described above. It is designed to increase proteome coverage without liquid-phase fractionation of a sample by selection of peptides of medium-to-low signal : noise ratio that are normally missed during a standard LC-MS/MS analysis where one typically selects parental ions via data-dependent methods from a wide m/z range; e.g. 400–1800 m/z (see *Fig. 3*, also available in the color section). Note that iterative analyses of a single, wide m/z range will result in an eventual plateau of proteins identified (10). For GPF of a bacterial sample, we typically analyze a sample from six unique m/z ranges as depicted in *Fig. 3*, e.g. 400–610, 600–810, 800–1010, 1000–1210, 1200–1510, and 1500–1810. The same amount of sample is injected for analysis over each of the narrower m/z ranges and each parental ion scan range slightly overlaps the next to avoid missing m/z values present at the interface of two regions. The only difference in each LC-MS/MS analysis is the range from which parental ions are observed and thus selected by the mass spectrometer's computer system for CID. In cases where we seek high proteome coverage of a sample or where the sample has high complexity (e.g. mammalian cells), we may select ions from as many 18 different m/z ranges and analyze each in triplicate. In this case, we usually select ions from 100 m/z-wide overlapping parental ion m/z ranges as: 400–510, 500–610, 600–710 ... 1300–1410, 1400–1610, and 1600–2000. Note that the last two ranges are wider because there are fewer ions at the high m/z range that show up when analyzing peptides from a tryptic digest. Of course, if a protease other than trypsin or if sample processing were used, then the peptides might be longer and thus this high m/z space would be more crowded. Although this method is simple to implement (see *Protocol 3*), it does require sufficient sample for multiple injections. We routinely use this method in the laboratory to define proteomes of varying complexity from bacterial to human samples.

Protocol 3
LC-MS/MS sample analysis by GPF

Equipment and Reagents
- Microcapillary HPLC column, e.g. binary HP1100 pump from Agilent Technologies
- Solvent A (see *Protocol 2*)
- Angiotensin (Sigma-Aldrich)
- Autosampler, e.g. Famos micro-autosampler from Dionex LC packings
- Data-dependent ESI tandem mass spectrometer, e.g. LCQ Deca ion trap mass spectrometer from Thermo Finnigan, or HCT ion trap from Bruker Daltonics, with integrated six-port switching valve pre-column (e.g. 100 µm internal diameter × 2.0 cm length)
- Analytical microcapillary column (e.g. 75 µm internal diameter × 10 cm), which also serves as an ESI emitter

Method
1. Equilibrate the microcapillary HPLC column with solvent A (see *Protocol 2*, step 2).
2. Check the chromatographic peak width and MS detection sensitivity by injection of a standard solution of angiotensin (10).
3. Inject the sample from *Protocol 2* and elute with a gradient that is compatible with the complexity of the sample (8).
4. Set the mass spectrometer parameters to select ions from a specific *m/z* range.
5. For thorough sample analysis, perform iterative analysis of the same sample over multiple overlapping *m/z* ranges: for example, in LC-MS/MS experiment 1, scan for parental ions from 400 to 510 *m/z*; in LC-MS/MS experiment 2, scan for parental ions from 500 to 610 *m/z*, etc.[a]
6. For each LC-MS/MS experiment, set the parameters for data-dependent ion selection such that ions are excluded from reselection for 1–3 min.

Note
[a] A basic configuration that allows samples to be analyzed in a shotgun proteomic experiment has been published (5). Here, we have presented just the basic steps for instrument set-up, which include GPF. The number of precursor ion ranges scanned in a GPF experiment depends directly on sample complexity. Thus, the *m/z* ranges used need to be adjusted accordingly.

4. DATABASE SEARCHING

Finally, with thousands of tandem mass spectra acquired, some of which are from peptides and some common contaminants, the process of converting this data into protein identifications must begin. Fortunately, peptides subjected to low-energy CID fragment in this manner can readily be interpreted in terms of amino acid sequence (see *Fig. 4*), as, under these physical conditions, the peptide amide bond is most likely to fragment. There are two basic steps:

1. Associating a tandem mass spectrum with a given amino acid sequence, typically in a database of sequences but also manually.
2. Proving that a given protein that contains the matched amino acid sequence is present in the sample, i.e. corroboration of the result by additional MS evidence or by an orthogonal method.

Confidence in the first step depends largely on generating high-quality data and in the second on either uniqueness of the single amino acid sequence for a peptide or parallel detection of other peptides from the same putative parental proteins.

For various reasons, the rate of efficiency for converting tandem mass spectra into protein identifications in shotgun proteomics is often as low as 20%. Practically, this means that if 20 000 tandem mass spectra are acquired, then only 4000 of these may be confidently matched to a peptide sequence. From the latter, a few hundred proteins may be identified, the actual number depending on many variables such as sample dynamic range and complexity. This low success rate drives the continued refinement of analytical methods and algorithm development. In fact, more than 35 algorithms for matching peptide tandem mass spectra to amino acid sequences have been described. Although only a handful are used routinely by our community, and many have a similar efficiency (11), the large number of different algorithms indicates the poor efficiency of the currently employed strategies in terms of overall success rate.

The two main strategies employed for protein identification are: (i) direct correlation of MS/MS spectra with sequences in a database, based on a best-fit approach; and (ii) *de novo* correlation of tandem mass spectra with sequences in the absence of sequence database information. Here again there are many differences in how the community chooses to approach protein identification from these two perspectives, with numerous algorithms devised to identify proteins by either approach (for a review, see 11). Given the numerous programs used to identify proteins, we will finish this chapter with some general thoughts on protein identification.

4.1 Matching peptide fragmentation patterns to amino acid sequence

There are two pieces of experimental data that are used primarily to match a peptide tandem mass spectrum with an amino acid sequence via an algorithm. The first is the measured precursor ion *m/z* value, which, in the absence of an amino acid sequence database, represents only a set of possible amino acid compositions. In *Fig. 4*, this *m/z* value was $[M + H^+]^{1+} = 1733$ for the peptide with one proton attached, but it was actually the $[M + 2H^+]^{2+}$ ion = 867 that was selected for CID. Note that the ion selected for CID will, under most circumstances, be fully destroyed in an ion trap and is thus not visible in *Fig. 4*. The second piece of experimental information is the measured *m/z* values of the fragment ions, which are related to the amino acid sequence, as shown in *Fig. 4*. Annotated as b and y ions (see *Table 1*, also available in the color section), the computer-calculated *m/z* values for each theoretical fragment ion are shown, along with a color-coded indication of which fragment ions were matched. Thus, a typical algorithm will take the precursor ion *m/z* value deconvoluted to molecular mass and ask the question: what sequences in a genomic sequence database could generate peptides with a molecular mass equal to the measured molecular mass? Next, a set of theoretical tandem mass spectra are generated and compared, often by cross-correlation, with the observed tandem mass spectrum.

Simplistically, the sequence pattern that matches best between the observed and the theoretical data has the highest possibility of being correct. It should be noted that these database sequence-matching algorithms will *always* produce a best-fit answer. If the sequence of the protein under interrogation in your sample is present in the database and the data quality of the tandem mass spectrum for a peptide from that protein is high, then the likelihood of a correct match is also very high. However, if the sequence database does not contain the protein in your sample for which you have real data in the computer, then the result will be a match to the theoretical tandem mass spectrum that most closely matches your observed tandem mass spectrum, and, of course, a false positive. Such a false

Table 1. List of all b and y ions for the peptide with amino acid sequence MFDFNDSMVSNAIIK[a] (see page xxiii for color version)

b$^+$ ions		Amino acid	y$^+$ and y^{2+} ions		
m/z value of b$^+$ ion	Ion number		m/z value of y$^+$ ions	m/z value of y^{2+} ions	Ion number
132.2005	1	M	–	–	15
279.3771	2	F	1601.8145	801.4112	14
394.4657	3	D	1454.6379	727.8229	13
541.6422	4	F	1339.5493	670.2786	12
655.7461	5	N	1192.3728	596.6904	11
770.8347	6	D	1078.2689	539.6384	10
857.9129	7	S	963.1803	482.0941	9
989.1054	8	M	876.1021	438.555	8
1088.238	9	V	744.9096	372.9588	7
1175.3162	10	S	645.777	323.3925	6
1289.42	11	N	558.6988	279.8534	5
1360.4988	12	A	444.595	222.8015	4
1473.6583	13	I	373.7162	187.2621	3
1586.8177	14	I	260.3567	130.6823	2
–	15	K	147.1973	74.1026	1

[a]The b and y ions detected experimentally are highlighted in red and blue, respectively, and correspond to those shown in *Fig. 4*.

match may also arise from a poor or low-quality tandem mass spectrum. Additionally, in the event that the gene/protein sequence for your protein is not known or is missing from the database but where a gene/protein of a closely related species is present, there is an added possibility of matching the tandem mass spectrum to the correct peptide sequence but incorrectly identifying the parental protein from a related species. Practically, this is not usually a problem, except perhaps for forensic or environmental studies, as one usually knows the identity of the species from which the sample was derived.

In most cases, however, protein identification has become considerably simpler, as essentially all possible protein sequences are represented in the sequence databases. At the time of writing, data for more than 2100 genomes are publicly available (www.genomesonline.org). Even though there are millions of genes in sequence databases, 'all-encompassing' databases are not typically used in search strategies, as one knows the organism and one can select a database representing an organism(s), compile a subdatabase from various large databases, or compile a large database from various smaller databases. A narrowed search (i.e. small database) greatly speeds up the search time and lowers the possibility of false matches.

There is considerable debate about the best practices for conducting and validating automated database searches (see also Chapters 11 and 12), but currently it is not possible to state agreed guidelines. However, it is appropriate to consider the protease used to process the protein sample prior to MS analysis. Generally, trypsin is the most commonly used enzyme to cleave proteins, as it cuts specifically at sites C-terminal to Lys or Arg (except at Lys-Pro/Arg-Pro sequences).

However, it can cleave, spuriously, at other amino acids and cleavage reactions may not be driven to completion due to various problems in a real biological sample such as intrinsic proteases, pH, and ionic strength. This tendency to fragment at residues other than those accepted as specific cleavage sites is a phenomenon characteristic of all proteases. Missed cleavage may occur often in which, in the case of trypsin, a Lys or Arg is 'missed' and left in place in the interior sequence of the peptide. This will become apparent from the results of the database search. The more contentious problem is nonspecific cleavage at amino acids other than those in the canonical recognition sequence for the protease used in the experiment. Such data are more difficult to interpret confidently and these tandem mass spectral matches to peptide sequences should be reviewed manually. In an initial first pass analysis, they may be ignored entirely by the algorithm by requiring that every peptide identified by the search algorithm contains an Arg or Lys at the C terminus, except for the C-terminal peptide proper. However, we argue in favor of a two-tiered approach where an initial database search is conducted with unrestricted enzyme specificity and the data sorted as desired by the analyst. Whilst the results from unrestricted database searches take longer to generate, they may provide novel matches to proteins that would be missed by a search restricted to only the known enzymatic specificity.

In *Fig. 4*, one can view the annotated peptide tandem mass spectrum for the tryptic peptide with amino acid sequence MFDFNDSMVSNAIIK and a molecular mass of 1733 Da. This peptide sequence was assigned via the database search process described above. *Table 1* shows the b, y and y^{2+} ions that the database search engine SEQUEST matched. Note that in both *Fig. 4* and *Table 1*, y ions are highlighted in blue and b ions in red. This tandem mass spectrum was acquired in a linear ion trap, where the *m/z* range that is scanned is a function of the precursor ion mass that was selected. With these instruments, the spectrum only begins at ~30% of the precursor ion *m/z* values (12). Note that this data loss does not occur with other types of tandem mass spectrometer, such as triple quadrupole or quadrupole time-of-flight mass spectrometers. Future software developments promise also to provide this low *m/z* region on ion traps.

5. CONCLUDING REMARKS

We have provided the reader with a detailed set of generic protocols for so-called shotgun proteomics to identify proteins from complex mixtures. In addition to the generation and separation of peptides and MS operations, which can be implemented and used routinely, the peptide tandem mass spectral matching to archived sequence in databases remains an issue of continued discussion, and developments are ongoing. As we shift from analysis of hundreds to thousands of protein identifications, greater care must be taken to generate the best-quality information. Successful proteomics is governed by three main factors: systematic, reproducible, and accurate analysis. When this is achieved, one can focus on the task of management and sharing of data. Ultimately, what is required is not a table with a list of proteins, but a humanly interpretable form of biological information. For this, we need the involvement and contribution of many types of scientist.

Acknowledgements

D.R.G. thanks the NIH for support from NCRR 1S10RR17262-01, NIEHS 5P30ES007033-10, and NIAID 1054 A157141-01 awards.

6. REFERENCES

★ 1. Aebersold R & Goodlett DR (2001) *Chem. Rev.* **101**, 269–295.
  2. Ideker T, Galitski T & Hood L (2001) *Annu. Rev. Genomics Hum. Genet.* **2**, 343–372.
  3. Ideker T, Thorsson V, Ranish JA, *et al.* (2001) *Science*, **292**, 929–934.
  4. Kitano H (2002) *Science*, **295**, 1662–1664.
★ 5. Yi EC, Lee H, Aebersold R & Goodlett DR (2003) *Rapid Commun. Mass Spectrom.* **17**, 2093–2098.
★ 6. Hunt DF, Henderson RA, Shabanowitz J, *et al.* (1992) *Science*, **255**, 1261–1263.
  7. Corthals GL, Aebersold R & Goodlett DR (2005) *Methods Enzymol.* **405**, 66–81.
  8. Lee H, Yi EC, Wen B, *et al.* (2004) *J. Chromatogr. B Analyt. Technol. Biomed. Life Sci.* **803**, 101–110.
★ 9. Yates JRI, Carmack E, Hays L, Link AJ & Eng JK (1998) In *Methods in Molecular Biology: 2-D Proteome Analysis Protocols*, pp. 553–569. Edited by AJ Link. Humana Press, Totowa, NJ.
  10. Yi EC, Marelli M, Lee H, *et al.* (2002) *Electrophoresis*, **23**, 3205–3216.
  11. Hernandez P, Muller M & Appel RD (2006) *Mass Spectrom. Rev.* **25**, 235–254.
  12. Corthals GL, Gygi SP, Aebersold R & Patterson SD (1999) In *Proteome Research: Two-dimensional Gel Electrophoresis and Detection Methods*, pp. 197–231. Edited by T Rabilloud. Springer, New York.

CHAPTER 6

Desorption electrospray ionization: proteomics studies by a method that bridges ESI and MALDI

Zoltán Takáts, Justin M. Wiseman, Demian R. Ifa, and
R. Graham Cooks

1. INTRODUCTION

Desorption ionization (DI) methods have played a key role in the development of mass spectrometry (MS)-based proteomics. For decades, the application of DI techniques (such as field desorption, plasma desorption (PD), fast atom bombardment (FAB), and secondary ion MS (SIMS)) were the only derivatization-free ways for the mass spectrometric investigation of large, nonvolatile, fragile molecules such as peptides, oligosaccharides, and nucleic acids (1). In the 1980s, the transformation of laser desorption (LD) into the widely useful and easily implemented method of matrix-assisted laser desorption/ionization (MALDI) (2) displaced most of the earlier DI methodologies, at least in proteomics applications, and extended MS to intact biomolecules of all sizes. MALDI, like most DI methods, is easy to automate and is an excellent tool for high-throughput analytics.

This sequence of developments provided a solution to the problem of analysis of intact proteins, provided that solid-phase samples were analyzed. A method that allowed intact solution-phase proteins to be examined was developed almost simultaneously, in the form of the spray ionization method of electrospray ionization (ESI) (3). This method has the advantages of compatibility with liquid chromatography and so allows on-line experiments involving purification by chromatography.

Desorption ESI (DESI), introduced in 2004 by Takats and co-workers (4), is a desorption ionization method by nature and, like MALDI, is used for the analysis of material present on a surface. DESI also includes features reminiscent of ESI in respect to both its instrumental and mechanistic aspects. However, the analyte in the DESI experiment is not in solution as in ESI. Instead, a microelectrospray ion source is used to produce charged droplets, ionic clusters, and/or gas-phase ions (depending on the chosen experimental conditions) and these are directed at the sample surface. The sample is present in the ambient environment. An electrical potential of several kilovolts is applied to the spray solution and pneumatic nebulization is used to assist in desolvation. Ionization of molecules present on

Proteomics: *Methods Express* (C.D. O'Connor and B.D. Hames, eds)
© Scion Publishing, 2008

the sample surface occurs on impact of the ESI-originated, charged particles with the surface. Surfaces include deposited samples on sample holder targets, as well as surfaces of natural objects such as biological tissues or minerals.

1.1 DESI instrumentation

A DESI ion source is a pneumatically assisted microelectrospray source equipped with a surface holder and positioning devices. The source comprises two main parts, a sprayer assembly and a surface assembly, both mounted on a source base (see *Fig. 1*). A high voltage (3–5 kV) is applied to the liquid junction on a stainless steel union or on the stainless steel syringe needle used to deliver the spray solvent. (*Caution: the high-voltage electrical connection should be isolated from the environment to prevent electrical shock.*) The sprayer itself is mounted onto a vertical rotating stage, which in turn is mounted onto a three-dimensional (3D) linear moving stage. The linear movement is used to change the sprayer-to-MS or sprayer-to-sample distance, and also to compensate for the different angles at which the sprayer is used. The rotating stage allows selection of charged droplet impact angles from 0 to 90°. The sample is placed on the surface holder, which is mounted onto a separate 3D moving stage. In one version, the surface holder can carry 5 cm × 1 cm large disposable surface slides, which lie in a thin piece of stainless steel embedded into a surface holder made of polytetrafluoroethylene (PTFE). The metal support is connected to an external high-voltage power supply to provide an appropriate surface potential. The range of surface potentials available is identical to the range for the ion source (0–6 kV). In a different surface holder design, the surface holder is a polyetheretherketone-coated aluminum block, which has a built-in heater cartridge, a Peltier cooling device, and a thermometer. This surface holder provides controlled surface temperatures in the range of −20 to 300°C. The newest DESI ion sources (Omni Spray Ion Source) feature two charge-coupled device cameras and a light source. For high-throughput applications, the surface assembly is replaced by a moving belt system or rotating disk system, as shown in *Fig. 2*. Sample targets using pre-deposited arrays of samples similar to those used in commercial MALDI systems are also feasible, but the moving belt and rotating disk devices are mostly used in DESI.

A recent development in DESI now allows for the direct analysis of microtiter well plates (e.g., N = 96, 384 etc.) in a high throughput format. This advance was brought about by the addition of a small pressure tight enclosure surrounding the sprayer and inlet into the mass spectrometer (*Fig. 3*). This allows for effective sampling of ions even when the spray is orthogonal to the sample surface and with an inlet capillary parallel to the sprayer. An embodiment of this configuration allows for direct analysis *inside* well plates where the well-wall naturally creates the enclosure. This technique, referred to as geometry independent DESI, promises to simplify DESI instrumentation and make it much easier to implement.

DESI ion sources can be coupled to practically any type of mass analyzer fitted with an atmospheric interface of the type used in electrospray or atmospheric pressure ionization sources. Most examples of spectra provided in this chapter were recorded using a linear ion trap mass spectrometer (LTQ; Thermo Scientific), although in some cases the data were recorded using a time-of-flight mass spectrometer.

INTRODUCTION 101

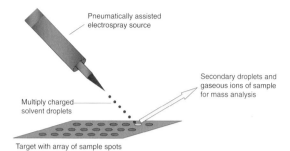

Figure 1. Concept of a DESI ion source.
The DESI ion source is a pneumatically assisted microelectrospray ion source capable of delivering a coaxial sheath gas at up to 350 m/s. High-voltage contact is applied at the liquid junction between solvent delivery lines or at the spray nozzle. The sample is in the ordinary ambient environment.

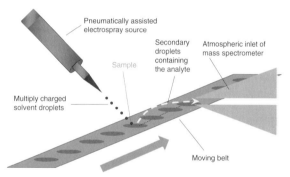

Figure 2. High-throughput DESI experiment.
Samples are spotted on a moving belt positioned in front of the inlet of the mass spectrometer.

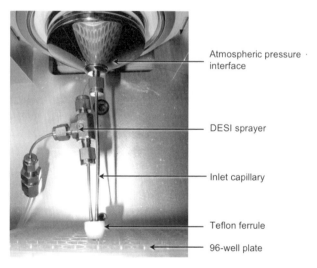

Figure 3. Picture of geometry independent DESI ion source on a Thermo Scientific LTQ mass spectrometer.

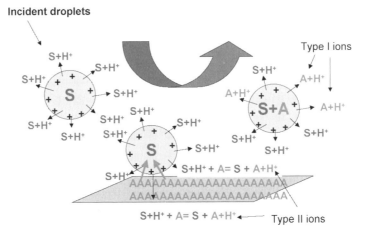

Figure 4. General scheme for two ion formation mechanisms in DESI-MS.
Type I ions are formed by charge (e.g. proton) transfer when primary ions or multiply charged incident droplets impact the surface and react with the analyte present at the surface. Type II ions are formed when charged droplets impact the surface, dissolve the analyte, and leave the surface as secondary charged microdroplets, which are desolvated in the mass spectrometer inlet to give gas phase ions. S, surface; A, analyte.

1.2 Ion formation

The relationship of DESI to the other DI methods is close, at least at the phenomenological level. All of the DI methods (PD, LDI, MALDI, SIMS, and FAB) involve condensed-phase samples being impacted by projectiles; these include photons (LD, including MALDI) and translationally excited atoms (FAB) and energetic ions (SIMS). The projectiles used in SIMS include polyatomic ions and it is well established that polyatomics are significantly more efficient than are atomic ions in sputtering molecules from surfaces. However, hitherto all DI experiments using particles as projectiles have been conducted in a high-vacuum environment: the fact that DESI is performed in air is the principal feature that distinguishes it from earlier DI methods. This also implies significant differences in the fundamental processes involved, as the momentum transfer collisions upon which traditional SIMS and other DI sputtering mechanisms are based are not applicable in the high-pressure environment of DESI, as the projectiles can have only low kinetic energies. Low-energetic collisions (1–100 eV in the laboratory frame of reference), however, can effect desorption/ionization of molecules from surfaces through a heterolytic process known as chemical sputtering (5). In this process, desorption/ionization occurs as a result of charge transfer (electron, proton, or other ionic fragment) from a low translational energy primary ion to a surface molecule.

The collisions of multiply charged aqueous droplets with a surface introduce an additional and fundamentally different mechanism of ion formation in DESI. The multiply charged primary droplets impacting the surface are scattered as secondary droplets that carry the analyte and a fraction of the charge and mass of the primary droplet. In this case, the 'desorption' event is essentially a phase transfer; the solid-phase analyte is adsorbed onto the impacting aqueous droplets and carried in solution to the mass spectrometer. In almost all cases, this is the primary desorption/ionization mechanism present in the DESI experiment (6). Subsequent ion formation from the secondary droplet closely resembles conventional ESI and proceeds either by ion emission (ion evaporation model) or by evaporation of neutral droplet molecules (charged residue model) to produce gaseous ions (see *Fig. 4*). Hence, DESI spectral characteristics are rather similar to those of ESI. DESI spectra of peptides and proteins feature a series of multiply charged molecular ions and almost completely lack ionization-related fragmentation (see *Fig. 5*).

2. METHODS AND APPROACHES

The application areas of DESI can be divided into two categories, based on the type of sample preparation method used.

The first category covers the investigation of pre-deposited samples, for example where solutions are deposited on substrates such as paper or plastic. One of the important features of this type of DESI experiment is its high-throughput character, which has been used in the analysis of biological samples as well as pharmaceutical formulations (7, 8). This type of sample and the high-throughput character of the application have potential value in the fields of industrial process monitoring, food safety, environmental analysis, and clinical diagnostics. High-throughput DESI analysis of biological fluid samples (e.g. urine) for metabolite identification is also receiving attention due to the potential value of information on the distribution of characteristic small molecules as biomarkers of disease. With limited sample preparation required, even for the analysis of raw urine, a few hundred samples per hour can be analyzed using DESI-MS when biological fluids are spotted onto paper or other suitable substrates. In work by Chen *et al.* (7), the detection of over 80 metabolites in urine without any sample preparation was achieved using DESI.

The second group of applications deals with the DESI-MS investigation of natural, unmodified surfaces. One of the most important applications of this type is the investigation of biological tissues. Thin tissue sections prepared for microscopy or native, freshly cut tissue surfaces are used for DESI-MS analysis and even – using a fine DESI probe beam – chemical imaging. Imaging MS based on MALDI and SIMS has become a powerful technique for analyzing histological sections of biological tissues (9, 10). These techniques require that the sample be confined to the high-vacuum region of the instrument, severely limiting any further chemical or physical manipulation of the sample, a limitation not inherent in DESI studies. The determination of the spatial distribution of natural tissue components such as membrane phospholipids has recently been demonstrated using DESI (11), and in other work an automated DESI source was used for two-dimensional (2D) imaging of tissue sections and other surfaces (12, 13). Other applications of DESI to natural surfaces occur in forensic analytics, especially the detection of explosives, toxins, and drugs of abuse on the surfaces of personal items.

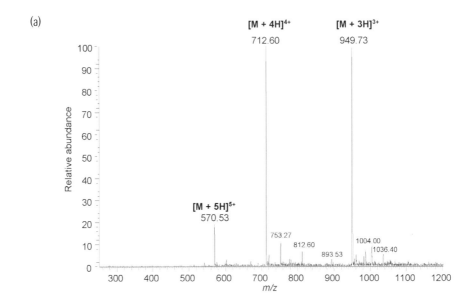

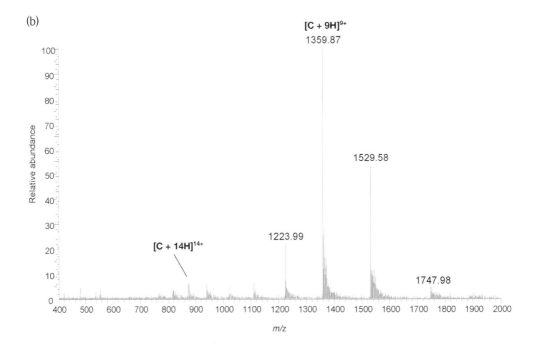

Figure 5. Examples of DESI mass spectra.
(*a*) DESI mass spectrum of 100 pg (total amount deposited on surface) of melittin (M) (molecular mass 2847 Da) deposited on a PVDF surface. (*b*) DESI mass spectrum of 1 pmol of cyctochrome c (C) (from equine heart) on a PTFE surface. Methanol : water (1 : 1) was used as the spray solvent at a flow rate of 5 μl/min. For more details, see *Protocol 1*.

DESI-MS is an emerging technique with great promise, but its application range is still being investigated. Therefore, the protocols presented below are general procedures used for the applications that have been investigated so far. Optimal ion source parameters and surface types may vary depending on the application.

2.1 Analysis of intact proteins

DESI is amenable to the study of intact proteins in complex mixtures, including blood or other biological media. Intact proteins can be desorbed and ionized from the surface under gentle conditions to produce compact conformations of the protein (14). In their report, Shin et al. (15) systematically investigated the detection of several proteins ranging in molecular mass from 12 to 66 kDa. The authors reported a reduction in the limits of detection with decreasing molecular mass of the proteins. Also reported was the detection of the intact bacteriophage MS2 capsid protein after the crude *Escherichia coli* host-cell suspension was passed through a 100 kDa molecular mass cut-off filter. These data demonstrate the ability of DESI-MS to detection proteins from biological matrices with minimal sample preparation and without chromatographic separation.

A procedure for DESI analysis of intact proteins and oligopeptides is described in *Protocol 1*. As an example of the type and quality of data obtained from intact pure proteins, *Fig. 6* shows the DESI mass spectrum of 1 pmol of lysozyme pre-

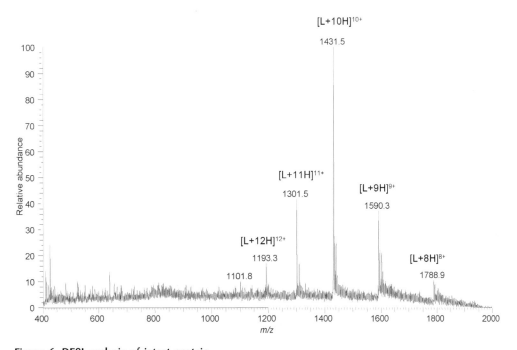

Figure 6. DESI analysis of intact protein.
DESI mass spectrum of 1 pmol of lysozyme (L) deposited onto a PTFE surface. Methanol : water (1 : 1) was used as the spray solvent at a flow rate of 5 µl/min. For more details, see *Protocol 1*.

pared by dissolution in methanol : water (1 : 1) solvent and deposition onto a PTFE surface. Both the composition of the solvent in the spray and that used to dissolve the analyte affect the appearance of the mass spectrum. For example, the use of acidified aqueous/organic solvents for deposition of the analyte produces results similar to the results obtained with ESI of the same sample where higher charge states appear in the mass spectrum indicating an unfolded or partially unfolded conformation of the protein. The mass spectra in *Fig. 7* show the data obtained when cytochrome c is dissolved in methanol : water : acetic acid (50 : 48 : 2) as opposed to methanol : water (50 : 50) as shown in *Fig. 5(b)*.

Protocol 1

DESI analysis of intact proteins/oligopeptides

Equipment and Reagents
- Protein of interest (1–100 ng/ml)
- Mass spectrometer equipped with a DESI ion source
- Surface slides (polymethyl methacrylate (PMMA), PTFE, or glass)
- Pipette tips
- Spray solvent: methanol/water or aqueous buffers[a]

Method
1. Deposit a 1 µl aliquot of a solution containing the protein of interest onto a PTFE, PMMA, or glass slide. An arbitrary solvent system can be used for deposition, as long as the protein is soluble in the selected solvent. Note that the unfolding and thus the charge state distribution of the protein ions depend on the composition of the solvent system used for deposition.
2. Allow the protein sample spot to dry.
3. Expose the dried sample to pneumatically assisted microelectrospray using a DESI ion source coupled to a mass spectrometer; typical parameters are given in *Table 1*[a].

Note
[a]Spray solvents include aqueous buffers (e.g. 10 mM NH_4CH_3COO), water/methanol and water/acetonitrile mixtures at various compositions. The use of nonvolatile buffers or high concentrations of acid (e.g. acetic acid) must be avoided.

2.2 Analysis of tryptic digests

The analytical utility of DESI is such that it can be applied to qualitative proteomics research in the same way as MALDI and ESI methods, although little work has yet been reported in this regard. As DESI is a surface analysis technique and easily automated, it can be implemented for high-throughput applications, which include the analysis of chromatographic fractions of digested proteins. The analysis of tryptic peptides follows the same protocols as in typical MALDI or ESI methods (see *Protocol 2*), except that the mixture is spotted directly onto an insulating surface, allowed to dry, and analyzed directly without adding matrix compounds

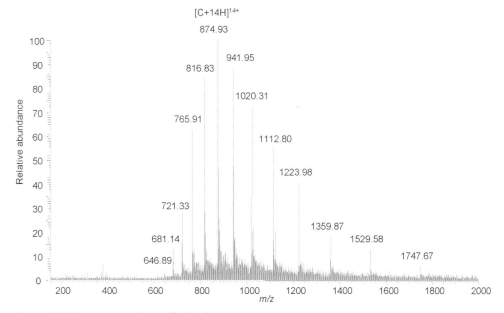

Figure 7. DESI mass spectrum of cytochrome c.
DESI mass spectrum of 1 pmol of cytochrome c (C) dissolved in methanol : water : acetic acid (50 : 48 : 2) and deposited onto a PTFE surface. Methanol : water (1 : 1) was used as the spray solvent at a flow rate of 5 μl/min.

Table 1. Typical parameters used for analysis of proteins and peptides

Parameter	Value
Spray tip-to-surface distance	1–2 mm
Incident angle of spray	70–80°
Spray tip-to-MS inlet distance	3–5 mm
Collection angle	0–5°
Solvent flow rate	1–5 μl/min
Nebulizing gas linear velocity	300–400 m/s
Spray high voltage	4–6 kV
Surface temperature	30–80°C

(as in the case of MALDI methods). The spectral characteristics are similar to those of ESI in that both singly and multiply charged analyte ions are detected. Spectra are highly similar to electrospray spectra of tryptic digests with regard to the overwhelming presence of multiply charged ions of peptides. DESI spectra show ion abundances that vary with time, similar to other DI methods. The signal can be stabilized by using higher spray tip-to-surface distances at lower incident angles. In optimal cases, spectra can be collected for 20–100 s without moving the surface in the case of deposited samples.

Recently, Creaser and co-workers (16) demonstrated the utility of the combination of DESI and ion mobility spectrometry/MS for the rapid analysis of tryptic peptides without sample pre-treatment or chromatographic separation.

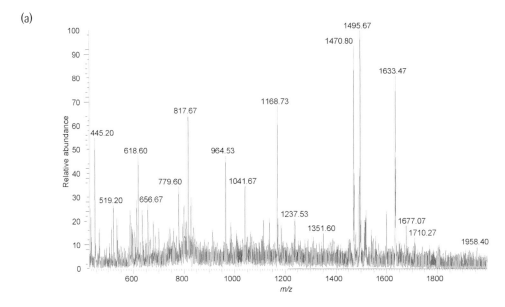
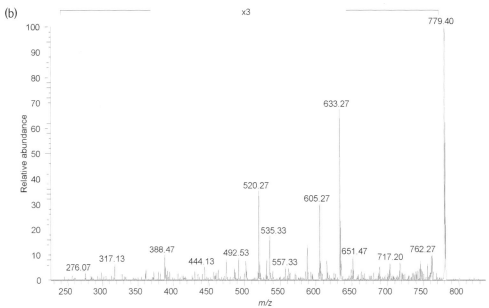

Figure 8. DESI mass spectra of a trypsin digest of cytochrome c.
(a) DESI mass spectrum of a trypsin digest of 100 ng solution of cytochrome c deposited onto a PMMA surface. Methanol : water (1 : 1) was used as the spray solvent at a flow rate of 5 µl/min.
(b) MS/MS spectrum of a tryptic fragment at m/z 779. For more details, see *Protocol 2*.

The results of their studies showed good agreement between DESI and ESI analysis of a tryptic digest of bovine serum albumin in terms of absolute signal intensity and signal-to-noise (S/N) ratio. Interestingly, their results also showed that in DESI there were more singly charged species detected than in the case of ESI, resulting in a less-complex spectrum in the mass range between m/z 650 and 800. Consequently, this resulted in better sequence coverage with higher confidence in peptide matches for protein identification. The analysis of a trypsin digest of cytochrome c is shown in *Fig. 8(a)* and the product ion MS/MS spectrum of one tryptic peptide is shown in *Fig. 8(b)*. This is one example of the type and quality of mass spectrum recorded using DESI-MS without chromatographic separation.

Protocol 2

DESI analysis of tryptic digests/peptides

Equipment and Reagents
- Tryptic digest/peptides (1–100 ng/ml)[a]
- Mass spectrometer equipped with DESI ion source
- Surface slides (PMMA or PTFE)
- Pipette tips
- Spray solvent: methanol/water or aqueous buffers[b]

Method
1. Deposit a 1 µl aliquot of a solution of tryptic digest onto a PMMA or PTFE slide; the use of other surface materials is not advised. The ideal surface concentration is 10 ng/mm^2; higher concentrations cause disadvantageous suppression effects[a].
2. Allow the peptide sample spot to dry.
3. Expose the dried sample spot to a pneumatically assisted electrospray using a DESI ion source coupled to mass spectrometer; typical parameters are given in *Table 2*.

Notes
[a] Aqueous buffers or buffered water/acetonitrile mixtures can be used for deposition. Methanol or methanol-containing solvent systems are not compatible with the PMMA surface.
[b] Commonly used spray solvents include aqueous buffers (e.g. 10 mM NH_4CH_3COO), water/methanol and water/acetonitrile mixtures at various compositions. Use of nonvolatile buffers or high concentrations of acid (e.g. acetic acid) must be avoided.

Table 2. Typical parameters used for analysis of tryptic digests/peptides

Parameter	Value
Spray tip-to-surface distance	2–3 mm
Incident angle of spray	60–80°
Spray tip-to-MS inlet distance	3–5 mm
Collection angle	0–5°
Solvent flow rate	1–2 µl/min
Nebulizing gas linear velocity	300–400 m/s
Spray high voltage	4–6 kV
Surface temperature	50–100°C

2.3 Coupling DESI with separation methods

The application of DESI in conjunction with planar separation methods is advantageous, as it provides a direct and rapid means of analysis of the separated components without the need for staining or other visualization procedures. In this respect, relevant considerations parallel those met in coupling MALDI to separation methods, except that DESI can be carried out more easily as a matrix is not required. There are few examples of the direct coupling of planar chromatography

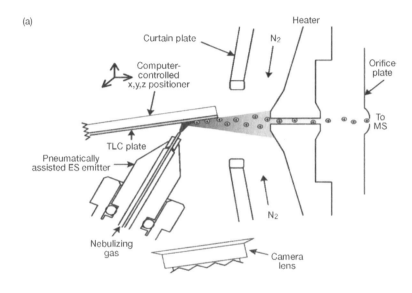

Figure 9. Example of a TLC/DESI-MS experiment (see page xxiv for color version). (*a*) Schematic illustration of the TLC/DESI-MS experimental set-up. (*b*) Color photograph of the DESI emitter and the TLC plate as viewed through the camera monitor during a TLC/DESI-MS experiment. Separate bands of rhodamine 6G (orange band), rhodamine B (pink band), and rhodamine 123 (yellow band) are observed on the RP C8 TLC plate. Reprinted with permission from *Anal. Chem.* (2005); **77**(5), 1207–1215. Copyright 2005, American Chemical Society.

methods with MS, primarily because of the physical constraints imposed by the vacuum system. For example, MALDI is the most common approach for MS analysis of thin-layer chromatography (TLC) plates but the need for extensive post-separation sample preparation and specialized plates has limited its widespread use (17, 18). As DESI operates at atmospheric pressure under ambient conditions, the physical constraints are lifted, allowing the native substrate to be analyzed *in situ*. The direct analysis of TLC plates using DESI-MS was demonstrated in an early publication by Van Berkel *et al.* (19). These authors used a computer-controlled motion stage to analyze different rhodamine dyes and various drug compounds separated on TLC plates (see *Fig. 9*, also available in the color section). This first combination of DESI-MS with planar chromatography showed the advantages of speed, versatility, and enhanced specificity through MS/MS studies.

Gel-based separations (e.g. 2D gel electrophoresis) have constituted the primary tool for global protein expression profiling in proteomics. In combination with MS, hundreds of proteins can be identified in this way. In the traditional approach, separated proteins are excised from the gel and digested using, for example, a serine protease. This approach is generally reagent- and time-intensive, involving destaining agents and extraction solvents. In some cases, it is advantageous to transfer the separated proteins onto an inert or chemically modified surface (e.g. immobilized trypsin on nitrocellulose membrane) for enzymatic digestion. The analysis of separated proteins or peptides from membrane substrates has been demonstrated using MALDI-MS (20). *Fig. 10(a)* shows the general concept of coupling electroblotting to DESI-MS; *Fig. 10(b)* shows the proof-of-

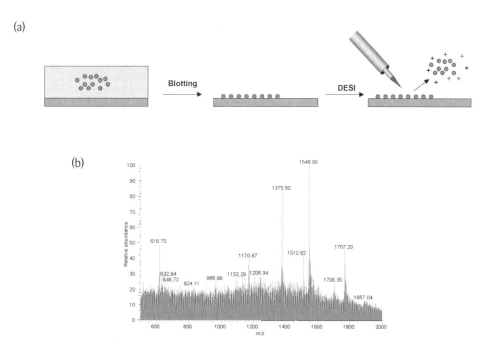

Figure 10. Analysis of electroblotted proteins or peptides using DESI-MS.
The concept of the procedure is shown in (*a*), whilst (*b*) shows a DESI mass spectrum of cytochrome c after SDS-PAGE separation and electroblotting onto PVDF membrane.

concept DESI mass spectrum of cytochrome c after sodium dodecyl sulfate polyacrylamide gel electrophoresis (SDS-PAGE) separation and electroblotting onto polyvinylidene difluoride (PVDF). As DESI is such a new technique, this application is also only in the early stages of development.

2.4 Reactive DESI

DESI can be operated as a traditional surface analysis technique capable of ionizing molecules present on surfaces of any type, including natural materials. DESI also offers an alternative mode of operation, termed 'reactive desorption', which is especially useful for the analysis of specific components in complex mixtures. In this experiment, a reagent is mixed into the primary electrospray solvent and undergoes a rapid chemical reaction with the analyte present on the surface. Examples include functional group-specific reactions involving new covalent bond formation and the formation of specific noncovalent complexes. Reactive DESI experiments can involve ion/molecule reactions, but these are quite different from traditional MS ion/molecule reactions in that the experiment is conducted using solution-based reagents in the atmosphere under ambient conditions. Reactions occur at the solution/solid interface, and the neutral analyte can be nonvolatile, each of these characteristics representing an extension to ion/molecule reaction methodology (21). Reactive desorption is generally applied to increase the sensitivity or specificity of the DESI-MS technique. However, the technique can also be applied to the study of small-molecule interactions with peptides or proteins for drug discovery or activity assays.

This unique analytical specificity of the *reactive* DESI experiment was recently demonstrated for the analysis of compounds containing *cis*-diol functionality (22). The specific recognition of aliphatic diols (carbohydrates and steroids) and aromatic diols (flavoids and catecholamines) was attempted using DESI by spraying a solution containing benzeneboronate anions (PhB(OH)$_3^-$ (see *Fig. 11*). These compounds pose a particular challenge in traditional ESI or MALDI analysis because their ionization is generally inefficient. *Fig. 12* shows the DESI mass spectra of the reaction of charged microdroplets containing PhB(OH)$_3^-$ impacting a surface (cotton swab tip) bearing D-fructose (*Fig. 12b*) and D-glucose (*Fig. 12c*). Cyclization reactions occur with the PhB(OH)$_3^-$ reagent under basic conditions and the characteristic complexation products were detected in the mass spectra, allowing specific detection of the *cis*-diol functionality. These reactions were also

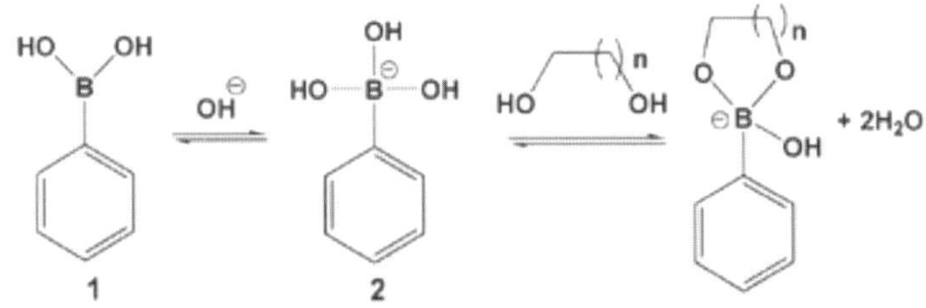

Figure 11. General equation for phenylboronic acid-*cis*-diol complexation.
(*1*) Phenylboronic acid (PhB(OH)$_2$); (*2*) arylboronate intermediate (PhB(OH)$_3^-$).
Ref. 22. Reproduced by permission of the Royal Society of Chemistry.

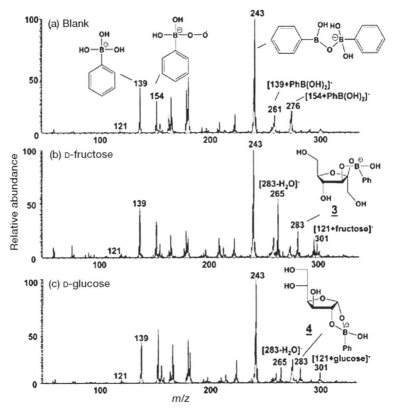

Figure 12. Reactive DESI mass spectra.
The spectra show the ionic species generated from PhB(OH)$_3^-$ anions upon interaction with (a) a blank surface, (b) D-fructose on the surface, and (c) D-glucose absorbed on a cotton tip. Ref. 22. Reproduced by permission of the Royal Society of Chemistry.

demonstrated for a number of other diols, including epinephrine, estriol, and quercetin, and application to glycopeptides is underway.

Another illustration of the potential of reactive DESI is the selective titration of amino acids. These experiments consist, for example, of monitoring covalent complexes between arginine residues of proteins and sulfonates such as 1-anilino-naphthalene-8-sulfonic acid (ANS) (23). The determination of ANS complexes is useful for probing folded structures, as some arginine residues are buried and not accessible to the solvent. These experiments are compatible with DESI and can lead to rapid structural screening of proteins and peptides, especially due to the high-throughput and ready exchange of solvents. *Fig. 13* shows a spectrum of lysozyme deposited on a Teflon surface and sprayed with 3.4 µM ANS; up to four ANS molecules were complexed.

2.5 *In situ* proteomics

DESI allows *in situ* analysis of biological tissues (11, 12). Wiseman *et al.* (11) analyzed adenocarcinoma tissue from human liver and determined the lipid

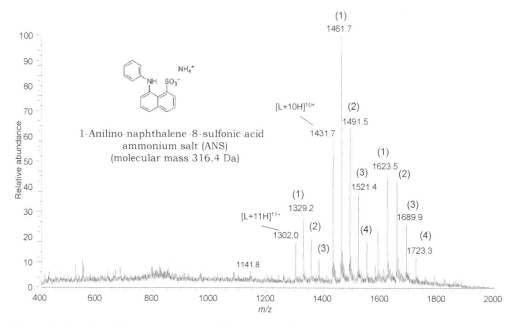

Figure 13. Reactive DESI mass spectrum of lysozyme–ANS complexes.
Lysozyme (1 pmol) was deposited on a PTFE surface and 3.4 μM ANS in methanol : water (1 : 1) was sprayed at the surface.

composition at a series of positions across the tissue. In these experiments, thin frozen tissue sections were analyzed directly in the laboratory environment without additional treatment. The results from these studies indicated that the abundance of specific lipid species was enhanced in the tumor portion of the tissue. *In situ* analysis of unprocessed tissue is typified by the DESI mass spectrum shown in *Fig. 14(a)*. In this case, an intact *Gallus gallus* (chicken) heart was analyzed using DESI. The most abundant ions present in the mass spectrum were phospholipids and other lipid species. *Fig. 14(b)* shows the DESI mass spectrum recorded on a dried blood spot on the *G. gallus* heart. The analysis of the blood residue on the chicken heart showed the presence of heme and hemoglobin chains, as would be expected. The analysis of less-abundant protein constituents within the tissue sample often requires the removal of lipid species prior to analysis, similar to the situation with MALDI (9). After removal of lipid constituents, the tissue can be treated with protease to degrade proteins present in the tissue. The tryptic products can be investigated directly from the tissue using DESI (see *Protocol 3*). The spectra obtained feature ions of tryptic fragments from abundant proteins present in the tissue sample. The digestion is usually not complete; hence, the presence of missed cleavage sites is typical in the peptides detected. The signal is more stable for longer times than in the case of deposited samples, so the recording of MS/MS data is simple in this case. The full potential of this application is currently being explored in detail.

(a)

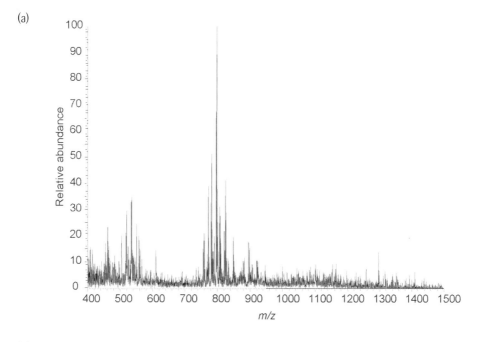

(b)

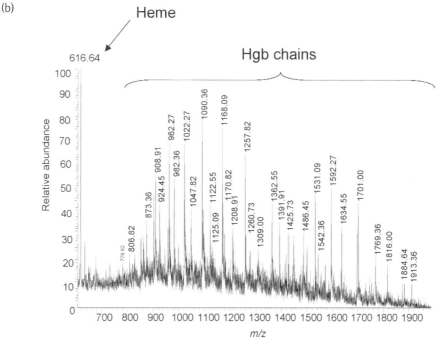

Figure 14. *In situ* DESI analysis.
(*a*) *In situ* DESI analysis of intact chicken (*Gallus gallus*) heart using ethanol/water spray solvent at 5 µl/min. (*b*) DESI mass spectrum of hemoglobin (Hgb) chains recorded on a dried blood spot on the chicken heart.

Protocol 3

In situ DESI analysis of tissue sections

Equipment and Reagents
- Mass spectrometer equipped with DESI ion source
- Spray chamber
- Vacuum desiccator
- Surface slides (PMMA, PTFE, or glass)
- Pipette tips
- Spray solvent: methanol/water or aqueous buffers[a]
- Chloroform : acetone (1 : 1)
- Dichloromethane
- Alkylated trypsin (10 mg/ml) in 10 mM ammonium bicarbonate (pH 8)

Method
1. Thaw mount a fresh or frozen tissue section onto a PMMA, PTFE, or glass slide and wash three times with chloroform : acetone (1 : 1).

2. Remove any traces of acetone by washing the sample in dichloromethane.

3. Spray deposit a 10 µg/ml solution of alkylated trypsin in 10 mM ammonium bicarbonate (pH 8) onto the tissue section to obtain a surface concentration of trypsin of 0.1–1 ng/mm^2.

4. Incubate the sample at 35–40°C for 1 h at 95–100 % relative humidity.

5. Dry the sample under vacuum for 3 h.

6. Expose the dried sample to DESI analysis; typical operating parameters are given in *Table 3*[a].

Note
[a]Typical spray solvents are acidified aqueous solvents, e.g. 0.1 % acetic acid in water.

Table 3. Typical parameters used for *in situ* analysis of tissue

Parameter	Value
Spray tip-to-surface distance	0.5–1 mm
Incident angle of spray	85°
Spray tip-to-MS inlet distance	3–5 mm
Collection angle	5°
Solvent flow rate	0.1–3 µl/min
Nebulizing gas linear velocity	300–400 m/s
Spray high voltage	4–6 kV
Surface temperature	80–100°C

3. TROUBLESHOOTING

- No surface-originated ions in the spectra
  A possible cause of this is that the sprayed droplets and ions are not reaching the surface due to the wrong spray parameters (i.e. the volumetric flow rate is

too low or there is no applied high voltage) or surface charging effects. Always check the spray pattern on the surface. If the spray is not visible on a glass surface, check to ensure that the solvent syringe pump is on and that there are no blockages in the solvent delivery line that would restrict flow.

- **No signal detected**
  There may be no spray present or the sprayed species are being deflected by the charged surface. Increase the solvent and gas flow rates. If there is no change, remove the surface and check whether the spray produces ions. Test the spray with 10 mg/ml bovine cytochrome c in 10 mM aqueous ammonium acetate. The test solution should give a narrow charge state distribution of the protein, with a main charge state of +7 or +8. If there are no ions, check whether the spray tip (or solvent line) is clogged. If the cytochrome c spectra are full of adducts, change the spray tip.

- **Excessive adduct formation**
  This may be caused by insufficient desolvation due to the presence of large droplets or contamination of the surface. Increase the gas flow rate and/or heat the surface. Check whether the sample contains salts at high concentration. If the deposited sample contains more than 100 mM inorganic, non-volatile salt, dilute it or remove the salts prior to deposition. If the PMMA surface is used in combination with methanol (as a solvent for the deposited sample), let the surface dry for an additional 10 min at 50°C.

- **Transient signal, not suitable for MS/MS**
  A possible cause of this is low surface concentration or lack of sample adhesion to the surface. Possible solutions are to try a different surface material or a roughened surface, increase the surface concentration of the sample, increase the spray tip-to-surface distance, or decrease the solvent flow rate.

- **Strong suppression effects and poor spectral resolution**
  Possible causes are high surface concentration or an inappropriate solvent system. Decrease the surface concentration of the sample or change the solvent composition.

4. REFERENCES

1. Pachuta SJ & Cooks RG (1987) *Chem. Rev.* **87**, 647–669.
2. Karas M & Hillenkamp F (1988) *Anal. Chem.* **60**, 2299–2301.
★ 3. Fenn JB, Mann M, Meng CK, Wong SF & Whitehouse CM (1989) *Science*, **246**, 64–71. – *The first description of ESI.*
★★★ 4. Takats Z, Wiseman JM, Gologan B & Cooks RG (2004) *Science*, **306**, 471–473. – *The original DESI reference, providing an excellent introduction to the technique.*
5. Vincenti M & Cooks RG (1988) *Org. Mass Spectrom.* **23**, 317–326.
6. Venter A, Sojka PE & Cooks RG (2006) *Anal. Chem.* **78**, 8549–8555.
7. Chen H, Talaty N, Takats Z & Cooks RG (2005) *Anal. Chem.* **77**, 6915–6927.
8. Weston DJ, Bateman R, Wilson ID, Wood TR & Creaser CS (2005) *Anal. Chem.* **77**, 7572–7580.
9. Caprioli RM, Farmer TB & Gile J (1997) *Anal. Chem.* **69**, 4751–4760.
10. Pacholski ML & Winograd N (1999) *Chem. Rev.* **99**, 2977–3006.
★★★ 11. Wiseman JM, Puolitaival SM, Takats Z, Cooks RG & Caprioli RM (2005) *Angew. Chem. Int. Ed. Engl.* **44**, 7094–7097. – *An example of DESI imaging of biological tissue.*

★★★ 12. Wiseman JM, Ifa DR, Song Q & Cooks RG (2006) *Angew. Chem. Int. Ed. Engl.* **45**, 7188–7192. – *An example of DESI imaging of biological tissue.*
13. Ifa DR, Wiseman JM, Song Q & Cooks RG (2007) *Int. J. Mass Spectrom.* **259**, 8–15.
14. Myung S, Wiseman JM, Valentine SJ, Takats Z, Cooks RG & Clemmer DE (2006) *J. Phys. Chem. B*, **110**, 5045–5051.
15. Shin YS, Drolet B, Mayer R, Dolence K & Basile F (2007) *Anal. Chem.* **79**, 3514–3518.
16. Kaur-Atwal G, Weston DJ, Green PS, Croland S, Bonner PLR & Creaser CS (2007) *Rapid Commun. Mass Spectrom.* **21**, 1131–1138.
17. Wilson ID (1999) *J. Chromatogr. A*, **856**, 429–442.
18. Busch KL (1995) *J. Chromatogr. A*, **692**, 275–290.
★ 19. Van Berkel GJ, Ford MJ & Deibel MA (2005) *Anal. Chem.* **77**, 1207–1215. – *An example of DESI combined with chromatography.*
20. Vestling MM & Fenselau C (1995) *Mass Spectrom. Rev.* **14**, 169–178.
21. Takats Z, Cotte-Rodriguez I, Talaty N, Chen HW & Cooks RG (2005) *Chem. Commun. (Camb.)*, **Issue 15**, 1950–1952.
22. Chen H, Cotte-Rodriguez I & Cooks RG (2006) *Chem. Commun. (Camb.)*, **Issue 6**, 597–599.
23. Friess SD & Zenobi R (2001) *J. Am. Soc. Mass Spectrom.* **12**, 810–818.

CHAPTER 7
Analysis of cellular protein complexes by affinity purification and mass spectrometry

Tilmann Bürckstümmer and Keiryn L. Bennett

1. INTRODUCTION

A major challenge of the post-genomic era lies in the necessity of understanding how individual gene products combine and conspire to fulfill the various biological processes (1–3). Cellular functions are nearly always the result of the coordinated action of several proteins in macromolecular assemblies and pathways (4). Protein complexes are highly ordered, dynamic structures translating biological information into function (5, 6). Protein complex composition is not immutable, but varies greatly in time and space to adapt to changing cellular requirements (7, 8). A particular order of assembly, conformational changes, and energy-dependent processes are often involved. Fundamental insights into protein interaction networks have been obtained through large-scale yeast two-hybrid experiments (9–12). Despite the formidable advantage that these interaction networks represent, it is not usually possible to reassemble molecular machines based exclusively on the analysis of these binary interactions because of inherent complex properties (8, 13). Protein complexes that form cellular machines are thus thought to be more than the sum of individual interactions. Affinity purification methods have been developed to retrieve macromolecules associated with a tagged bait protein and subsequently to identify the individual components by mass spectrometry (MS) (14). Such approaches have been applied on a large scale to prokaryotic and eukaryotic cells (15–20). The ability to recover protein complexes and reassemble pathways has contributed to a major shift in perspective, particularly when addressing pathological processes and in drug target discovery strategies (21, 22). In our experience and in the experience of others, tandem affinity purification (TAP) approaches have proved to be particularly efficient in the elucidation of protein complexes, protein pathways, and protein networks. The approach is especially amenable to MS analysis, as a significant purification of the bait and the interacting proteins from a contaminating cellular 'soup' is achieved. Additionally, the high-affinity first step of the strategy followed by mild enzymatic elution

Proteomics: *Methods Express* (C.D. O'Connor and B.D. Hames, eds)
© Scion Publishing, 2008

warrants purification under close-to-physiological conditions (16, 17, 23, 24). Several two-step (tandem) methods have been described in the literature and all appear to result in successful complex characterization. Slightly different variations are available from commercial sources (e.g. Stratagene, Invitrogen). In this chapter, we concentrate on the original method developed by Bertrand Séraphin and colleagues at the EMBL (14), adapted according to Veraksa et al. (25) and Brajenovic et al. (26). TAP has also been used successfully in *Saccharomyces cerevisiae*, *Schizosaccharomyces pombe*, *Candida albicans*, plants, *Escherichia coli*, trypanosomes, *Caenorhabditis elegans*, *Drosophila*, mouse, and human cells (15–17, 19, 25–31). Important information on TAP can be obtained from the web pages of Bertrand Séraphin (http://www-db.embl-heidelberg.de/jss/servlet/de.embl.bk.wwwTools.GroupLeftEMBL/ExternalInfo/seraphin/TAP.html). TAP vectors and cassettes are available from EUROSCARF (http://web.uni-frankfurt.de/fb15/mikro/euroscarf/cz_plas.html).

2. METHODS AND APPROACHES

2.1 TAP compared with other experimental approaches

TAP relies on two affinity purification steps that are applied sequentially. It requires the fusion of the TAP tag to the protein of interest. The TAP-tag comprises two IgG-binding units of protein A derived from *Staphylococcus aureus* and the calmodulin-binding peptide (CBP), separated by a protease cleavage site derived from tobacco etch virus (TEV). During the first purification step, protein A binds to immunoglobulin beads and bound proteins are eluted by TEV protease-mediated cleavage. The second step involves the interaction of calmodulin-binding peptide with calmodulin in the presence of calcium ions. Bound proteins are eluted by calcium withdrawal (using EGTA) or by boiling in sodium dodecyl sulfate (SDS) sample buffer.

Several other tags have been used successfully for the analysis of protein complex formation. These include the poly-His tag, biotin, GST, and a number of epitopes that are recognized by specific antibodies (e.g. FLAG, c-myc, hemagglutinin) (32, 33). In many instances, however, the affinity that the tags have for the cognate ligands is not high enough to ensure sufficient purification efficiency and background reduction of the sample for comprehensive MS analysis.

TAP is a powerful approach because the purification of proteins expressed at close-to-physiological level is possible. This is achieved by:

1. Sequestering a large proportion of the TAP-tagged protein in the first purification step due to the high affinity of the protein A–IgG interaction (K_d in the nanomolar range).
2. Applying a second purification step to remove contaminants that can potentially hinder identification of low-abundance proteins.

In quantitative terms, the number of proteins that co-purify with a given bait is reduced at least threefold when comparing the single-step procedure with the tandem purification procedure (T. Bürckstümmer, unpublished observation). This suggests that a substantial proportion of the proteins that are found to associate with the bait during the first purification step may in fact be nonspecific binders or weak interactors. Another major advantage of the TAP procedure is the fact that it is operational under native conditions, which is a prerequisite for the purification of protein complexes formed *in vivo*. Moreover, TAP is an unbiased approach in which protein complex formation is limited only by the proteins that are naturally expressed in a given cell line.

There are at least three major differences between TAP and the yeast two-hybrid approach. Firstly, TAP is operational in cells of human origin. Given the fact that the human proteome differs quite substantially from the yeast or the murine proteome, the importance of identifying the relevant protein complex cannot be overestimated. Secondly, TAP provides a means of detecting protein–protein interactions directly rather than through a genetic ploy involving transcriptional activation. Moreover, in contrast to the yeast two-hybrid approach, it is applicable to proteins involved in gene regulation. Thirdly, and most importantly, TAP enables multi-protein complex formation. Protein complexes are not just the sum of binary interactions. Many proteins only assemble because another protein is already present that triggers a conformational change or a post-translational modification. In turn, the binding of yet another component is enabled. Such proteins are not facultative components in a complex, but rather are part of the fundamental principles by which biological matter is organized to fulfill biological tasks.

There are also some major drawbacks of the TAP-MS procedure. Firstly, TAP ideally requires the establishment of cell lines that stably express the TAP-tagged protein of interest. Moreover, the cells must proliferate and grow to large quantities. As a consequence, the utilization of primary cells is prohibitive under current experimental restrictions. The second disadvantage of TAP is that it may fail to detect transient interactions. Although there are examples that document the successful identification of transient protein–protein interactions (17), the TAP protocol usually results in the detection of the more 'sturdy' interactors. The obtained datasets are therefore best considered to be the 'hard-wired' portion of a protein–protein interaction network. Lastly, TAP-MS requires a larger investment and commitment of resources compared with the yeast two-hybrid approach. Proficient MS requires expensive equipment and highly skilled personnel, supplemented by a well-established and integrated bioinformatic infrastructure.

2.2 Methodology

The overall TAP process is illustrated in *Fig. 1* (also available in the color section). A total period of 6 weeks is required, as depicted in the timeline shown in *Fig. 2*.

For certain cases transient transfection of target cells may be suitable; however, ideally TAP requires the establishment of cell lines that stably express the protein of interest fused to the TAP tag. Stable cell lines can be generated by standard procedures, such as transfection or infection. In any case, it is advisable to include a selection marker that allows the selection of successfully transduced cells.

CHAPTER 7: ANALYSIS OF CELLULAR PROTEIN COMPLEXES

Expression of TAP-tagged entry point in target cells

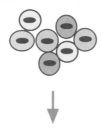

Tandem affinity purification (TAP)

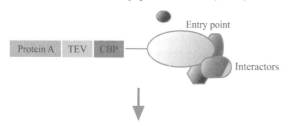

Mass spectrometric protein identification by LC-MS/MS

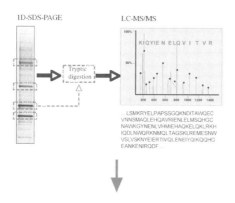

Physical network

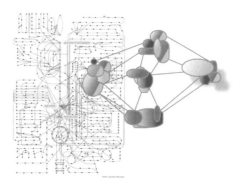

Figure 1. Overview of the TAP procedure (see page xxv for color version). The gene of interest is expressed as a TAP-tagged fusion protein in the target cell line. TAP-tagged proteins are purified by TAP and co-purified proteins are identified by SDS-PAGE and subsequent MS analysis. Information gathered by multiple TAPs is integrated to generate a physical map of the pathway of interest.

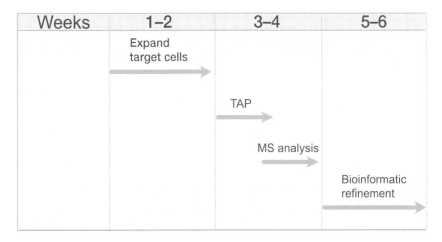

Figure 2. Time line of the TAP procedure.
The TAP procedure from expansion of the target cells to bioinformatic refinement requires an estimated period of 6 weeks, as depicted in the time line.

Once it has been verified that the target cells express the protein of interest, the cells need to be expanded so that TAP can be applied. The final number of cells that are required for a single TAP purification is dependent on the type of target cells. For adherent cell lines such as HEK293, approximately 40 confluent 15 cm plates of cells are required. This corresponds to an estimated total number of 4×10^8 cells. For suspension cells such as the human monocytic cell line U937, approximately 80 dense 15 cm plates are required. This corresponds to a total volume of 1.6 l of cell suspension or 1×10^9 cells. The difference between adherent cells and suspension cells arises from the fact that suspension cells are usually smaller and therefore the same number of cells will contain less total protein. The cells are harvested as described in *Protocol 1*.

Protocol 1

Harvesting cells

Equipment and Reagents
- Centrifuge bottle
- Cells (approx. 80 dense 15 cm plates or 1.6 l of cell suspension equivalent to 1×10^9 cells)
- PBS
- PBS/EDTA (PBS containing 1 mM EDTA)

Method
Adherent cells

1. Remove the medium from the 15 cm plates by aspiration and replace it with PBS/EDTA.

2. After incubating the cells in PBS/EDTA for 5 min, detach the cells by pipetting.
3. Transfer the cell suspension to a centrifuge bottle.
4. Once all of the plates have been harvested, centrifuge the cell suspension at 300 *g* for 3 min. Aspirate the supernatant and wash the pellet with 50 ml of PBS.
5. Centrifuge the cell suspension at 300 *g* for 3 min and remove the supernatant. Freeze the cell pellet immediately in liquid nitrogen and store at −80°C.

Suspension cells

1. Pellet the cells by centrifugation at 300 *g* for 3 min.
2. Resuspend the cells in PBS by pipetting.
3. Transfer the cell suspension to a centrifuge bottle.
4. Continue as described in steps 4–6 for adherent cells.

The conditions for preparation of the cell extract are highly dependent on the nature of the protein of interest. *Protocol 2* describes a method used for cytosolic proteins. TAP can also be performed from subcellular fractions (e.g. nuclear or plasma membrane), as described elsewhere (34).

Protocol 2

Preparation of cell extract

Equipment and Reagents
- Cell pellet (from *Protocol 1*)
- Lysis buffer (50 mM Tris/HCl, pH 7.5, 125 mM NaCl, 5% glycerol, 0.2% NP-40, 1.5 mM $MgCl_2$, 1 mM DTT, 25 mM NaF, 1 mM Na_3VO_4, 1 mM EDTA, protease inhibitors) (25)
- Ultracentrifuge tube
- Falcon tube

Method
1. Thaw the cell pellet on ice and resuspend the cells in 15 ml of lysis buffer. As soon as the lysis buffer has been added to the cells, all subsequent steps should be performed on ice with pre-cooled reagents and material. Incubate the cell lysate on ice for 30 min to enable complete lysis.
2. Centrifuge the sample at 15 000 *g* for 15 min.
3. Transfer the supernatant to an ultracentrifuge tube and centrifuge at 100 000 *g* for 1 h at 4°C.
4. Transfer the supernatant containing the cell extract to a Falcon tube. Remove an aliquot (100 µl) for later analyses. It is preferential to continue immediately to *Protocol 3* at this point, but, if necessary, the cell extract can be frozen in liquid nitrogen and stored at −80°C.

Protocol 3

TAP (adapted from 23, 32)

Equipment and Reagents
- Lysis buffer (see *Protocol 2*)
- IgG–Sepharose beads (Amersham)
- Calmodulin–Sepharose beads (Amersham)
- Filter columns (e.g. MobiTec Mobicol M1002, Bio-Rad Poly-Prep)
- Cleavage buffer (modified from 23) (10 mM Tris/HCl, pH 7.5, 150 mM NaCl, 0.5 mM EDTA, 0.2% NP-40)
- Cleavage buffer containing 250 units/ml TEV protease (Gibco)
- Cleavage buffer containing 2 mM $CaCl_2$
- 1 M $CaCl_2$
- 4× Laemmli sample buffer

Method
1. Wash the IgG–Sepharose beads twice with 10 ml of lysis buffer.
2. Add the cell extract to 200 μl of IgG–Sepharose (50% slurry in lysis buffer) and rotate the sample for 2 h at 4°C.
3. Centrifuge at 300 *g* for 3 min. Remove an aliquot (100 μl) of the supernatant for Western blot analysis.
4. Transfer the IgG–Sepharose to a small filter column. Remove the upper plug from the column and attach a 10 ml syringe. Add 10 ml of lysis buffer and remove the lower plug. Allow the column to drain by gravity flow.
5. Wash the IgG–Sepharose with 5 ml of cleavage buffer[a].
6. Once the cleavage buffer has drained from the column by gravity flow, replace the lower plug on the column. Resuspend the IgG–Sepharose in 400 μl of cleavage buffer containing 100 units of TEV protease. Replace the upper plug on the column and rotate for 1 h at 16°C.
7. Allow the column to drain by gravity flow.
8. Resuspend the IgG–Sepharose in 400 μl of cleavage buffer. Centrifuge the column at 300 *g* for 3 min at 4°C. Remove the supernatant and combine with the initial eluate. Remove an aliquot (20 μl) for later analyses.
9. Add 2 μl of 1 M $CaCl_2$ to the combined eluates (final concentration >2 mM).
10. Wash the calmodulin–Sepharose beads three times with 10 ml of cleavage buffer supplemented with 2 mM $CaCl_2$.
11. Add 200 μl of calmodulin–Sepharose beads (50% slurry in cleavage buffer supplemented with 2 mM $CaCl_2$) to the eluate from step 9. Rotate the beads for 1 h at 4°C.
12. Centrifuge the beads at 300 *g* for 3 min. Remove an aliquot (20 μl) of the supernatant for later analyses.
13. Transfer the calmodulin–Sepharose beads and supernatant to a small filter column. Remove the upper plug from the column, then remove the lower plug and allow the column to drain by gravity flow.
14. Attach a 10 ml syringe and add 10 ml of cleavage buffer supplemented with 2 mM $CaCl_2$. Allow the column to drain by gravity flow.

15. Elute the bound proteins from the Sepharose by boiling in 50 µl of 4× Laemmli sample buffer. Cool the samples to room temperature.

16. Transfer the column to a new Eppendorf tube. Remove the lower plug from the column and centrifuge at 300 *g* for 1 min to retrieve the final TAP eluate.

> **Note**
>
> ^aThe presence of protease inhibitors is detrimental at this stage, as the inhibitors may prevent TEV cleavage. In addition, exclude reducing agents (such as DTT or β-mercaptoethanol), as these will result in reduction and subsequent dissociation of the immunoglobulin chains. Peptides from IgG are a major contaminant in MS analysis of the final samples.

Analysis of the proteins purified by TAP is carried out by SDS-polyacrylamide gel electrophoresis (PAGE) as described in *Protocol 4*.

Protocol 4
One-dimensional SDS-PAGE (pre-cast gels)

Equipment and Reagents
- Eluted proteins in 4× Laemmli sample buffer (from *Protocol 3*)
- Iodacetamide stock solution in water (133 mg/ml)
- Pre-cast SDS-PAGE gels, e.g. 1.0 mm NuPAGE Novex 4–12% Bis/Tris gels with ten wells (Invitrogen) (25) and electrophoresis equipment
- Ultrahigh-quality (UHQ) water
- 20× SDS running buffer, e.g. NuPAGE MOPS SDS running buffer (for Bis/Tris Gels only) (Invitrogen)
- 1× MOPS SDS running buffer (500 ml required for each gel box by mixing 25 ml of NuPAGE MOPS SDS running buffer with 475 ml of UHQ water)
- 200 ml of 1× MOPS SDS running buffer containing 500 ml of NuPAGE antioxidant (Invitrogen)

Method
1. Add 7 µl of iodoacetamide stock solution per 50 µl of reduced and denatured sample.

2. Rinse the electrophoresis chamber with water.

3. Open the gel wrapping of the pre-cast SDS-PAGE gel, discard the liquid, and rinse the gel with 1× MOPS SDS running buffer.

4. Remove the white strip. Place two of the gels in the holder (so that the anode part of the gel, which was covered with the white strip, faces the outer side of the box). Lightly press the holder down.

5. Fill the internal chamber with 1× MOPS SDS running buffer containing antioxidant and check for leakage.

6. If the internal chamber is sealed, fill the outer chamber with 1× MOPS SDS running buffer.

7. Remove the plastic combs from the gel. Carefully wash the wells by adding 200 ml of 1× MOPS SDS running buffer.

8. Use thin gel-loading tips to load the sample into the pockets. When releasing the sample from the pipette, pull the tip slowly upward in the sample well.
9. Load the empty sample wells with 10 µl of 4× Laemmli sample buffer.
10. Place the electrodes on the top of the two chambers and connect them to the electric power supply; switch on the main power.
11. Check that the gel sample wells are covered with 1× MOPS SDS running buffer and that the bottom part of the gel is a minimum of 3 cm below the buffer surface.
12. Adjust the power to 120 mA, 200 V, for 55 min to start the electrophoresis. When the run is complete, unplug the electrodes and discard the solutions.

Following SDS-PAGE, the gel should be silver-stained to detect the separated proteins. A suitable method is given in *Protocol 5*.

Protocol 5

Silver staining

Equipment and Reagents
- 96% Ethanol (Merck)
- 99.9% Glacial acetic acid (Merck)
- $Na_2S_2O_3$ (anhydrous) (Sigma)
- $AgNO_3$ (Sigma)
- Na_2CO_3 (Merck)
- Formaldehyde (approx. 35%; Sigma)
- UHQ water
- Fixing solution (375 ml of ethanol, 435 ml of water, 90 ml of glacial acetic acid)
- Wash solution (564 ml of ethanol, 1236 ml of water)
- Sensitizing solution (0.06 g of $Na_2S_2O_3$ dissolved in 300 ml of water; prepare just before use)
- Silver nitrate solution (0.6 g of $AgNO_3$, 60 µl of formaldehyde made up to 300 ml with water; prepare just before use and store in a refrigerator)
- Developer (18 g of Na_2CO_3, 300 µl of formaldehyde, made up to 600 ml with water; prepare just before use)
- Quenching solution (45 ml of acetic acid, 855 ml of water)
- Petri dish

Method
Note: Perform all steps in a fume hood, with gels placed on a rocking shaker during incubation, with sufficient solution to cover the gel completely at each stage.

1. Fix the gel for 1 h with fixing solution.
2. Wash the gel twice for 20 min with wash solution.
3. Wash the gel for 20 min with UHQ water.
4. Sensitize the gel for 1 min with sensitizing solution.
5. Wash the gel three times for 20 s each with UHQ water.

6. Incubate the gel for 20 min with cold silver nitrate solution.
7. Wash the gel three times for 20 s with UHQ water. After the last rinse, transfer the gel to a clean Petri dish.
8. Develop the gel with developer. Add sufficient developer to the gel in the Petri dish to cover the gel. Rock the Petri dish until the solution changes to a yellow color. Remove the liquid and add new developing solution.
9. When all the bands are visible on the gel, quench the developer by removing the liquid from the Petri dish and adding quenching solution. Leave the gel in the quenching solution for a minimum of 5 min, or place the covered gel in the refrigerator overnight.

Tryptic digestion of proteins in gel plugs can be carried out *in situ* as described in Protocol 6.

Protocol 6

In situ tryptic digestion (adapted from 35)

Equipment and reagents
- Thermofast 96-well microtitre plate with a hole in each well, 0.4–0.5 mm in diameter
- Thermofast 96-well microtitre plate without holes
- NH_4HCO_3 (Sigma)
- Dithiothreitol (DTT; Sigma)
- Iodoacetamide (Sigma)
- Absolute ethanol (Merck)
- Sequencing-grade modified trypsin, frozen (Promega)
- UHQ water
- 50% 50 mM NH_4CO_3, 50% ethanol
- Reduction buffer (10 mM DTT in 50 mM NH_4HCO_3; prepare using 0.5 M DTT stock (0.3855 g of DTT dissolved in 5 ml of UHQ water) and store in aliquots of 110 and 210 µl at −20°C)
- Alkylation buffer (55 mM iodoacetamide in 50 mM NH_4HCO_3; prepare just before use and protect from the light by wrapping in aluminum foil)
- 50 mM NH_4HCO_3 in UHQ water
- Trypsin solution (12.5 ng/µl in 50 mM NH_4HCO_3; prepare just before use)
- 10% Trifluoroacetic acid (TFA)[a]
- 5% Formic acid

Method
1. Excise the gel plugs and place each plug in individual wells of a 96-well microtitre plate containing holes[b].
2. Wash the stained gel plugs in 150 µl of 50% 50 mM NH_4CO_3, 50% ethanol for 20 min on a mixer.
3. Centrifuge the plates at 1500 r.p.m. for 2–3 min and discard the solution after the wash[c].
4. Repeat step 2.

5. Repeat step 2[d].
6. Dehydrate the gel plug with 150 µl of absolute ethanol for 10 min at room temperature. Discard any liquid by centrifuging the plates at 1500 r.p.m. for 2–3 min.
7. Reduce disulfide bonds by adding a volume of reduction buffer sufficient to cover the gel plug (~50 µl). Cover the plate with plastic film. Incubate for 45 min at 56°C. Discard the liquid by centrifuging the plates at 1500 r.p.m. for 2–3 min.
8. Alkylate cysteine residues by adding the same volume (~50 µl) of alkylation buffer. Wrap the plate in foil to protect it from the light. Incubate at room temperature for 30 min on the shaker. Discard the liquid by centrifuging the plates at 1500 r.p.m. for 2–3 min.
9. Wash with 150 µl of 50 mM NH_4HCO_3 for 10 min. Discard the liquid by centrifuging the plates at 1500 r.p.m. for 2–3 min.
10. Dehydrate with 150 µl absolute ethanol for 10 min. Discard any liquid by centrifuging the plates at 1500 r.p.m. for 2–3 min.
11. Repeat steps 8 and 9.
12. Dehydrate with 150 µl of absolute ethanol for 10 min. Discard the solution and leave the gel at room temperature for 5 min.
13. Add 10 µl of trypsin solution and incubate at room temperature for 10 min.
14. Add 20 µl of 50 mM NH_4HCO_3 (without trypsin). This is added to keep the gel plugs wet during digestion.
15. Cover the plate with thermosealing tape. Take care not to press down on the sealing tape with force, as this may force some of the liquid out of the holes in the bottom of the plates. Place the plate on top of a second plate containing 80 µl of 50 mM NH_4HCO_3 in each well. Digest by incubating at 37°C for 4 h or overnight[e].
16. Place a fresh 96-well microtitre plate (without holes) under the plate containing the gel plugs.
17. Stop digestion by adding 3 µl of 10% TFA[a]. Mix well and shake for 5 min.
18. Centrifuge the liquid into the ThermoFast 96-well microtitre plate without holes. Keep this solution as it contains the tryptic peptides.
19. Extract the peptides by adding 20 µl of 5% formic acid to the gel plug. Extract for 30 min and centrifuge the liquid into the plate together with the liquid from the first centrifugation.

Notes

[a]TFA-containing solutions should be made on the day of use. Extreme care should be taken when using TFA as it is an extremely hazardous chemical and the concentrated acid should only be used in a fume hood and dispensed wearing suitable protective clothing. Consult the safety datasheet for handling details and disposal.

[b]If the gel plugs are large, it may be necessary to cut them in half. Excise plugs as close to the staining margin as possible. This aids penetration of the trypsin into the gel plug and avoids dilution effects.

[c]Adjust the speed of the centrifuge as required.

[d]Omit this wash if the gel is silver-stained or if the Coomassie blue-stained gel is destained after the first two washes.

[e]It may be necessary to add larger volumes of 50 mM NH_4HCO_3 (without trypsin) if the gel plugs are large, but avoid adding too much liquid as this will dilute the trypsin and decrease the likelihood of it penetrating the gel plug. Check the digestion after approx. 2 h to ensure that all gel plugs are still wet. If some of the plugs have absorbed all the liquid, add more 50 mM NH_4HCO_3.

Protocol 7

Stage tip purification (adapted from 36)

Equipment and Reagents
- Empore high-performance extraction disks (3M)
- 17-gauge KF metal needle (Hamilton)
- Pipette tips
- Liquid chromatography LC-MS-grade methanol (Sigma)
- Glacial acetic acid (Merck)
- TFA (Sigma)[a]
- Heptafluorobutyric acid (Sigma)
- LC-MS-grade acetonitrile (Fisher)
- Solution A (0.5% acetic acid, 2% TFA)[a]
- Solution B (0.005% heptafluorobutyric acid, 0.4% acetic acid, 90% acetonitrile)
- Solution C (0.005% heptafluorobutyric acid, 0.4% acetic acid)
- High-speed vacuum centrifuge

Method
1. Excise small discs of C18 reverse-phase beads embedded in Teflon mesh (Empore disks) using a 17-gauge KF metal needle.
2. Insert the metal needle into a pipette tip (200 µl preferred).
3. Release the excised disc into the tapered end of the tip by pushing on the insert to produce a completed stage tip. Ensure that the Teflon disc is firmly pushed into the end of the pipette tip, otherwise the liquid containing the peptides will run through the tip and the peptides will not bind to the C18 material.
4. Activate the stage tips with 50 µl of methanol.
5. Equilibrate the stage tip with 50 µl of solution A.
6. Load the *in situ*-acidified tryptically digested samples (from *Protocol 6*).
7. Wash the stage tip with 50 µl of solution A and centrifuge at 1500 r.p.m. for 2–3 min.
8. Elute the samples directly via centrifugation into a ThermoFast 96-well microtitre plate with 30 µl of solution B. For thick, heavily stained bands, collect the initial flow-through and re-purify through a second stage tip. Pool the eluents.
9. Remove the acetonitrile using a high-speed vacuum centrifuge. Take care that the samples are not dried to completeness (leave ~3 µl).
10. Reconstitute the samples with ~7 µl of solution C so that the final volume is 10 µl[b].

Notes
[a]TFA-containing solutions should be made on the day of use. Extreme care should be taken when using TFA as it is an extremely hazardous chemical and the concentrated acid should only be used in a fume hood and dispensed wearing suitable protective clothing. Consult the safety datasheet for handling details and disposal.
[b]Depending on the sensitivity of the mass spectrometer, it may be necessary to dilute the samples further with 10–20 µl of solution C.

Protocol 8

Analysis by LC-MS/MS

Equipment and Reagents
- Tryptically digested sample (from *Protocol 7*)
- Heptafluorobutyric acid (Sigma)
- Glacial acetic acid (Merck)
- LC-MS-grade acetonitrile (Merck)
- LiChrosolv water (Merck)
- NanoLC system (Agilent Technologies)
- Quadrupole time-of-flight mass spectrometer (Waters)
- Solvent A (0.005% heptafluorobutyric acid, 0.4% acetic acid)
- Solvent B (0.005% heptafluorobutyric acid, 0.4% acetic acid, 90% acetonitrile)

Method

Liquid chromatography[a,b]

1. Inject 8 µl of the tryptically digested sample and load at 0.8 µl/min for 20 min onto the analytical column with 3% solvent B.
2. Decrease the flow rate to 250 nl/min within 1 min.
3. Switch to bypass mode and maintain the flow rate at 250 nl/min with 3% solvent B for 2 min[c].
4. Increase to 10% solvent B within 1 min.
5. Over 25 min, increase the composition to 35% solvent B.
6. Within 3 min, increase to 50% solvent B and in the subsequent 3 min, increase solvent B to 100%.
7. Increase the flow rate to 0.8 µl/min and wash the column for 5 min.
8. Within 1 min, return to the starting conditions and re-equilibrate the analytical column.

Mass spectrometry[d]

9. Acquire an MS survey spectrum from m/z 350 to 1300 for 1 s.
10. Switch to MS/MS mode when the intensity of an eluting precursor ion exceeds 25–40 counts/s.
11. Acquire an MS/MS spectrum for the three most abundant precursor ions from m/z 350 to 1300 for 2 s each.
12. Once an MS/MS spectrum has been acquired, exclude the precursor ion from selection for 60 s.
13. Return to the MS survey spectrum mode.

Notes
[a]The LC system is configured with an analytical column only. No pre-column is included.
[b]To minimize sample carry-over, the loop is washed once with acetonitrile, once with water, once with solvent A, and once with 3% solvent B.
[c]This enables stabilization of the flow rate in bypass mode.
[d]MS acquisition is from 30 to 65 min in positive-ion mode.

Data analysis is carried out using the Mascot search engine (MatrixScience) and EPICENTER (Proxeon Biosystems). Data analysis is covered in Chapter 11.

3. TROUBLESHOOTING

A sample TAP purification is shown in *Fig. 3*. In order to troubleshoot the process of TAP, it is essential to implement a series of check points during the purification (as suggested in the protocols).

- IgG binding is inefficient
  If binding in the first step is inefficient, the binding of TAP-tagged proteins to IgG–Sepharose can be improved by: (i) increasing the amount of IgG–Sepharose in the first binding step (up to 400 µl of 50% slurry) or (ii) by increasing the incubation time to 4 h. It is important to bear in mind, however, that complex formation (and dissociation) may be affected by increased incubation times.

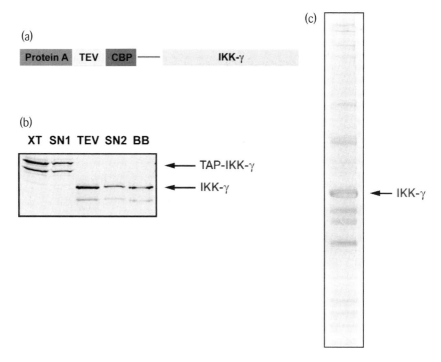

Figure 3. Example of a TAP purification.
(*a*) Schematic representation of the N-terminally TAP-tagged IKK-γ. (*b*) TAP-tagged IKK-γ was purified by TAP. Samples were taken at different steps of the purification protocol: XT, cell extract (70 µg of total protein); SN1, supernatant after IgG–Sepharose incubation (70 µg of total protein); TEV, eluate after TEV protease cleavage (5% of total); SN2, supernatant after calmodulin–Sepharose incubation (5% of total); BB, final eluate (5% of total). Purification was monitored by Western blotting using an IKK-γ-specific antiserum. (*c*) The final eluate was analyzed by silver staining.

- The bait is not detected in the TEV eluate
  This indicates that TEV cleavage is not particularly efficient. In order to improve the cleavage efficiency, the amount of TEV that is used should be increased. Alternatively, the cleavage time may be prolonged.
- Calmodulin binding is inefficient
  Unfortunately, recovery rates in the second purification step are sometimes unsatisfactory. Attempt to alleviate this problem by increasing the amount of calmodulin-Sepharose and/or the incubation time as recommended above for IgG-Sepharose.

Acknowledgements

Gregor Schütze is acknowledged for provision of *Protocol 4* and for critical reading of the manuscript. We thank Marc Brehme for providing *Fig. 1*. Work in our laboratory is supported by the Austrian Academy of Sciences and the Austrian National Bank. T.B. is supported by a research fellowship (BU 2180/1-1) from the German Research Foundation (DFG).

4. REFERENCES

1. Hood L, Heath JR, Phelps ME & Lin B (2004) *Science*, **306**, 640-643.
2. Hood L & Perlmutter RM (2004) *Nat. Biotechnol.* **22**, 1215-1217.
3. Weston AD & Hood L (2004) *J. Proteome Res.* **3**, 179-196.
4. Alberts B (1998) *Cell*, **92**, 291-294.
5. Cramer P (2002) *Curr. Opin. Struct. Biol.* **12**, 89-97.
6. Cramer P (2004) *Curr. Opin. Genet. Dev.* **14**, 218-226.
7. de Lichtenberg U, Jensen LJ, Brunak S & Bork P (2005) *Science*, **307**, 724-727.
8. Aloy P, Böttcher B, Ceulemans H, *et al.* (2004) *Science*, **303**, 2026-2029.
9. Li S, Armstrong CM, Bertin N, *et al.* (2004) *Science*, **303**, 540-543.
10. Rual JF, Venkatesan K, Hao T, *et al.* (2005) *Nature*, **437**, 1173-1178.
11. Hazbun TR & Fields S (2001) *Proc. Natl. Acad. Sci. U.S.A.* **98**, 4277-4278.
12. Uetz P, Giot L, Cagney G, *et al.* (2000) *Nature*, **403**, 623-627.
13. von Mering C, Krause R, Snel B, *et al.* (2002) *Nature*, **417**, 399-403.
14. Rigaut G, Shevchenko A, Rutz B, Wilm M, Mann M & Seraphin B (1999) *Nat. Biotechnol.* **17**, 1030-1032.
15. Butland G, Peregrín-Alvarez JM, Li J, *et al.* (2005) *Nature*, **433**, 531-537.
16. Gavin AC, Bösche M, Krause R, *et al.* (2002) *Nature*, **415**, 141-147.
17. Bouwmeester T, Bauch A, Ruffner H, *et al.* (2004) *Nat. Cell Biol.* **6**, 97-105.
18. Zhao R, Davey M, Hsu YC, *et al.* (2005) *Cell*, **120**, 715-727.
19. Roguev A, Shevchenko A, Schaft D, Thomas H & Stewart AF (2004) *Mol. Cell. Proteomics*, **3**, 125-132.
20. Ho Y, Gruhler A, Heilbut A, *et al.* (2002) *Nature*, **415**, 180-183.
21. Fishman MC & Porter JA (2005) *Nature*, **437**, 491-493.
22. Brown D & Superti-Furga G (2003) *Drug Discov. Today*, **8**, 1067-1077.
23. Puig O, Caspary F, Rigaut G, *et al.* (2001) *Methods*, **24**, 218-229.
24. Dziembowski A & Seraphin B (2004) *FEBS Lett.* **556**, 1-6.
25. Veraksa A, Bauer A & Artavanis-Tsakonas S (2005) *Dev. Dyn.* **232**, 827-834.
26. Brajenovic M, Joberty G, Kuster B, Bouwmeester T & Drewes G (2004) *J. Biol. Chem.* **279**, 12804-12811.
27. Rohila JS, Chen M, Cerny R & Fromm ME (2004) *Plant J.* **38**, 172-181.
28. Kaneko A, Umeyama T, Hanaoka N, Monk BC, Uehara Y & Niimi M (2004) *Yeast*, **21**, 1025-1033.

29. Westermarck J, Weiss C, Saffrich R, *et al.* (2002) *EMBO J.* **21**, 451–460.
30. Walgraffe D, Devaux S, Lecordier L, *et al.* (2005) *Mol. Biochem. Parasitol.* **139**, 249–260.
31. Gottschalk A, Almedom RB, Schedletzky T, Anderson SD, Yates JR III & Schafer WR (2005) *EMBO J.* **24**, 2566–2578.
32. Monti M, Orru S, Pagnozzi D & Pucci P (2005) *Clin. Chim. Acta* **357**, 140–150.
33. Terpe K (2003) *Appl. Microbiol. Biotechnol.* **60**, 523–533.
34. Dignam JD, Lebovitz RM & Roeder RG (1983) *Nucleic Acids Res.* **11**, 1475–1489.
35. Shevchenko A, Wilm M, Vorm O & Mann M (1996) *Anal. Chem.* **68**, 850–858.
36. Rappsilber J, Ishihama Y & Mann M (2003) *Anal. Chem.* **75**, 663–670.

CHAPTER 8
Clinical proteomic profiling and disease signatures

Rosamonde E. Banks, David A. Cairns, David N. Perkins, and Jennifer H. Barrett

1. INTRODUCTION

Most diseases have a high degree of inter-individual heterogeneity due to the involvement of multiple genetic pathways, genetic polymorphisms, and epigenetic influences. Consequently, and analogous to the use of microarrays for transcriptomic analyses, the simultaneous examination of multiple proteins in parallel should ultimately provide highly individualized information, generating marker profiles for diagnosis, prognosis, or treatment selection and monitoring. The concept of using matrix-assisted laser desorption/ionization (MALDI) mass spectrometry (MS)-based approaches to generate mass profiles of mixtures of intact proteins or protein fragments present in biological fluids and tissue is now being explored widely. In many cases, the identities of the peaks are unknown, with the profiles or 'molecular signatures' *per se* being used in computational models to discriminate between groups. This is exemplified by the discrimination between cancerous and normal lung tissue (1) (see *Fig. 1*, also available in the color section). This is extended further in the technically highly challenging approach of 'imaging MS', which spatially 'maps' specific areas of tissue sections in terms of the biomolecules present, with a lateral resolution of 30–50 mm (2). As an example, S100β has been shown to be expressed at higher levels in high-grade gliomas compared with low-grade samples and localized to areas with higher concentrations of astrocytes (3).

The majority of MS profiling studies to date have used surface-enhanced laser desorption/ionization (SELDI) MS. Based on the principles of MALDI, SELDI utilizes sample arrays ('ProteinChips') of differing chromatographic surfaces as the target, selectively capturing proteins/peptides on the basis of their chemical properties (4). Biologically or clinically relevant results have been shown in many studies including the identification of α-defensins as the CD8 T-lymphocyte-derived anti-human immunodeficiency virus type 1 activity (5), profiling of different β-amyloid forms in Alzheimer's disease (6, 7), and the development of many computer-based algorithms generating impressive predictive diagnostic tests in many cancers and other diseases such as trypanosomiasis (e.g. 8–11). These have yet to

136 CHAPTER 8: CLINICAL PROTEOMIC PROFILING AND DISEASE SIGNATURES

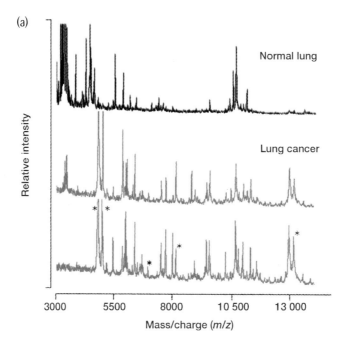

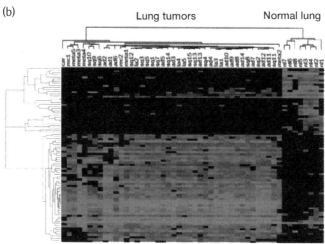

Figure 1. Detection of optimum discriminatory biomarker sets in lung tumors (see page xxvi for color version).
(a) Representative MALDI-TOF-MS spectra obtained from frozen tissue sections showing the relative intensities of the peaks of different molecular masses (m/z values; asterisks indicate examples of discriminatory peaks). (b) Hierarchical clustering analysis of 42 lung tumors and eight normal lung tissues in the training cohort based on 82 MS signals. Each row represents an individual signal and each column an individual sample with the dendrogram at the top showing the similarity in protein expression profiles of the samples. Substantially raised expression is indicated in red. Reproduced with permission from (1).

be validated in long-term studies, and translation of such findings into longer-term clinical utility will require further studies addressing both pre-analytical and analytical issues (12–16). The subsequent identification of the discriminant peaks in the profiles may increase the biological value of the data and also allow the subsequent generation of more-robust and quantitative multiplex assays. This chapter describes specifically the use of SELDI in clinical proteomic profiling and illustrates some of the critical methodological issues, many of which apply equally to other profiling approaches, mass spectrometric or otherwise.

2. METHODS AND APPROACHES

A typical clinical proteomic profiling study has several recognizable stages, the exact details of each varying slightly depending on the technological platform used, as exemplified in the flow chart in *Fig. 2* using SELDI as the model system. The following section describes some critical aspects of the methods and approaches adopted.

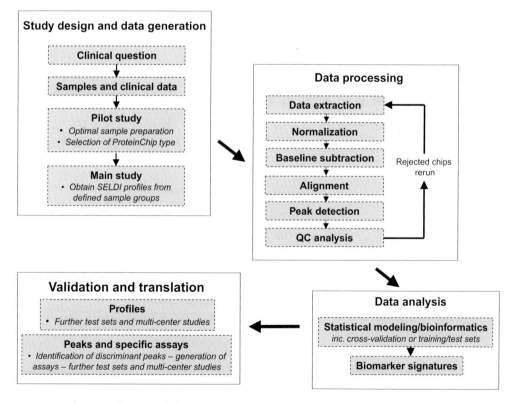

Figure 2. Schematic diagram of the various stages involved in a typical clinical proteomic profiling study for biomarker discovery.
The specific details are those using SELDI as the model system, but the overall stages and progression are common to many different profiling approaches.

2.1 Principles of MALDI and SELDI

The principles underlying MS and its many applications in proteomics have been reviewed recently (17). Mass spectrometers essentially consist of an ion source that generates ionized analyte molecules in the gas phase that 'fly' upon application of an accelerating voltage, a mass analyzer that separates the ions on the basis of their mass-to-charge (m/z) ratios, and a detector that records the number of ions for each m/z species as mass spectra. MALDI is one of the two most common ionization techniques, with laser energy being used to ionize the sample from a solid-phase mixture of sample and energy-absorbing matrix on the MALDI target. MALDI is most often coupled to a time-of-flight (TOF) analyzer, which resolves ions according to their m/z ratios on the basis of the time taken to travel along a flight tube with heavier ions being slowest; this has played a pivotal role in peptide mass fingerprinting of tryptic peptides for identification. With very high levels of mass accuracy and resolution, improved ionization, and the wider mass ranges of current instruments, the use of MALDI-TOF for profiling-based analyses of polypeptides or intact proteins has also now become realistic.

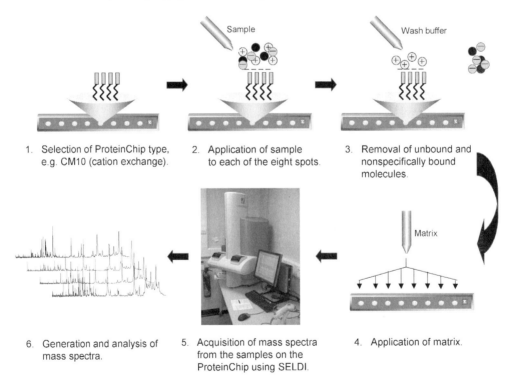

Figure 3. Illustration of the principles of the SELDI approach.
ProteinChips with specific surface chemistries are selected and the sample, diluted in appropriate binding buffer, is applied to each sample spot on the chip surface. After incubation, unbound and nonspecifically bound molecules are removed using wash buffer and energy absorbing matrix is applied to each sample spot. After drying, mass spectra are acquired using a SELDI instrument (in this case, the Enterprise 4000 is shown) and are then analyzed using Ciphergen software or exported for further manipulation.

SELDI-TOF incorporates an initial 'selection' step through the derivatization of the target surface (18). The commercially available SELDI system introduced in 1999 (Ciphergen Biosystems) utilizes a ProteinChip target (most commonly containing eight sample spots) with specific chemically derivatized surfaces – normal phase, reverse phase, ion exchange, or immobilized metal affinity. These selectively immobilize particular molecular species present in a sample prior to profiling and permit removal of contaminating substances such as salts (see *Fig. 3*). ProteinChips with pre-activated surfaces reacting with primary amines (PS20) are also available, allowing the immobilization of antibodies, receptors, or peptides, for example. The original PBS II machines have evolved, with the PBS IIc having increased sensitivity with the incorporation of a mass deflector to prevent detector saturation with matrix signal and an autoloader option for higher throughput. The new-generation Enterprise 4000 SELDI platform introduced in 2004 has an increased mass range (0–400 kDa) and significantly improved sensitivity and resolution (2.5–40-fold more sensitive, particularly in the higher mass regions, and able to detect 0.5–2.5 fmol of several peptides/proteins tested in the range of 2–150 kDa).

2.2 Choice of sample type

Analysis of tissues gives a direct insight into disease-related protein changes. There are relatively few SELDI-based tissue studies, with several incorporating laser-based microdissection techniques upstream to enrich for the cells of interest, generating profiles from as few as ten cells, but more often from 500–2000 cells (e.g. 19–26) and in some cases identifying discriminant peaks. Methods vary considerably and no systematic examination of the various parameters is available and so no protocols will be provided here. However, from our preliminary results and the optimization studies undertaken for MALDI-based tissue profiling/imaging (although solubilization and compatibility with ProteinChip chemistries are not issues, several issues will be relevant), factors identified as critical include initial sample freezing and storage, the use of embedding media, histological stain and fixative, the choice of extraction buffer/detergents, the amount of material, and the matrix and solvent (27–29).

Biological fluids representing many diseases have been profiled using SELDI, particularly serum and plasma but also urine (12, 30–32), cerebrospinal fluid (33–35), nipple aspirate fluid (36–38), and pancreatic juice (39). Fluids are often more accessible than tissue, contain proteins in soluble form, and may contain not only proteins and protein fragments shed by the diseased tissue but also those reflecting secondary systemic changes. Particular challenges with serum or plasma, for example, are the dominance of 10–12 proteins that make up 90% of the total protein, the wide range of concentrations ranging from 50–100 g/l down to levels of less than 1 ng/l, and the dilution effects for disease-derived proteins (40–42). Multi-dimensional proteomic approaches and pre-fractionation strategies are proving promising, detecting 'classical' circulating blood proteins but also many lower-abundance molecules such as receptors, transcription factors, and even some cytokines/growth factors (41, 42). SELDI has the advantage of its inherent target-based chemical selectivity, but further pre-fractionation will ultimately increase potential. Many proteins or fragments of biological interest are carried by other proteins such as albumin and strategies are now being developed to examine these (43, 44).

2.3 Study design, and pre-analytical and analytical issues

Both the requirements of *in vitro* diagnostic tests and the importance of study design in avoiding bias in biomarker discovery have been reviewed recently (45, 46). Many proteomic profiling studies are essentially observational rather than experimental in nature and the principles of study design appropriate to epidemiological studies apply (47). Pre-analytical influences on many aspects of clinical laboratory measurements are widely recognized (48) but often overlooked in proteomic studies. Similarly, the importance of examining any biomarkers or profiles in the context of an 'index of individuality', i.e. taking into account pre-analytical, analytical, inter-individual, and intra-individual variation in the overall interpretation of the results, has been emphasized (49). Many of the published studies have not addressed the factors identified below and examples of possible spurious differences have been highlighted (14, 15, 50, 51).

Using as a paradigm a study with the aim of comparing two or more different groups, the following general principles should be adhered to:

- There should be no systematic differences in the ways in which samples from different groups or centers are processed. Several studies have now identified differences depending on sample source and how samples are processed, for example elapsed time post-venepuncture and type of blood tube are critical (16, 52–56). Many of the low-molecular-mass peaks in serum and several of those in plasma depending on the type and handling arise from *ex vivo* events such as the proteolytic coagulation cascade and other clotting events leading to cellular release of peptides (16, 52, 57). These and other issues are discussed further in the protocols.
- Samples must be analyzed at least in duplicate to partially remove, and provide some measure of, intra-sample variability. A quality control (QC) sample should be included on each chip (the location varying with each chip).
- Samples, including replicates, must be assigned to spots on chips entirely at random to eliminate the potential for bias due to variability in chip or machine performance.
- Sample groups must be matched for any major confounding factors, including as a minimum age and sex; depending on context, other variables such as smoking or diet may need to be included.
- To produce results that can be generalized, the cases and controls should be representative samples from a clearly defined population (e.g. incident cases of bladder cancer in a geographical region) and the inclusion criteria defined.

Samples numbers in SELDI studies are typically in the order of 25–100 subjects per group, but it is currently difficult to provide a good theoretical justification for any choice, particularly without knowing the extent of variation between and within groups and the sort of differences expected. Studies should be sufficiently powered to detect differences between groups, allowing multiple testing, and many of the ideas arising from consideration of this issue in the context of the high-dimensional data generated by gene expression microarrays (58, 59) are applicable.

Several key factors can be identified at the analytical level, although most have not been investigated systematically. Choice of chip type and buffer conditions have to be determined empirically and therefore vary among studies. Appropriate QC samples should be included as part of the overall experimental design, as discussed above, and a pooled sample formed from multiple samples representing all groups may be most appropriate to monitor the experimental run. A separate type of QC sample run daily can also monitor the whole process over the long term, including SELDI performance and any liquid-handling robotics – the latter generally leads to improved reproducibility, but QC measures such as dye titration tests should be performed routinely. From a consideration of chromatographic principles and experimental data (12), similar amounts of protein should be profiled for each sample. Matrix application is critical in terms of ensuring good crystallization and coverage, with both composition and timing being important (60, 61; R.E. Banks *et al.*, unpublished data), and, similarly, machine settings should be optimized (all discussed further in the protocols).

2.4 Data processing

Data processing tasks may be undertaken either with the proprietary Ciphergen software, or following export of spectra as text files containing *m/z* and intensity values (~1 Mb/file for a mass range of 200 kDa), which allows greater flexibility and the application of less-conservative approaches. The Ciphergen PROTEINCHIP software controls the SELDI and analyses the data, the data files being stored on the local disk in a proprietary binary format. The Ciphergen EXPRESS software supplied with the Enterprise 4000 system is based on a Structured Query Language (SQL) database, with all data stored along with experimental and operator-entered information and analysis results in virtual directories. Typically, the SQL database is run on a server running Microsoft WINDOWS 2000, with client software used for data analysis and operating the SELDI run on a separate computer. The Enterprise 4000 SELDI itself has an embedded PC running a version of REDHAT LINUX. These three computers all communicate with each other via a network connection. The spectra data consist of the raw TOF values along with the parameters required to produce the appropriate *m/z* and intensity values. If required, data can be imported into the database from the older PROTEINCHIP software via an intermediate file format.

The 'typical' analysis of SELDI data has several stages (see *Fig. 2*). Firstly, the spectra are normalized if required, to adjust for slight variations in machine performance, for example between samples. Most often this is performed by linearly scaling the intensities of a set of spectra to an average total ion current. The data are then de-noised by applying a smoothing algorithm and baseline subtraction.

A variety of smoothing algorithms have been used including Gaussian smoothers (62) or Fourier transforms (63). Baseline subtraction is used to remove the electronic and chemical noise contributed by the matrix from a spectrum and, in the case of the Ciphergen software, a modified convex piece-wise hull algorithm is used. Alternative methods include the use of locally weighted regression (LOESS) (64, 65), in some cases using different parameters for different regions of the spectrum (66), the use of linear regression on peak-devoid areas of spectra with subsequent subtraction of the fitted line from the whole spectra (67, 68), or the use of various types of running means and medians (63, 69). A unified approach to baseline subtraction and spectral alignment and peak resolution has also been developed using a method of optimal smoothing and target filtering by time series analysis (70).

The alignment of spectra to correct for machine drift during the course of an experiment is critical. Many methods of peak detection use, in a manner similar to the proprietary software, a simple mass window of width ±0.1%, for example, to align peaks (12, 71), with alternatives including one-dimensional hierarchical clustering (65), alignment based on improved internal calibration using the quadratic calibration equation or cubic splines (72), or the use of an insertion/deletion heuristic algorithm based on the comparison of all spectra with a mean or reference spectrum (73). In the PROTEINCHIP software, peaks are defined on the basis of identifying changes in intensity above a specified threshold value and are then 'clustered' across multiple spectra. Other methods of peak detection use sliding windows of varying sizes to define peaks as local maxima and then filter on the basis of signal-to-noise (S/N) ratio (13, 71). Methods using wavelet transformations that will de-noise the spectra, perform baseline subtraction, and detect a greater number of peaks than standard methods have also been proposed (74).

A stringent data QC strategy must be applied, monitoring machine and chip performance over the long term and data quality during the course of an experiment. Few studies appear to have implemented well-defined processes, but increasing attention is being given to this area (13, 75–77).

The relative merits of many computational methods relating to data processing, including the ones above, have recently been reviewed in the context of comparative proteomic profiling using liquid chromatography-tandem MS (LC-MS/MS) (78) but are relevant to many profiling approaches and should be considered.

2.5 Data analysis

The analysis of proteomic profiles obtained from SELDI-TOF-MS broadly falls into three categories dependent on the study aims:

1. The determination of specific peaks that are expressed differentially in two or more groups of samples and subsequent statistical modeling (described in detail in the protocols given in this chapter).
2. The use of profiles as predictors in classification algorithms to separate different clinical or biological states. Usually developed using detected peaks (typi-

cally numbering several hundred) as predictors, the alternative approach of using the raw data output as predictors (typically tens of thousands of intensity measures corresponding to different m/z values) has also been explored. The classification algorithms used have included methods from classical statistics including principal components analysis (79), logistic regression (68), and linear and quadratic discriminant analysis (15). Machine-learning methods such as neural networks (12, 80), random forests of classification trees (81; described in section 2.7.6), and support vector machines (66) have also been used. These methods generally use a training set, where the correct classification is known, to educate the classification algorithm on the patterns in the profiles. The algorithm is then evaluated by applying it to an independent test set, and the performance is measured by the sensitivity and specificity or other similar indicators.

3. The use of various types of cluster analysis in an unsupervised context (i.e. blind to their actual disease status), which may uncover previously unknown classes. Hierarchical clustering has been used for a vivid display of the effect of blood sample processing on the SELDI profiles obtained (16) and the separation of lung cancer tissue samples from normal lung tissue on the basis of MALDI profiles (1).

The relative merits of different approaches and a discussion of many of the relevant issues pertaining to SELDI profile analysis are covered in greater detail in section 2.7. Additionally, this area has been reviewed recently in terms of utility in the area of LC-MS/MS-based profiling (78), but many of the issues are applicable to other approaches such as SELDI.

2.6 Validation and translation

Whether studies adopt a profile-based approach or identification of specific discriminant peaks, validation of the findings is necessary. For profiles, this often takes the form of a blind 'test' set to test the model developed using a training set (although usually run at the same time, the sets should be selected prior to profiling), and ultimately the model should be tested further with additional samples over time with eventual expansion to multiple centers. Few studies have tested their models beyond the initial main sample set, but results tend to be poorer with reduced sensitivity and specificity. Lack of reproducibility can be caused by the initial algorithm overfitting the data (82), but can also be due to variability in the performance of the technology over time (12). With improved QC measures, our recent study profiling urine samples from patients with bladder cancer has shown that, using random forest classification, similar sensitivity and specificity of approximately 70% obtained with a training and test set could be achieved in a late blind test set 6 months later (N.P. Munro et al., unpublished data). Clearly improvements in sensitivity and specificity are needed, but this is promising in terms of long-term reproducibility. Recent results from preliminary multi-center studies indicate that, with appropriate standardization of protocols and machines, promising levels of correlation can be achieved between centers (83, 84).

In several studies, specific discriminant peaks have been identified. These include transthyretin, cystatin C, 7B2, VGF, chromogranin B, HIP/PAP-I, α-defensins, β-defensins, parvalbumin, haptoglobin, SAA, C3a(desArg), and apolipoprotein A-II (5, 33, 34, 37, 39, 84–91), either as mature proteins, fragments, or post-translationally modified forms. Such identification is challenging and has been achieved by SELDI-guided enrichment/purification by chromatographic or gel-based methods followed by tryptic digestion, either on-spot or followed by LC-MS/MS sequencing (33, 34, 84, 85, 89–91) or by prediction of identity from databases on the basis of mass with confirmation using antibody-based precipitation (37, 39, 86, 87). With the introduction of a ProteinChip interface for specific tandem mass spectrometers, it has been possible to obtain sequence data directly from the chip analyzed by SELDI for peaks of <6 kDa (5, 34, 84), although the sensitivity means that enrichment is still needed for many of the peaks. In several cases, further subsequent validation of the findings has been undertaken using various approaches such as enzyme-linked immunosorbant assays or immunohistochemistry. At present, it seems likely that whatever the proteomic discovery technology, the validation and translation stages are likely to use complementary approaches, which are by necessity robust, quantitative, and feasible to use in clinical chemistry laboratories, such as immunoassays and possibly quantitative MS or antibody-based arrays (92).

2.7 Methodology

The protocols described here act as a starting point for generating profiles using the SELDI system. The examples given are those optimized for serum and plasma but work equally well for other sample types with modification of parameters such as sample load. A basic knowledge of operation of the Ciphergen instruments and proprietary Ciphergen software is assumed, with only basic outline guidelines to appropriate settings for some key aspects of data collection, processing, and analysis being provided. For further detail, consultation of the Ciphergen manuals or technical support is recommended. For groups with bioinformatic and statistical expertise, we also provide alternative strategies that we have found useful, or are developing, to illustrate the possible approaches. Space precludes including detailed protocols for purification methods for subsequent peak identification (and anyway these will be dependent on the nature of the protein/peptide), but this area has been referenced extensively in section 2.6, with many of the references providing detailed methodology of the various approaches for several different molecules and fluids.

2.7.1 Blood samples for serum or plasma banking

Protocol 1 describes the processing of blood for serum or plasma banking. These recommendations, given in step 1, are based on a recent study (16). Heparinized or citrated plasma may also be used – indeed citrated plasma may be superior in terms of ensuring platelet stability, but care must be taken to ensure accurate filling of the blood tube to achieve the correct ratio of blood to liquid anticoagulant in this case to avoid effects on sample dilution and this is not always possible in routine hospital clinics. It is still not clear whether plasma or serum provides the optimum fluid for biomarker discovery, with many of the peaks seen in serum potentially arising from clotting events, and so protocols are provided for both.

Protocol 1

Processing of blood samples for serum or plasma banking

Equipment and Reagents
- Centrifuge (controlled temperature)
- Labeled 0.5 ml Eppendorf tubes
- 5 ml Bijoux for pooling aspirates
- Narrow-tipped Pastettes
- EDTA anticoagulant tubes
- Plain tubes containing silica activator[a]

Method

1. Collect 5–10 ml[b] of venous blood into EDTA anticoagulated tubes (plasma) and 5–10 ml of blood into plain tubes containing silica activator (serum).

2. Centrifuge the samples for 10 min at 2000 g at 20°C. The different tube types will need to be centrifuged separately, as samples for serum (red top tubes) should be allowed to clot for 1 h before centrifugation, whereas samples for plasma should be centrifuged within 30 min of venepuncture[c].

3. Following centrifugation of the EDTA sample, remove the plasma from the sample using a narrow-tipped Pastette and place in a pooling tube. Only remove the upper two-thirds of the plasma to avoid contamination with buffy coat (the white cell/platelet layer above the red blood cells).

4. Using a second Pastette, divide the plasma among ten labeled Eppendorf tubes. Store the sample aliquots at −80°C.

5. Following centrifugation of the plain (silica-coated) serum tube, remove the serum from the sample using a narrow-tipped Pastette and place in a pooling tube. Remove as much of the serum as possible without disturbing the red cells.

6. Using a second Pastette, divide the serum among ten labeled Eppendorf tubes. Store the sample aliquots at −80°C and avoid freeze-thawing.

Notes

[a]Avoid tubes containing gel-based activator/separator.
[b]Smaller or larger volumes may be taken, the exact volume depending on the size of available tubes – the number of stored aliquots should be adjusted accordingly, but we would advise a minimum volume of ~200 µl per aliquot.
[c]Record any deviations from the protocol times or any specific observations regarding the sample, e.g. hemolyzed or lipaemic. No studies have yet looked at the effects of posture, time of day, etc. on the sample profiles and it may be useful to record such information if available (although it is recognized that, particularly in the experimental discovery phase, standardization of such factors is likely to be impossible).

2.7.2 ProteinChip sample loading

Protocol 2 describes ProteinChip sample loading. The binding buffers described in *Protocol 2* for each ProteinChip type are starting points only that we have found work well with serum and plasma, and can be modified as needed to alter the stringency of the binding conditions. With the NP20 chip, the use of increasing salt (50–500 mM), detergent (0.01–0.1%, v/v), organic solvent (0–50%, v/v), or decreasing the pH will increase the selectivity. For the H50 chip, inclusion of salts such as 50–100 mM NaCl or organic solvents such as 10–30% acetonitrile will increase the hydrophobic interactions and detergents will have the converse effect. For the ion exchange chips CM10 and Q10, selectivity can be altered by varying the pH of the binding buffers used – as buffer pH is increased above the isoelectric point (pI) of the protein, the protein will bear a net negative charge and the converse for pH below the pI. For the IMAC30 ProteinChip, alternative metal ions that can be used to prime the ProteinChip include nickel and gallium, which may alter the profile of bound proteins. Denaturing buffers containing acetonitrile or chaotropes such as urea have also been used (e.g. 42, 84). Examples of profiles obtained are provided in *Fig. 4*.

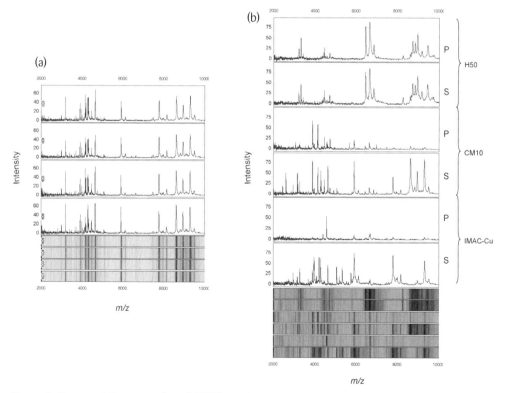

Figure 4. Representative examples of SELDI spectra.
(*a*) Spectra acquired from a single serum sample profiled on four separate sample spots, loaded as described in *Protocols 2* and *3* on a CM10 chip with HEPES buffer (pH 7.6), to show the typical reproducibility expected between different runs of the same sample (*b*) Typical appearances of EDTA plasma (P) and serum (S) profiled on three different chip types (H50, CM10, and IMAC-Cu) using the conditions described in *Protocols 2* and *3*. In both cases, the corresponding 'gel' view of the spectra are presented below.

Protocol 2

ProteinChip sample loading

Equipment and Reagents
- Milli-Q purified water or equivalent for all solutions[a]
- Calibrants: bovine insulin (M_r 5733.6), bovine ubiquitin (M_r 8564.8), cytochrome c (M_r 12 230.9), equine myoglobin (M_r 16 951.5), horseradish peroxidase (M_r 43 240), bovine serum albumin (M_r 66 433), and bovine IgG (M_r 147 300) (Ciphergen)
- ProteinChips (Ciphergen): NP20 (normal phase), H50 (hydrophobic), CM10 (weak cation exchange), IMAC30 (immobilized metal affinity), and Q10 (strong anion exchange)[b]
- Biomek 2000 liquid handling robot or Gilson pipettes
- Saturated sinapinic acid (SPA; Fluka) matrix solution[c] (prepare by addition of 240 µl of acetonitrile and 160 µl of 1% (v/v) trifluoroacetic acid (TFA)[c] to 5 mg of SPA)
- CM10 diluent/wash buffer (50 mM HEPES, pH 7.5)
- Q10 diluent/wash buffer (10 mM Tris/HCl, pH 8.0)
- 20% (v/v) Acetonitrile, 0.1 % (v/v) TFA[d]
- H50 diluent (0.1 M sodium phosphate, 0.5 M NaCl, 0.1% (v/v) TFA, pH 7.0)
- H50 wash buffer (10% (v/v) acetonitrile, 0.1% (v/v) TFA)[d]
- 0.1 M Copper sulfate
- 0.1 M Sodium acetate (pH 4.0)
- IMAC30 diluent/wash buffer (0.1 M sodium phosphate, 0.5 M NaCl, pH 7.0)

Method

Calibrant chip

1. Dilute 2 µl of each of bovine insulin, bovine ubiquitin, and cytochrome c (all 10 nmol/ml) with 34 µl of SPA to provide a 'low' calibrant.
2. Dilute 2 µl of each of equine myoglobin (10 nmol/ml), horseradish peroxidase (20 nmol/ml), bovine serum albumin (20 nmol/ml), and bovine IgG (50 nmol/ml) with 32 µl of SPA to provide a 'high' calibrant.
3. Apply 2 µl of each to the surfaces of an NP20 ProteinChip.
4. Allow to air dry in the dark for at least 30 min and use for calibrating SELDI.

Sample preparation

1. Immediately before use, allow plasma or serum samples to thaw, mix gently, and place on ice.
2. Microfuge at 13 000 r.p.m. (15 000 g) for 5 min at 4°C[e].
3. Remove the required amount of plasma or serum and dilute 1 : 10[f] with the appropriate diluent depending on the ProteinChip type.
4. Mix gently, place in 96-well plates if using the robotic liquid handler and load samples onto the ProteinChips using the following protocols[g].

Q10 and CM10 ProteinChips

1. Wash each sample spot with 5 µl of Q10 diluent/wash buffer or CM10 diluent/wash buffer as appropriate, twice for 1 min each.
2. Remove the solution from each sample spot, and apply 5 µl of diluted sample to the ProteinChip surface, taking care not to allow the sample spots to dry.
3. Incubate at room temperature for 30 min.
4. Aspirate the sample from each spot and wash three times with repeated aspirations/dispensing of Q10 or CM10 wash buffer as appropriate.

5. Wash three times with repeated aspirations/dispensing of water.
6. Remove the final wash and allow to dry for 3 min.
7. Add two applications of 1 ml of SPA matrix solution[h] to each sample spot with 5–6 min drying time between applications.
8. Air dry for at least 30 min before analyzing using SELDI.

H50 ProteinChip

1. Pre-treat each sample spot with 5 µl of 20% (v/v) acetonitrile, 0.1 % (v/v) TFA.
2. Remove the solvent from each sample spot and apply 5 µl of diluted sample to the ProteinChip surface, taking care not to allow the sample spots to dry.
3. Proceed as in step 3 onwards in the Q10 and CM10 ProteinChip methodology but using H50 wash buffer.

IMAC30 ProteinChip

1. Add 5 µl of 0.1 M copper sulfate solution to the sample spots and incubate for 10 min at room temperature.
2. Remove the solution and wash twice with water for 2 min each.
3. Aspirate the water and wash with 0.1 M sodium acetate solution (pH 4.0) for 5 min.
4. Proceed as in step 2 onwards in the Q10 and CM10 ProteinChip methodology, but using IMAC 30 wash buffer.

Notes

[a]For solutions, all chemicals should be of at least Analar grade or equivalent (sequencing grade for the solvents) unless otherwise indicated and nonautoclaved tips should be used throughout.

[b]Care should be taken to avoid touching the surface of ProteinChips with the pipette tips, whether manual or robotic. In addition, powder-free gloves should be worn, as spurious peaks can be generated by powder contamination.

[c]Matrix solutions should be prepared fresh each day and stored in the dark, microfuging immediately before use. The energy-absorbing molecule described here for the sample matrix is SPA, which is most suited for a wider range of molecular masses, particularly those >5 kDa. If a region of <5 kDa is of particular interest, an alternative energy-absorbing molecule is CHCA (α-cyano-4-hydroxy-cinnamic acid). The time of drying of ProteinChips before matrix application is critical to reproducibility and should be optimized in each laboratory. Similarly, the volume and solvent content of matrix should be optimized to produce optimal crystallization and coverage of the sample.

[d]TFA-containing solutions should be made on the day of use. Extreme care should be taken when using TFA as it is an extremely hazardous chemical and the concentrated acid should only be used in a fume hood and dispensed wearing suitable protective clothing. Consult the safety datasheet for handling details and disposal.

[e]Microfuging is primarily to remove any insoluble aggregates formed during storage. Centrifugation at 15 000 g for 15 min may also be used to remove lipids (as an upper layer), although whether effects on other proteins/peptides in the profiles would also result has not been investigated.

[f]For meaningful comparison, the protein load on each sample spot should be the same for all samples if the surface is not saturated. At a dilution of 1 : 10 for plasma or serum, the surface is saturated (i.e. the sample is in excess, over a range of total protein concentrations found in health and disease), thereby obviating the need for measurement of protein concentration. If pre-fractionation steps are incorporated, the protocol will need to be reoptimized accordingly.

[g]Inclusion of QC samples on each chip and randomization of sample replicates should be adhered to, as described in earlier sections outlining the methods.

[h]The exact volume of the matrix should be optimized for either manual pipetting or liquid robotics, and also depending on the solvent composition.

2.7.3 Generation of SELDI profiles

SELDI profiles can be generated as described in *Protocol 3*, both from the calibration chips and the sample chips.

Protocol 3
Generation of profiles using SELDI

Equipment
- SELDI PBS II or PBS IIc with autoloader (Ciphergen)

Method
1. Obtain SELDI profiles from the high-mass and low-mass calibration chips using the optimum mass range settings for each, which should be based on those intended to be used for the experimental samples[a]. Collect a minimum of 50 transients across the surface[b].
2. Using spare sample spots, optimize the laser and detector settings iteratively by manually firing the laser several times for each setting, ensuring that the settings do not result in saturation of any peaks in the mass range and that the S/N ratio is maximized. For plasma or serum loaded as above, initial settings for the low-mass range are likely to be in the range of laser intensity 170–200 at a detector sensitivity of 9 for example, but will vary between machines.
3. Construct appropriate spot and chip protocols incorporating the above optimized settings and including warm-up shots.
4. Calibrate the SELDI using the appropriate low-mass calibrants (singly and doubly charged).
5. Obtain SELDI profiles from each of the three QC chips, reading each chip at the optimized parameters for one selected mass range before progressing to the next, making sure that the software protocols are designed so that readings are offset in their positions on the sample spot.
6. Construct calibration equations using the singly and doubly charged peaks for the appropriate higher-mass calibrants for each mass range examined and apply the resultant calibration to the spectra generated at these settings for those mass ranges.
7. Analyze the data for the QC samples over the different mass ranges as described below and, if acceptable, proceed to prepare and profile the samples.
8. Obtain SELDI profiles from each of the sample chips, ensuring that all chips are stored for approximately equal lengths of time in the dark after drying of the matrix before reading[c]. Read each chip at the optimized parameters for one selected mass range before progressing to the next, making sure that the software protocols are designed so that readings are offset in their positions on the sample spot.
9. Using the appropriate calibration equations, ensure that the spectra are all calibrated as above and proceed to process that data.

Notes
[a]Unless only interested in one particular mass region, we would recommend examining a low-mass range of, for example, <10 kDa, followed by 10–20 kDa and 20–100 kDa, with the width of the different ranges depending on the areas of particular interest. Increasing laser intensities and center

masses (and hence extraction delay times) should be selected with increasing mass to optimize performance. Generally, laser intensities will be in the region of 180–220 with a detector sensitivity of 9 or 10. The most informative regions are often <30 kDa due to the higher resolution in this area and greater ease of ionization compared with larger molecules.

[b]It is a good idea to also rerun the calibrant chips during and after long experimental runs to check for any systematic drift by examining the peak masses. New calibration equations can be generated if necessary, although calibration drift to this extent should not normally occur unless, for example, major changes in ambient temperature are experienced.

[c]In practice, our current data indicate that the exact timing of this step after matrix drying is not critical, with little change in signal over at least the first 2 h post-drying, but this may be highly dependent on ambient conditions and hence delays or variation should be avoided.

2.7.4 Processing of raw SELDI data

This essentially consists of sequential application of the different processes indicated in *Fig. 2*. The process is described in outline for the PROTEINCHIP software with BIOMARKER WIZARD (version 3.2.1), using menus for clarity rather than tabs, although similar methods are used with the Ciphergen EXPRESS software.

1. *Normalization.* The most common approach is to normalize by total ion current (TIC).
   - Select the spectra to be normalized and follow Experiment → Normalize.
   - Select the dialog options such as the mass region that you wish to normalize spectra over if using TIC, or choose an external normalization coefficient or specific common peak if this is appropriate and normalize the spectra.
2. *Baseline subtraction.* This can be done as an integral part of the above process or following normalization.
   - Select spectra and select Options → Analysis Protocol → Baseline. There are options to smooth the spectra before fitting the baseline and vary the width of the window used for this smoothing, but generally start with the default settings.
   - View the data and adjust the settings iteratively accordingly.
3. *Alignment of spectra.* This is not an automated process in the PROTEINCHIP software.
   - Check spectra visually, which may be easier in the gel view, and realign using internal peaks/calibrants if used, essentially by correcting the calibration for individual spectra or all spectra if deemed necessary. In the PROTEINEXPRESS software, there is a discrete process that can be used to align peaks.
4. *Peak detection.* Adjust the noise setting.
   - Follow Options → Analysis Protocol → Noise and adjust the mass window of the spectra that you want to set the noise over (generally one setting is not appropriate for the whole spectrum and typically the matrix region is excluded).
   - Using the default values for the other noise parameters, inspect the spectra and noise plots and adjust other settings iteratively until the best fit is obtained. Peaks can be detected under the Analysis Protocol option,

varying the settings in terms of the S/N ratio and minimum valley depth, or may be more easily detected and clustered using the BIOMARKER WIZARD cluster tool.
- Select the Find New Clusters tab, define the S/N ratio for the first pass for peak detection (which is often set initially at 5), the minimum percentage of spectra that must contain the peak for it to be defined as a cluster, the cluster mass window (i.e. how close a peak has to be to be part of the same cluster; this is dependent on the mass region of the spectrum and must be determined empirically in each laboratory by determination of the mass accuracy of repeat measurements, for example using the PBS II at 0-10 kDa, settings of ±0.05-0.1% may be appropriate) and the S/N ratio for peaks to be detected in the second pass, which is often 2 or 3.
- Change these values iteratively and inspect the spectra visually to check the results.

Alternative methods that we are exploring include the following:

1. *Normalization.* The use of TIC is really only appropriate for very similar samples, as otherwise real differences in data may be masked. The possibility of using 'spiked' peaks as internal standards is attractive, being added with the matrix to avoid issues of competitive binding, but potential effects of signal suppression need to be explored.
2. *Baseline subtraction.* A modified two-stage LOESS algorithm can be used, which is a regression technique that fits simple models to localized sections of the data with parameters being dependent on the mass range being analyzed (66, 93). After the first baseline estimation, those intensity values that deviate by greater than 1 standard deviation are temporarily replaced by the baseline value and the smoothing technique is reapplied. LOESS has the advantage of being a well-understood statistical method and is relatively easy to implement.
3. *Noise reduction.* Each spectrum is smoothed using a moving window average with window lengths of 3-21 points.
4. *Peak detection, alignment, and detection of common peaks.* Peaks of the appropriate width are identified using a varying size window. For a data point to be identified as a peak, there must be a given number of progressively lower values on each side, with the number of points dependent on the varying size window; for example, with a smoothing window size of five points, peaks of a width of five points are counted. This is done using an in-house algorithm (12). After peak detection, a 'seed' spectrum is used to produce a list of common peaks. For each mass point in the 'seed' spectrum, the number of peaks detected in all of the other spectra within a defined mass window is counted. The use of a mass window means that spectra alignment is implicit in the technique. Internal calibrants, which can be used to increase the mass accuracy, may also assist in the alignment. After this process, for each mass point there is an integer value representing the number of peaks detected for that value. These values are then smoothed using a moving window average and those

values that are above a defined threshold value are defined as a common peak. This method of peak detection is less conservative than the Ciphergen software and in particular improves the detection of lower-intensity peaks.

2.7.5 Analysis of QC data

QC data comes from three sources: (i) data collected over a long period of time to monitor machine performance, (ii) every spot on the first three chips of an experimental data collection run, and (iii) from one spot of each chip used in the experimental sample run.

The data collected routinely over a long period of time are from a pooled serum QC sample, the spectra of which are examined as described below for the various parameters to determine whether any variation is within acceptable limits and whether systematic changes are apparent. The first three chips of an experiment are loaded with a pooled QC sample of the type being investigated and are used as a basis to define reference values for subsequent analysis of chip QC samples and also the assessment of the agreement of sample replicates.

Spectra are assessed for the following parameters:

- Total ion current, i.e. the total area under the spectrum
- The contribution of the subtracted baseline to the total ion current
- The number of detected peaks
- The intensity of the detected peaks
- The number and intensity of common peaks across all spectra

Examination of these parameters graphically for the QC spots present on sample chips with simple 'flagging' of values greater than 2 standard deviations from the mean of the values obtained for the first three chips give an indication that a specific sample chip should be investigated further, and any spectra of samples that fall outside the appropriate limits may need to be excluded.

The assessment of the agreement between duplicates can be undertaken by comparing the coefficient of variation (CV) for each set of duplicates for the above parameters with the distribution of the CVs between all possible pairs of QC samples measured on the first three chips of the experiment. Any duplicate pair of samples is 'flagged' for further examination if the CV obtained is greater than the 95th percentile of the distribution of QC CVs.

Viewing the transformation of peak intensities from QC samples in principal component space can also give a visual method of examining any changes in the machine over time, chip effects, or other changes in the QC sample (see *Fig. 5*, also available in the color section).

2.7.6 Analysis of SELDI data

Analysis of peaks to detect differences between groups

There are a variety of ways of identifying which peaks are expressed differentially in two or more groups of samples. For simplicity, only two-level comparisons such as case–control comparisons have been considered here. Both versions of the

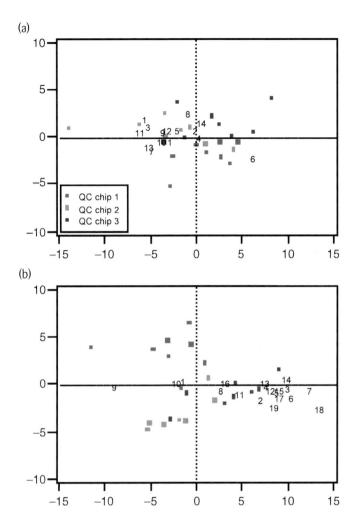

Figure 5. Examples of the use of plots of the first two principal components in QC (see page xxvi for color version).
These examples are based on the method outlined by Coombes et al. (13). Routinely, three chips with a pooled QC sample are run at the beginning of the experiment (indicated by colored squares). Principal components are calculated on the correlation matrix of the identified peak intensities for these 24 samples. Subsequently, one spot on each chip contains the same pooled QC sample (indicated by 1 for QC from chip 1, 2 for QC spot from chip 2, etc.). Peak intensities are projected into the principal component space of the original 24 samples, the Mahalanobis distance from the center of the space is calculated, and the QC spot is assessed by comparing this with χ^2_2 distribution. (*a*) An example of acceptable QC. None of the sample chip QC spots are very far from the origin and the spread of values is generally within or very close to those found on the QC chips. As the spot from chip 6 is at the edge of the space occupied by the QC samples, the spectra from this chip should be examined further. (*b*) An example of QC flagging unacceptable results. The QC spots from sample chips are on the edge of the space occupied by the original QC spots, indicating that the spectra from these chips ought to be investigated further. The QC spots from chips 7 and 18 are identified as being unusual, as they have statistically significant large values of Mahalanobis distance, i.e. these QC spots cannot be said to be the same as those from the first three QC chips. Subsequently, a machine fault was identified. Note also the marked chip effects between the three QC chips, also observed to a much lesser extent in (*a*).

Ciphergen software (and any statistical software) can also allow the analysis of more than two groups using, for example, analysis of variance or a nonparametric equivalent such as the Kruskall–Wallis test. The *lme* (see below) can also be adapted to estimate effects for multi-level factors of interest.

In the simplest situation of comparison of two groups with no replicates, a *t*-test may be appropriate, if distributional assumptions are satisfied (possibly after transformation)[a]. This can be done in the Ciphergen PROTEINCHIP software BIOMARKER WIZARD module following peak clustering using the Biomarker Wizard → Sample Group Statistics menu and selecting the groups to be compared. The Ciphergen EXPRESS software allows further statistical tests, notably a *t*-test with replicate averaging capability, cluster analysis, and principal components analysis. A further Ciphergen software package that can be purchased is the BIOMARKER PATTERNS software, which allows the creation of classification and regression trees using binary recursive partitioning.

Exporting of the data and independent processing allows further flexibility in the possible analysis methods, which may be more appropriate given the complexity of the data. The linear mixed effects model (*lme*, also referred to as random effects regression) is a flexible extension of the *t*-test that can accommodate more than one source of random variation and is valid with different numbers of replicates for different samples. Differentially expressed peaks can be identified, and the method also quantifies the level of agreement or discordance among replicates by the intra-class correlation coefficient. The model is defined mathematically as:

$$y_{ij} = \alpha + \beta x_i + \gamma z_i + \eta_i + \varepsilon_{ij}$$

where y_{ij} is the intensity value at one peak cluster for individual *i* and replicate *j*. The explanatory variables are x_i (an indicator variable for case–control status, or more generally indicating which group the sample comes from) and z_i (a vector of covariates such as age and sex, which need to be corrected for). These terms in the model are the fixed effects. There is a significant difference in peak intensities between cases and controls when the coefficient β is significantly different from 0. To account for the correlation among replicates of the same sample, a random effect term η_i for sample *i* is included in the model, where η_i are assumed to be normally distributed

---

Note

[a]Some statistical methods assume normality of the distribution of peak intensities over samples. Common transformations to achieve approximate normality include the natural logarithm (e.g. 81) and various power transformations such as the cube root (13). When a peak is absent in a particular sample, one approach is to impute a non-zero value (e.g. the mean). Otherwise, if the peak here has intensity zero, then the distribution of intensity values may follow a mixture distribution consisting of a (log)-normal distribution and an atom of probability at zero. Likelihood ratio tests can then be used to identify differentially expressed peaks, identifying whether these differences are apparent in the proportion of zeros in cases and controls or the mean intensities in the peaks present in these groups.

with mean 0 ($N(0, \sigma_\eta^2)$). The variance σ_η^2 is thus a measure of the variability among patients in the study. The final term describes the residual variation ($\varepsilon_{ij} \sim N(0, \sigma_\varepsilon^2)$), assumed to be uncorrelated with the other terms. The intra-class correlation coefficient is calculated as $\sigma_\eta^2/(\sigma_\eta^2 + \sigma_\varepsilon^2)$, with values close to 0 indicating poor agreement between replicates and values close to 1 indicating good agreement. This analysis can be carried out using standard statistical software such as STATA (Stata Statistical Software, release 9; StataCorp LP) or R (http://www.R-project.org).

When performing many tests simultaneously, as here where many peak clusters are considered, it is important to consider the effect that multiple testing has on the interpretation of statistical significance. When applying a 5% significance level, on average one in 20 peaks will be significantly differentially expressed, even if no peaks truly differ between the groups (type I error). In this situation (using either the *t*-test or the *lme*), a useful approach is to estimate the false discovery rate (FDR). The FDR is the proportion of the peaks that are called significant that in fact do not differ between groups. In this type of study, our aim is to keep this level acceptably low, whilst not applying such a stringent test that true differences are undetected. Applying a Bonferroni correction (dividing the significance level by the number of tests) is one way of adjusting the significance level, but this will generally be unacceptably conservative as the tests are not independent. An alternative is to use the significance analysis of microarrays procedure (94), which can be implemented easily in an EXCEL interface or in R. Permuted samples from the data are used to estimate the null distribution of the statistics (such as the *t*-statistics from each comparison of peaks); the user can decide on a level of differential expression that gives an acceptable FDR (see *Fig. 6*, also available in the color section).

Classification of samples

Another common use of peaks from SELDI profiles is as predictors in classification algorithms. A method that performs well is based on random forests (80, 95). A classification tree is a set of decision rules based on a set of predictors that successively splits the data set until all of the samples in the same node are in the same class. Random Forest (96), which can be implemented in R or Fortran, is a method of classification based on growing an ensemble of many such trees. From a data set (SELDI profiles) of n samples averaged over replicates, a large number (S) of random samples of size n are drawn from the data set, sampling with replacement (bootstrap samples). Hence, the bootstrap samples will contain single or multiple replicates of some SELDI profiles and other profiles will not be represented in each sample drawn out. From each such sample, a classification tree is developed, resulting in S trees (typically tens of thousands). At each branching of each tree, m input variables (peaks) are randomly selected as the predictors on which the data set is split. (Here, m is chosen to be much smaller than the total number of peaks.) The random selection of predictors reduces the correlation between the trees in the forest. In order to classify a new sample, each tree in the forest is applied to the sample, resulting in a classification into one of the possible groups. Each tree thus yields a 'vote' for a particular classification of the sample. The forest

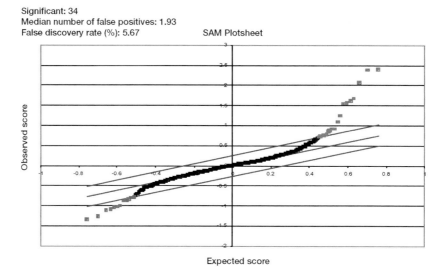

Figure 6. Comparison of MS profiles of two sample groups using the significance analysis of microarrays (SAM) method (94) (see page xxvii for color version).
The x-axis shows the expected score (in this case a two-sample t-statistic) for each of 510 peaks, ordered from lowest to highest, under the null hypothesis of no difference between groups. This is obtained by randomly permuting the group labels (e.g. case–control status), calculating the t-statistic for each peak on this permuted data set, and ordering these 510 statistics. The mean of each ordered statistic is then obtained on the basis of 1000 such permuted samples. The y-axis shows the observed scores for the 510 peaks, again ordered by magnitude. Deviation greater than a certain distance from the line $y=x$ indicates a significant result. In Microsoft EXCEL, a slider tool allows the adjustment of the width of the lines parallel to $y=x$. This allows the interactive selection of the number of significant results with an acceptable FDR, which is indicated in the top left-hand corner of the graph.

chooses the classification having the most votes (over all of the trees in the forest), and the proportion of trees choosing that classification provides a corresponding measure of certainty about the classification.

Although it is usual to partition the data into training and test sets, in this method there is no need for a separate test set to get an unbiased estimate of performance. In drawing a bootstrap sample of size n from the data, approximately one-third of the original data will be absent from each sample and thus not used in the construction of the corresponding tree. For each sample in the data set, there will thus be approximately $S/3$ trees that have been developed without using the sample, and these trees can be used to classify the sample. Sensitivity and specificity measured on these 'out-of-bag' samples provide unbiased estimates of the performance of the method overall.

3. TROUBLESHOOTING

- **Unexpected peaks present in samples**
  This may be due to interference from, for example, detergents or latex gloves.
- **Poor or little signal**
  Check the SELDI performance with a previous QC chip – if adequate, then the most likely reason is problems with sample loading such as pipetting of sample, buffers of wrong composition, or inadequate washing of contaminants such as salts causing ion suppression.
- **Variable signal generated across the chip spot surface**
  This is most likely due to poor/uneven matrix application and crystallization, or an inadequate number of warming shots, or both.
- **Declining signal over time and/or poor reproducibility**
  Check the QC of any robotics used for chip loading and if this is fine, then the likely cause of the problem is a change in machine performance, e.g. declining laser performance. Contact Ciphergen.
- **Sudden change in spectra obtained**
  Check that this has not coincided with a new batch of buffer, new batch of chips, etc.
- **Differences between the groups in age, sex, or other factors that might influence the peak profiles**
  These can be adjusted for by including covariates in the statistical analysis if analyzed externally. If the distribution of the factor is very different between groups, it may only be possible to eliminate its confounding effect by restricting the analysis to a subset of the data matched retrospectively on the factor.
- **The baseline plots do not look appropriate for all samples**
  If using the Ciphergen-supplied software, it is imperative that the baseline for each spectrum is examined manually, as the default settings will not be appropriate for all spectra – it is occasionally necessary to adjust the width of the smoothing window manually to get the baseline correction to function optimally.
- **Normalizing by TIC has marked effects on some spectra**
  This may be due to quite major differences in samples, either due to poor technique or inherent sample differences. If the former, the reproducibility of the approach should be explored. If the latter, then normalization by TIC is inappropriate.

4. REFERENCES

★★ 1. Yanagisawa K, Shyr Y, Xu BJ, *et al.* (2003) *Lancet*, **362**, 433–439. – *A good illustration of how hierarchical clustering can be applied to MALDI data with clinical diagnostic utility.*
2. Stoeckli M, Chaurand P, Hallahan DE & Caprioli RM (2001) *Nature Med.* **7**, 493–496.
3. Chaurand P, Schwartz SA, Reyzer ML & Caprioli RM (2005) *Toxicol. Pathol.* **33**, 92–101.
4. Merchant M & Weinberger SR (2000) *Electrophoresis*, **21**, 1164–1177.
5. Zhang L, Yu W, He T, *et al.* (2002) *Science*, **298**, 995–1000.
6. Lewczuk P, Esselmann H, Groemer TW, *et al.* (2004) *Biol. Psychiatry*, **55**, 524–530.

7. Goldstein LE, Muffat JA, Cherny RA, et al. (2003) *Lancet*, **361**, 1258–1265.
8. Adam BL, Qu Y, Davis JW, et al. (2002) *Cancer Res.* **62**, 3609–3614.
9. Petricoin EF, Ardekani AM, Hitt BA, et al. (2002) *Lancet*, **359**, 572–577.
10. Li J, Zhang Z, Rosenzweig J, Wang YY & Chan DW (2002) *Clin. Chem.* **48**, 1296–1304.
11. Papadopoulos MC, Abel PM, Agranoff D, et al. (2004) *Lancet*, **363**, 1358–1363.
★ 12. Rogers MA, Clarke P, Noble J, et al. (2003) *Cancer Res.* **63**, 6971–6983. – One of the first studies to show the potential issues over long-term robustness of profiling and some critical issues in urine processing.
★★★ 13. Coombes KR, Fritsche HA Jr, Clarke C, et al. (2003) *Clin. Chem.* **49**, 1615–1623. – A good early example of a thorough and statistically sound QC paper with a case study.
14. Diamandis EP (2004) *Mol. Cell. Proteomics*, **3**, 367–378.
15. Sorace JM & Zhan M (2003) *BMC Bioinformatics*, **4**, 24.
★★ 16. Banks RE, Stanley AJ, Cairns DA, et al. (2005) *Clin. Chem.* **51**, 1637–1649. – One of the first studies to show the importance of pre-analytical variables, in this case sample processing, in clinical proteomic profiling.
★ 17. Aebersold R & Mann M (2003) *Nature*, **422**, 198–207. – A good introductory review of MS.
★★ 18. Hutchens TW & Yip TT (1993) *Rapid Commun. Mass Spectrom.* **7**, 576–580. – The initial 'invention' of the SELDI concept.
19. Paweletz CP Gillespie JW, Ornstein DK, et al. (2000) *Drug Dev. Rev.* **49**, 34–42.
20. Von Eggeling F, Davies H, Lomas L, et al. (2000) *Biotechniques*, **29**, 1066–1070.
★ 21. Wright GW Jr, Cazares LH, Leung SM, et al. (1999) *Prostate Cancer Prostatic Dis.* **2**, 264–276.
22. Melle C, Ernst G, Schimmel B, et al. (2006) *Int. J. Oncol.* **28**, 195–200.
23. Wellmann A, Wollscheid V, Lu H, et al. (2002) *Int. J. Mol. Med.* **9**, 341–347.
24. Wong YF, Cheung TH, Lo KW, et al. (2004) *Cancer Lett.* **211**, 227–234.
25. Kwapiszewska G, Meyer M, Bogumil R, et al. (2004) *BMC Biotechnol.* **4**, 30.
26. Cheung PK, Woolcock B, Adomat H, et al. (2004) *Cancer Res.* **64**, 5929–5933.
27. Xu BJ, Caprioli RM, Sanders ME & Jensen RA (2002). *J. Am. Mass Spectrom. Soc.* **13**, 1292–1297.
★★ 28. Chaurand P, Schwartz SA, Billheimer D, Xu BJ, Crecelius A & Caprioli RM (2004) *Anal. Chem.* **76**, 1145–1155. – A useful starting point for illustrating the critical issues involved in sample preparation of tissue for MS imaging and profiling.
29. Craven R & Banks RE (2004) In *Current Protocols in Protein Science*, Suppl. 31, 22.3.1–22.3.27. John Wiley & Sons.
30. Liu W, Guan M, Wu D, et al. (2005) *Eur. Urol.* **47**, 456–462.
★ 31. Schaub S, Wilkins J, Weiler T, Sangster K, Rush D & Nickerson P (2004) *Kidney Int.* **65**, 323–332. – A description of some important issues in urine processing.
32. Clarke W, Silverman BC, Zhang Z, Chan DW, Klein AS & Molmenti EP (2003) *Ann. Surg.* **237**, 660–664.
33. Ranganathan S, Williams E, Ganchev P, et al. (2005) *J. Neurochem.* **95**, 1461–1471.
★ 34. Ruetschi U, Zetterberg H, Podust VN, et al. (2005) *Exp. Neurol.* **196**, 273–281. – Examples of the complementary use of chromatographic purification and gel electrophoresis for purification and sequencing of discriminate peptides and of direct sequencing using the ProteinChip interface.
35. Carrette O, Demalte I, Scherl A, et al. (2003) *Proteomics*, **3**, 1486–1494.
36. Pawlik TM, Fritsche H, Coombes KR, et al. (2005) *Breast Cancer Res. Treat.* **89**, 149–157.
37. Li J, Zhao J & Yu X (2005) *Clin. Cancer Res.* **11**, 8312–8320.
38. Sauter ER, Shan S, Hewett JE, Speckman P & Du Bois GC (2005) *Int. J. Cancer*, **114**, 791–796.
★★ 39. Rosty C, Christa L, Kuzdal S, et al. (2002) *Cancer Res.* **62**, 1868–1875. – One of the earliest examples of antibody-based identification of a discriminant peak in a SELDI profile.
★★★ 40. Anderson NK & Anderson NG (2002) *Mol. Cell. Proteomics*, **1**, 845–867. – An excellent review of the history of plasma proteomics with additional useful information.
★★★ 41. Anderson NL, Polanski M, Pieper R, et al. (2004) *Mol. Cell. Proteomics*, **3**, 311–326. – A compilation of the extensive results from four studies adopting different approaches to mining the plasma proteome.

42. Tirumalai RS, Chan KC, Preito DA, Issaq HJ, Conrads TP & Veenstra TD (2003) *Mol. Cell. Proteomics*, **2**, 1096–1103.
43. Zhou M, Lucas DA, Chan KC, *et al.* (2004) *Electrophoresis*, **25**, 1289–1298.
★★ 44. Lowenthal MS, Mehta AI, Frogale K, *et al.* (2005) *Clin. Chem.* **51**, 1933–1945. – A qualitative study illustrating the diversity of proteins and fragments 'carried' by proteins such as albumin.
45. Vitzthum F, Behrens F, Anderson NL & Shaw JH (2005) *J. Proteome Res.* **4**, 1086–1097.
★★★ 46. Ransohoff DF (2005) *Nat. Rev. Cancer*, **5**, 142–149. – Review of the importance of various sources of bias in biomarker discovery studies.
47. Potter JD (2003) *Trends Genet.* **19**, 690–695.
48. Guder WG, Narayanan S, Wisser H & Zawta B (2003) *Samples: From the Patient to the Laboratory. The Impact of Preanalytical Variables on the Quality of Laboratory Results*, 3rd edn. Wiley VCH, Weinheim, Germany.
49. Lundblad RL (2006) *The Evolution from Protein Chemistry to Proteomics. Basic Science to Clinical Application.* CRC Press, Boca Raton.
50. Hu J, Coombes KR, Morris JS & Baggerly KA (2005) *Brief. Funct. Genomic. Proteomic.* **3**, 322–331.
51. Baggerley KA, Morris JS & Coombes KR (2004) *Bioinformatics*, **20**, 777–785.
★★ 52. Tammen H, Schulte I, Hess R, *et al.* (2005) *Proteomics*, **5**, 3414–3422. – A qualitative examination and identification of peptides present in plasma and serum.
★ 53. Rai AJ, Gelfand CA, Haywood BC, *et al.* (2005) *Proteomics*, **5**, 3262–3277. – A consensus view of the Human Proteome Organization on blood sample handling and effects on proteomic findings.
54. Marshall J, Kupchak P, Zhu W, *et al.* (2003) *J. Proteome. Res.* **2**, 361–72.
55. Villanueva J, Philip J, Chaparro C, *et al.* (2005) *J. Proteome Res.* **4**, 1060–1072.
★ 56. Karsan A, Eigl BJ, Flibotte S, *et al.* (2005) *Clin. Chem.* **51**, 1525–1528. – A brief report on critical effects of pre-analytical factors (e.g. blood sample handling) and analytical factors on profiling results.
★★ 57. Koomen JM, Li D, Xiao LC, *et al.* (2005) *J. Proteome Res.* **4**, 972–981. – A qualitative examination and identification of peptides in plasma and serum.
58. Lee ML & Whitmore GA (2002) *Stat. Med.* **21**, 3543–3570.
59. Gadbury GL, Page GP, Edwards J, *et al.* (2004) *Stat. Methods Med. Res.* **13**, 325–338.
60. Aivado M, Spentzos D, Alterovitz G, *et al.* (2005) *Clin. Chem. Lab. Med.* **43**, 133–140.
61. Jock CA, Paulauskis JD, Baker D, *et al.* (2004) *Biotechniques*, **37**, 30–32.
62. Wang MZ, Howard B, Campa MJ, Patz EF Jr & Fitzgerald MC (2003) *Proteomics*, **3**, 1661–1666.
63. Baggerley KA, Morris JS, Wang J, Gold D, Xiao LC & Coombes KR (2003) *Proteomics*, **3**, 1667–1672.
64. Wu B, Abbot T & Fishman D (2003) *Bioinformatics*, **19**, 1636–1643.
★★ 65. Tibshirani R, Hastie T, Narassimhan B, *et al.* (2004) *Bioinformatics*, **20**, 3034–3044. – A thorough method for the analysis of MS profiles, introducing an innovative method of spectral alignment.
66. Wagner M, Naik D & Pothen A (2003) *Proteomics*, **3**, 1692–1698.
67. Purohit PV & Rocke DM (2003) *Proteomics*, **3**, 1699–1703.
68. Neville P, Tan PY, Mann G & Wolfinger R (2003) *Proteomics*, **3**, 1710–1715.
69. Hilario M, Kalousis A, Müller M & Pellegrini C (2003) *Proteomics*, **3**, 1706–1719.
★ 70. Malyarenko DI, Cooke WE, Adam BL, *et al.* (2005) *Clin. Chem.* **51**, 65–74. – A unified approach to peak alignment and resolution using time series analysis.
71. Yasui Y, Pepe M, Thompson ML, *et al.* (2003) *Biostatistics*, **4**, 449–463.
72. Jeffries N (2005) *Bioinformatics*, **21**, 3066–3073.
73. Wong JWH, Cagney G & Cartwright HM (2005) *Bioinformatics*, **21**, 2088–2090.
74. Morris JS, Coombes KR, Koomen J, Baggerley KA & Kobayashi R (2005) *Bioinformatics*, **21**, 1764–1775.
75. White CN, Zhang Z & Chan DW (2005) *Clin. Chem Lab. Med.* **43**, 125–126.
76. Bons JA, de Boer D, van Dieijen-Visser MP & Wodzig WK (2005) *Clin. Chim. Acta*, **366**, 249–256.
77. Hong H, Dragan Y, Epstein J, *et al.* (2005) *BMC Bioinformatics*, **6** (Suppl. 2), S5.

★★★ 78. Listgarten J & Emili A (2005) *Mol. Cell. Proteomics*, **4**, 419–434. – *A comprehensive review of many aspects of data processing and statistical analysis for LC-MS/MS-based profiling with many concepts applicable to other profiling approaches.*
79. Lee KR, Lin X, Park DC & Eslava S (2003) *Proteomics*, **3**, 1680–1686.
80. Poon TC, Yip TT, Chan AT, *et al.* (2003) *Clin. Chem.* **49**, 752–760.
★★ 81. Izmirlian G (2004) *Ann. New York Acad. Sci.* **1020**, 154–174. – *The use of an effective classification method (random forests) explored and evaluated using SELDI profiles and simulation.*
82. Tibshirani R, Hong WJ, Warnke R, *et al.* (2005) *New Engl. J. Med.* **352**, 1496–1497.
★ 83. Semmes OJ, Feng Z, Adam BL, *et al.* (2005) *Clin. Chem.* **51**, 102–112 – *Promising results of inter-center SELDI performance.*
84. Rai AJ Stemmer PM, Zhang Z, *et al.* (2005) *Proteomics*, **5**, 3467–3474.
85. Dare TO, Davies HA, Turton JA, Lomas L, Williams TC & York MJ (2002) *Electrophoresis*, **23**, 3241–3251.
86. Vlahou A, Schellhammer PF, Mendrinos S, *et al.* (2002) *Am. J. Pathol.* **158**, 1491–1502.
87. Diamond DL, Kimball JR, Krisanaprakornkit S, Ganz T & Dale BA (2001) *J. Immunol. Methods*, **256**, 65–76. – *One of the earliest examples of antibody-based identification of a discriminant peak in a SELDI profile.*
88. Ye B, Cramer DW, Skates SJ, *et al.* (2003) *Clin Cancer Res.* **9**, 2904–2911.
89. Tolson J, Bogumil R, Brunst E, *et al.* (2004) *Lab. Invest.* **84**, 845–856. – *An example of purification and MS/MS sequencing of discriminant peaks showing multiple truncated isoforms.*
90. Li J, Orlandi R, White CN, *et al.* (2005a) *Clin. Chem.* **51**, 2229–2235.
91. Malik G, Ward MD, Gupta SK, *et al.* (2005) *Clin. Cancer Res.* **11**, 1073–1085.
92. Anderson NL (2005) *Mol. Cell. Proteomics*, **4**, 1441–1444.
93. Cleveland WS & Devlin SJ (1988) *J. Am. Stat. Assoc.* **83**, 596–610.
★★ 94. Tusher V, Tibshirani R & Chu C (2001) *Proc. Natl. Acad. Sci. U.S.A.* **98**, 5116–5121. – *An intuitive and easily implemented method of identifying differential expression with a strategy for controlling FDR.*
95. Tong W, Xie Q, Hong H, *et al.* (2004) *Environ. Health Perspect.* **112**, 1622–1627.
96. Breiman L (2001) *Mach. Learn.* **45**, 5–32.

# CHAPTER 9

Characterization of post-translational modifications: undertaking the phosphoproteome

W. Andy Tao, Bernd Bodenmiller, and Ruedi Aebersold

1. INTRODUCTION

Proteomics started with the great ambition of characterizing all of the proteins expressed in a cell or tissue. This aim has proved highly challenging, in spite of rapid technical and instrumental advances. Although the number of genes in the human genome is considerably smaller than originally thought, the number of proteins in a proteome has remained high. Many proteins are post-translationally modified or processed so that one gene can generate multiple polypeptide species (1). Furthermore, the dynamic range in biological samples may differ by seven to ten orders of magnitude, and the relatively low-abundance proteins are usually buried in a number of proteins of higher abundance. Therefore, the analysis of subproteomes (i.e. subsets of a proteome that have common functional or structural properties) has been more successful than the analysis of complete proteomes (2). This is particularly true for post-translationally modified proteins, which are frequently expressed at low abundance.

Post-translational modifications (PTMs) regulate the activity, localization, and stability of proteins and their interactions with other proteins (3). Whilst the biological significance of PTMs is well understood, their study has been complicated by a number of factors: proteins are frequently modified at a low stoichiometry and the degree of modification is highly dynamic. In addition, many PTMs are linked to the polypeptide via labile chemical bonds that can result in loss of PTM during sample preparation. In fact, PTMs have often been discovered by accident during studies of individual proteins using standard analytical techniques. Direct targeted analysis of PTMs has usually been a slow and laborious process. Along with its increasing ability in protein chemistry in general, mass spectrometry (MS) has also become the natural choice for the detection and identification of PTMs (4). PTMs usually lead to a mass increment or deficit compared with the mass of the unmodified residue and are therefore easily detected by MS. However, MS analysis of PTMs also has specific

Proteomics: *Methods Express* (C.D. O'Connor and B.D. Hames, eds)
© Scion Publishing, 2008

challenges (5). First, as it is not known *a priori* which residue of a protein might be modified, all peptides of the targeted protein need to be detected and analyzed. This is in contrast to protein identification where one or a few peptides are sufficient to declare the identity of a protein. Secondly, the presence of a PTM in a peptide can alter its ionization and/or fragmentation behavior, complicating the analysis of the generated fragment ion species. Finally, the multitude of possible PTMs complicates the assignment of the generated spectra to sequence databases.

Among all PTMs, protein phosphorylation has received the highest level of attention (6). Reversible phosphorylation of proteins plays a vital role in controlling many complex biological processes including cellular growth, cell division, apoptosis, and signaling. Aberrations in normal phosphorylation-dependent signaling systems relate to many human diseases including cancer, immune diseases, and diabetes. It is estimated that 30% of all research funding expended on drug development is focused on kinases, enzymes that transfer phosphate groups from high-energy donor molecules, such as ATP, to specified substrates or target molecules and their substrates. A comprehensive study of protein phosphorylation involves the identification of phosphoproteins and sites of phosphorylation, the identification of the kinase(s) and phosphatase(s) responsible for reversible phosphorylation and dephosphorylation, and an understanding of the biological consequence of the observed phosphorylation events. As a field of discovery science-based proteomic approaches (7), phosphoprotein profiling, resolved to the level of individual phosphorylation sites (5, 8, 9), is of high significance. There are four different types of phosphorylation that have been reported (10). The first and most common type is *O*-phosphorylation on serine, threonine, and tyrosine residues. The second type is *N*-phosphorylation, which is far less reported. *N*-phosphates occur mostly on histidine and lysine. The last two types of phosphorylation are *S*-phosphorylation, which occurs on cysteine, and acyl-phosphorylation, which occurs on both aspartate and glutamate. In this chapter, we will focus on *O*-phosphorylation.

Numerous factors complicate the analysis of protein phosphorylation in complex mixtures by MS-based methods. First, as with other PTMs, the stoichiometry of phosphorylation is frequently low, and only a fraction of an expressed protein may be phosphorylated at any given time. Secondly, a specific phosphoprotein may exist in several different phosphorylated forms, and the state of phosphorylation may be dynamically changing with changing states of the cell. In particular, many signaling molecules are present at extremely low abundance. Thirdly, with the exception of tyrosine phosphate, the phosphate bonds range from labile to highly labile. Therefore, specific precautions have to be taken to prevent the elimination of these phosphates during sample preparation. Specifically, *N*-phosphorylation and acyl-phosphorylation are extremely acid labile, whilst *O*-phosphorylation is relatively base labile. Lastly, chemical lability of the phosphate group on amino acid residues induced by collisions in the gas phase also has a negative influence on database-searching-based protein/peptide identification.

Whilst specific proteomic strategies for the investigation of protein phosphorylation keep evolving, there is a consensus on the general approach to studying protein phosphorylation in complex samples. The strategy is illustrated schematically in *Fig. 1*. In the first step, phosphoproteins are enriched. This step is useful but not absolutely necessary unless proteins of very low abundance are being

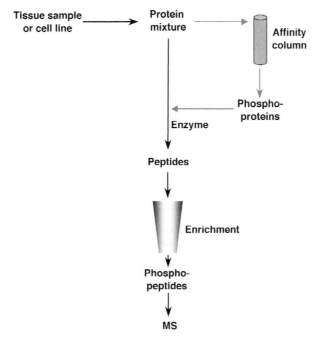

Figure 1. A generic protocol for phosphoproteomic analysis.

analyzed. In the second step, peptides are generated by chemical or enzymatic cleavage of the protein sample. In the third step, phosphopeptides are selectively enriched or isolated to identify sites of phosphorylation from complex samples. Finally, the phosphopeptides are analyzed by MS using specific data acquisition and database search protocols optimized for sequence and phosphorylation-site determination. In the following section, we describe the experimental approaches involved in each of these four stages.

2. METHODS AND APPROACHES

2.1 Enrichment of phosphoproteins

As protein phosphorylation is dependent on and an indicator of the cellular state, it is highly dynamic. For the isolation of phosphoproteins, it is therefore essential that the state of the cell at the time of isolation is considered. It may be possible to increase the overall level of phosphorylation (e.g. by the application of specific stimuli or, less specifically, the addition of phosphatase inhibitors), but such increased sites of phosphorylation may be of little biological significance.

Phosphoproteins are routinely enriched by affinity approaches. The most successful method uses anti-phosphotyrosine antibodies to enrich tyrosine phosphoproteins from complex protein mixtures such as cell lysates (see *Protocol 1*) (11–13). In several phosphoproteomics studies, effective enrichment of low-abundance tyrosine-phosphorylated proteins was used as the first fractionation step (14–16). A more specific approach takes advantage of the fact that most tyrosine-phosphorylated proteins provide high-affinity docking sites for specific protein interaction modules, such as Src homology 2 (SH2) domains and phosphotyrosine-binding domains (8, 17, 18). Using tyrosine-phosphorylated affinity substrates is functionally relevant because biologically significant interactors can be isolated and identified. There are several commercially available epitope-tagged SH2 fusion proteins that can be bound to affinity beads to provide a qualitative and quantitative fingerprint of the overall state of tyrosine phosphorylation. In contrast, antibodies for phosphoserine and phosphothreonine are generally not thought to have sufficient specificity or affinity for immunoprecipitation and therefore enrichment of serine- or threonine-phosphorylated proteins by immunoprecipitation with these antibodies has been reported in only a limited number of studies (19, 20).

Another popular affinity approach for enrichment of phosphoproteins is immobilized metal ion affinity chromatography (IMAC), which is based on the high affinity of phosphate groups toward a metal-chelated stationary phase, mainly Fe^{3+} or Ga^{3+} (21, 22). The major limitation of this method is that the specificity of this procedure is variable because of the affinity of acidic groups (aspartic and glutamic acid) and electron donors (e.g. histidine) for the metal chelate. The affinity capture of phosphoproteins followed by elution using alkaline phosphatase improves the specificity, but at the same time also loses the information on sites of phosphorylation (23).

Protocol 1

Enrichment of tyrosine phosphoproteins from whole cell extract (15)

Equipment and Reagents
- Agarose-conjugated 4G10 anti-phosphotyrosine monoclonal antibody (Upstate Biotechnology, Inc.)
- Glass homogenizer
- Lysis buffer (50 mM Tris/HCl, pH 7.8, 150 mM NaCl, 1% Triton X-100)
- Protease and phosphatase inhibitor cocktail (Roche)
- 50 mM Phenylphosphate
- Microsep (10K) centrifugal filter device (Pall Filtron Co.)

Method
1. Suspend the cells in lysis buffer containing protease and phosphatase inhibitor cocktail. Homogenize the cells using a glass homogenizer.
2. Transfer the lysate to Eppendorf tubes and spin down the nuclear debris at 2800 *g* at 4°C for 10 min.
3. Add agarose-conjugated 4G10 monoclonal antibody to the supernatant. Incubate at 37°C for 2 h.
4. Wash the beads extensively with lysis buffer.
5. Elute the bound proteins with 50 mM phenylphosphate.
6. Concentrate the eluted proteins using a Microsep (10K) centrifugal filter device.

2.2 Enrichment of phosphopeptides

The enrichment of phosphoproteins is preferred but by itself is not sufficient to identify sites of phosphorylation for complex mixtures. After the enrichment of phosphoproteins, enzymatic digestion is subsequently performed to generate peptides that are suitable for MS analysis (see *Protocol 2*).

Protocol 2

Protein digestion and peptide desalting

Equipment and Reagents
- Urea (Pierce)
- Tris(2-carboxyethyl)phosphine hydrochloride (TCEP; Pierce)
- 1 M Tris/HCl (pH 8.3; Sigma)
- Iodoacetamide (Sigma)
- Sequencing-grade trypsin (Promega)
- Sep-Pak desalting cartridge (Waters)
- 80% Acetonitrile (Aldrich), 0.1% trifluoroacetic acid (TFA; Aldrich)[a]
- SpeedVac (Thermo Electron)

Method
1. Add solid urea to the protein solution to a final concentration of 6–8 M and TCEP to 5 mM. Incubate at 37°C for 30 min.
2. Adjust the pH of the solution to around pH 8 using 1 M Tris/HCl (pH 8.3).
3. Add iodoacetamide to a final concentration of 15 mM. Incubate at room temperature in the dark for 1 h.
4. Dilute the solution using 10 mM Tris/HCl (pH 8.3) to bring the urea final concentration below 2.0 M.
5. Add trypsin to give a ratio of protein to trypsin of 25–100 : 1. Incubate at 37°C overnight.
6. Remove the urea and salts using a Sep-Pak cartridge. Elute peptides from the cartridge with 80% acetonitrile, 0.1% TFA.
7. Remove the solvent using a SpeedVac.

Note
[a]TFA-containing solutions should be made on the day of use. Extreme care should be taken when using TFA as it is an extremely hazardous chemical and the concentrated acid should only be used in a fume hood and dispensed wearing suitable protective clothing. Consult the safety datasheet for handling details and disposal.

In the resulting peptide mixtures, nonphosphopeptides are the dominant species. Therefore, overall enrichment of phosphopeptides is required. There are a larger number of methods for the enrichment of phosphopeptides than for phosphoproteins. They can be in classified into three groups: affinity enrichment, chemical derivatization, and charge-based fractionation. A recent study indicated that these methods are complementary (24).

2.2.1 Affinity methods

Antibodies have been used to enrich phosphopeptides but the efficiency of such affinity steps was relatively low (25, 26). Therefore, IMAC developed into the leading method of affinity enrichment of phosphopeptides (see *Protocol 3*) (26–32). The main problem with IMAC has been the nonspecific binding of non-phosphopeptides containing residues with acidic side chains. This limitation has been partially overcome by the employment of a different protease (e.g. Glu-C protease) that minimizes the occurrence of acidic residues on the peptides (33), and of methyl esterification to block the carboxylic group (28). However, the method appears to be highly dependent on the type of resin and pH conditions for binding and elution, and prefers peptides with multiple phosphorylation sites (34).

Protocol 3

Isolation of phosphopeptides by IMAC (32)

Equipment and Reagents
- Sample peptides (from *Protocol 2*)
- IMAC resin, e.g. POROS 20 MC (Applied Biosystems) or IDA–agarose (Sigma-Aldrich)
- Activating solution: 100 mM $GaCl_3$ or $FeCl_3$ (Aldrich)
- Loading/washing buffer: 100 mM NaCl in acetonitrile (Aldrich) : water : acetic acid (25 : 74 : 1, v/v/v)
- 50 mM Na_2HPO_4 (pH 9.0; Aldrich)

Method
1. Activate the IMAC resin with 100 mM $FeCl_3$ or $GaCl_3$.
2. Wash the resin with loading/washing buffer to remove free $FeCl_3$ or $GaCl_3$.
3. Dissolve the sample peptides in the loading/washing buffer. Incubate the peptides with the activated IMAC resin at room temperature for 30 min.
4. Wash the resin extensively with the loading/washing buffer to remove nonspecifically bound peptides.
5. Elute the bound peptides with 50 mM Na_2HPO_4 (pH 9.0).

Recently, a number of groups have successfully employed the affinity of titanium dioxide (TiO_2) for phosphate groups to enrich phosphopeptides from complex mixtures (see *Protocol 4*) (35–38). Progress has continuously been made with regard to particle size (39), washing and elution conditions (40), different metal oxide (41), and online automation (38) to improve the sensitivity and specificity of the method.

Protocol 4

Isolation of phosphopeptides using TiO$_2$ (24)

Equipment and Reagents
- TiO$_2$ (GL Science)
- Peptide mixture (from *Protocol 2*)
- Acetonitrile (Aldrich)
- TFA[a]
- Loading/washing buffer: 200 mg of 2,5-dihydroxybenzoic acid (Sigma-Aldrich) in acetonitrile : water : TFA[a] (80 : 17.5 : 2.5, v/v/v)
- Mobicol spin column (MoBiTec)
- 0.3 M NH$_4$OH

Method
1. Equilibrate the TiO$_2$ in the loading buffer.
2. Dissolve the peptide mixture in the loading buffer.
3. Incubate with TiO$_2$ in the spin column at room temperature for 15 min with end-over-end rotation.
4. Wash the TiO$_2$ extensively with the loading/washing buffer to remove nonspecifically bound peptides.
5. Wash twice with 80% acetonitrile and twice with 0.1% TFA.
6. Elute with 0.3 M NH$_4$OH.

Note
[a]TFA-containing solutions should be made on the day of use. Extreme care should be taken when using TFA as it is an extremely hazardous chemical and the concentrated acid should only be used in a fume hood and dispensed wearing suitable protective clothing. Consult the safety datasheet for handling details and disposal.

2.2.2 Chemical derivatization

Chemical methods to isolate phosphopeptides are based on the reactivity or lability of the phosphate group on amino acid residues. The first method (see *Protocol 5*) makes use of a β-elimination reaction that occurs when phosphoserine and phosphothreonine residues are exposed to strongly alkaline conditions (42, 43). A nucleophile reagent (e.g. 1,2-ethanedithiol or dithiothreitol (DTT), which provides a new reactive thiol group) is added to serve as a linker to be captured by an affinity tag such as biotin or via covalent attachment, respectively (see *Fig. 2*). However, the method has a number of undesired side reactions. First, the cysteine residue is reactive under the same conditions (42). To overcome this problem, the sample is first treated with performic acid to oxidize cysteine and methionine residues, thereby inactivating them. Secondly, under strong alkaline conditions, β-elimination may also take place for some unmodified serine residues (43). Although the yield of the undesirable reaction is low due to the high content of serine residues in

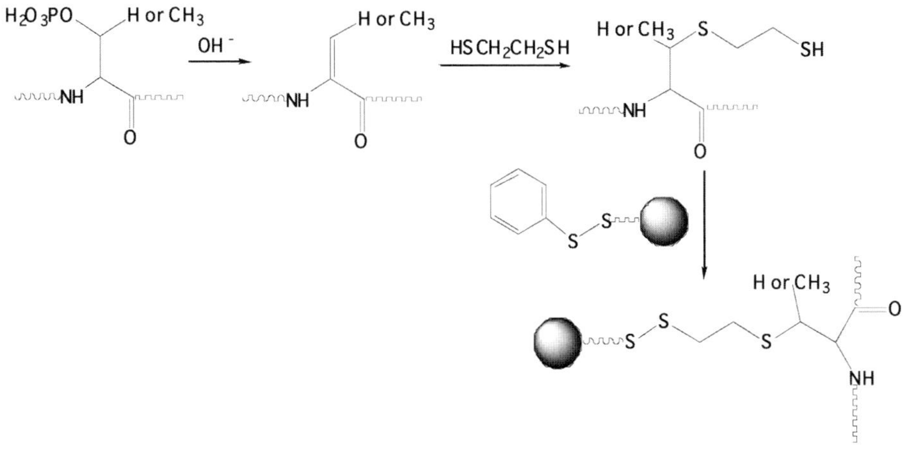

Figure 2. Chemical derivatization and affinity purification of phosphoserine- or phosphothreonine-containing species based on β-elimination and Michael addition.

proteins, this side reaction constitutes a serious problem. In addition, β-elimination is not applicable to tyrosine phosphorylation. Finally, other modifications, including glycosylation, are eliminated under the conditions used, leading to potentially erroneous assignment of phosphorylation sites.

Protocol 5

Isolation of phosphopeptides by β-elimination/Michael addition (43)

Equipment and Reagents
- Protein or peptide sample
- Performic acid solution (prepare by mixing mix 30% H_2O_2 (Aldrich) with 95% formic acid (Aldrich) (1 : 9, v/v) and allowing the solution to stand at room temperature for 2 h)
- SpeedVac (Thermo Electron)
- Sequencing-grade trypsin (Promega) in 50 mM NH_4HCO_3 buffer (pH 7.8)
- Ba^{2+}/NaOH/DTT reaction mixture: 20 mM DTT (Pierce), 65 mM NaOH, 100 mM $Ba(OH)_2$ (Aldrich)
- 5% TFA[a] (Aldrich)
- POROS R2 resin (Applied Biosystems)
- TE buffer (50 mM Tris/HCl, pH 8.0, 1 mM EDTA)
- Thiol affinity resin (Sigma)

- 60% Acetonitrile (Aldrich) in water
- 10 mM DTT in TE buffer

Method

1. For oxidation, incubate the dried peptide or protein sample with performic acid at room temperature for 2 h.

2. Remove residual performic acid using a SpeedVac.

3. For digestion of protein samples, incubate the proteins with trypsin (25 : 1) in 50 mM NH_4HCO_3 buffer at 37°C for 4 h.

4. Completely remove the solvent and buffer using a SpeedVac.

5. For β-elimination and Michael addition, incubate the peptide mixtures with Ba^{2+}/NaOH/DTT reaction mixture at 37°C for 1 h. Quench the reaction with 5% TFA.

6. For capture, first remove excess reagents using POROS R2 resin. Then reconstitute the peptides in TE buffer and incubate with activated thiol affinity resin for 1 h with gentle mixing. Place the resin in a pulled gel-loading tip for subsequent washing. Wash the resin with TE buffer and followed by 60% acetonitrile in water.

7. Elute peptides from the thiol affinity resin by incubating the resin with 10 mM DTT in TE buffer at room temperature for 30 min.

8. Acidify the eluted fractions with 5% TFA, and dilute with 0.1% TFA for MS analysis.

Note

[a]TFA-containing solutions should be made on the day of use. Extreme care should be taken when using TFA as it is an extremely hazardous chemical and the concentrated acid should only be used in a fume hood and dispensed wearing suitable protective clothing. Consult the safety datasheet for handling details and disposal.

The second chemical method employs the derivatization of phosphate to a phosphoramidate group (44), a reaction commonly used for the immobilization of oligonucleotides (45). The method is applicable to phosphotyrosine-containing peptides, as well as to those containing phosphoserine and phosphothreonine. The reaction is activated by water-soluble carbodiimide (e.g. N-(3-dimethylaminopropyl)-N'-ethylcarbodiimide (EDC)) to add cystamine to phosphate moieties. This allows the purification of phosphopeptides on a solid phase containing immobilized iodoacetyl groups. However, the amino groups and carboxyl groups of the peptides have to be protected to avoid unwanted reactions. Elution of phosphopeptides is performed by cleavage of phosphoramidate bonds by TFA, a step that also regenerates the amino groups.

As is generally the case with chemical-modification-based approaches, they require several reaction and purification steps prior to MS analysis, which can lead to substantial losses. As a result, the above methods require large amounts of sample with the result that only abundant proteins can be easily identified.

A variation of the phosphoramidate method that promises to be general and simple has been described recently (see *Protocol 5*) (15). It is based on a solution polymer support, in this case an amino-functionalized, generation 5, polyamidoamine dendrimer as the solid-phase capture reagent. In a single step, methylated phosphopeptides were captured directly on the dendrimer via a carbodiimide-activated reaction (see *Fig. 3*). Covalently bound phosphopeptides were readily

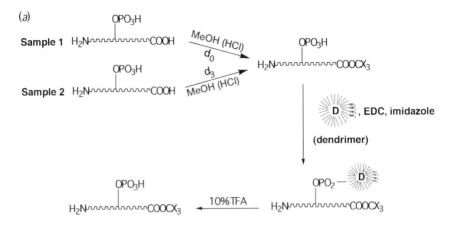

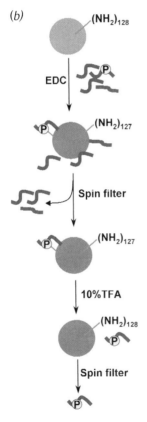

Figure 3. Schematic illustrations of isolation of phosphopeptides based on dendrimer conjugation and the soluble polymer-based phosphopeptide enrichment procedure.
(*a*) Isolation of phosphopeptides based on dendrimer conjugation. Peptides from samples 1 and 2 are d_0- and d_3-methylated, respectively. Methylated peptides are combined and subjected to a dendrimer conjugation catalyzed by EDC and imidazole. Phosphopeptides are captured on the dendrimers, whereas nonphosphopeptides are removed by extensive washing steps. Finally, methylated phosphopeptides are released from the dendrimer via acid hydrolysis using 10% TFA. X is H or D. Adapted from (15). (*b*) The soluble polymer-based phosphopeptide enrichment procedure. Dendrimer G5 containing 128 free amino groups captures any phosphopeptides. The resulting complexes are separated from unbound peptide using a molecular mass cut-off filter. Finally, the phosphopeptides are separated from the polymers by acidification and filtration, and the sequences and phosphorylation sites of the phosphopeptides are determined by MS.

isolated from unbound nonphosphopeptides using size-selective methods such as a simple membrane-based filter device. Phosphopeptides were detached from the dendrimer through acid hydrolysis of the phosphoramidate bonds and isolated using the same membrane-based filter device. The advantage of this strategy is that the utility of soluble polyamines such as dendrimers allows a homogeneous reaction with adequate amino groups in the solution. Large reagent excesses can still be used to drive reactions to completion. Noncovalently associated molecules and excess reagents can be removed by size-selective methods such as size-exclusion chromatography or a molecular mass cut-off filter device. As demonstrated in *Fig. 3*, quantitative measurement can be conveniently achieved with this method when peptides are d_0- or d_3-methylated. The method also demonstrated that the utility of functionalized synthetic soluble polymers such as dendrimers is not limited to the reaction but is applicable to any reaction where a solid-phase format cannot be achieved.

Protocol 6

Isolating phosphopeptides based on dendrimer conjugation (15)

Equipment and Reagents
- Methanolic HCl (prepare by dropwise addition of 160 µl of acetyl chloride into 1 ml of pre-cooled anhydrous methanol (Aldrich) while stirring/shaking[a]
- SpeedVac (Thermo Electron)
- Reaction solution (100 mM EDC, 100 mM imidazole, 100 mM 2-(*N*-morpholino)ethanesulfonic acid and 1 M PAMAM dendrimer G5; all from Aldrich. Make sure the pH is ~5.5)
- Biomax filtering device (molecular mass cut-off of 5000; Millipore)
- 2 M NaCl in 50% methanol
- 50% Methanol in water
- 5% TFA (Aldrich)[b]

Method
1. For methylation, incubate the dried peptide samples with 75 µl of methanolic HCl at 12°C for 90 min. Then remove the solvent using a SpeedVac.

2. For capture, dissolve the peptide methyl esters from step 1 in 40 µl of reaction solution. Incubate at room temperature with strong shaking for 20–25 h.

3. To wash, transfer the reaction solution to a Biomax membrane filter device (molecular mass cut-off of 5000). Make sure the membrane side is inward or perpendicular. Wash three times with 2 M NaCl in 50% methanol and three times with 50% methanol (for all steps, vortex or mix well). Discard the filtrate to remove nonspecifically bound nonphosphopeptides.

4. For acid hydrolysis, add 5% TFA and incubate with the dendrimer for 1 h to recover the phosphopeptides. Spin down and collect the eluate. Add 50% methanol in water, mix well, and spin down to collect the eluate. Pool the filtrates together. Remove the solvent using a SpeedVac.

Notes
[a]Take care: strong heat development.
[b]TFA-containing solutions should be made on the day of use. Extreme care should be taken when using TFA as it is an extremely hazardous chemical and the concentrated acid should only be used in a fume hood and dispensed wearing suitable protective clothing. Consult the safety datasheet for handling details and disposal.

2.2.3 Charge-based fractionation

The negative charge carried by phosphate groups on phosphopeptides has also been employed to fractionate phosphopeptides (46). This strategy reasons that tryptic phosphopeptides can be enriched by a differential net solution charge state. At low pH, most tryptic peptides carry a net solution charge state of 2+. For example, in the human protein database, over two-thirds of tryptic peptides are predicted to have a net charge of 2+. As a phosphate group maintains a negative charge at acidic pH values, the net charge state of a tryptic phosphopeptide is generally only 1+. Therefore, using strong cation exchange (SCX) column chromatography, many phosphopeptides elute before nonphosphopeptides, which are multiply charged. A suitable method is described in *Protocol 7*. It is obvious that this strategy does not have high specificity and also cannot recover all phosphopeptides, but it has the attraction that it can be fully automated and is particularly suitable for large-scale experiments.

Protocol 7

Fractionation of phosphopeptides using SCX column chromatography (46)

Equipment and Reagents
- 5–15% SDS-PAGE gradient slab gel (15 cm × 15 cm × 0.15 cm) plus SDS-PAGE apparatus and electrophoresis buffer
- 50 and 100% Acetonitrile
- 20 and 100 mM NH_4HCO_3
- 10 mM DTT (Pierce) in 100 mM NH_4HCO_3
- 55 mM Iodoacetamide in 100 mM NH_4HCO_3
- SpeedVac (Thermo Electron)
- Trypsin (12.5 ng/ml in 50 mM NH_4HCO_3)
- 5% Acetic acid
- SCX column: 3 mm × 20 cm column (PolyLC) packed with 5 μm polysulfoethyl aspartamide beads with a 200 Å pore size
- SCX solvent A: 5 mM KH_2PO_4, 30% acetonitrile (pH 2.7)
- SCX solvent B: solvent A with 350 mM KCl
- SCX solvent C: 0.1 M Tris/HCl (pH 7.0), 0.5 M KCl
- Surveyor high-performance liquid chromatography (HPLC) and photodiode array (PDA) detector (Thermo Electron)

Method
1. Fractionate the proteins by SDS-PAGE using a preparative 5–15% SDS-PAGE gradient gel. Stop the gel when the buffer front migrates 4 cm into the gel and stain with Coomassie blue.
2. Cut the entire gel into ten regions (~4 mm × 150 mm) for in-gel digestion as described in the following steps.
3. Wash the gel pieces with water followed by 50% acetonitrile.
4. Incubate for 15 min each with 100% acetonitrile and 100 mM NH_4HCO_3, respectively, to remove the Coomassie dye. Dry the gel pieces.

5. Cover the gel pieces with 10 mM DTT in 100 mM NH_4HCO_3. Reduce the proteins for 1 h at 56°C.

6. Cool to room temperature, remove the DTT solution, and add an equal volume of 55 mM iodoacetamide in 100 mM NH_4HCO_3. Incubate for 45 min in the dark at room temperature.

7. Wash the gel pieces with 100 mM NH_4HCO_3 for 5 min.

8. Add an equal volume of pure acetonitrile for 15 min. Remove the liquid phase and dry in a SpeedVac for 15–30 min.

9. Re-swell the gel pieces at 4°C for 45 min in trypsin solution (approx. 5 µl/mm² gel). The gel pieces should *just* be covered. Digest overnight at 37°C.

10. Centrifuge the gel pieces and collect the supernatant.

11. Further extract the peptides using 20 mM NH_4HCO_3, 5% acetic acid, and 80% acetonitrile in water, respectively. Combine all the supernatants and dry down in a Speedvac until the desired volume has been reached.

12. For SCX chromatography, dissolve the peptides in 500 µl of SCX solvent A. Load the tryptic peptides onto the SCX column (PolyLC) and separate the peptides at a flow rate of 350 µl/min. Set the UV detector at 220 and 280 nm. A suitable elution gradient would be 5 min at 100% solvent A, 15 min gradient to 15% solvent B, 1 min gradient to 100% solvent B, 15 min at 100% solvent B, 15 min at 100% solvent C, and 20 min at 100% solvent A. Collect fractions every 2 min. Collect the first four fractions and desalt off-line for MS analysis.

2.3 MS data acquisition, phosphopeptide identification, and determination of sites of phosphorylation

Peptides are commonly identified via the generation of tandem mass (MS/MS) spectra of individual peptides and searching the fragment ion species against protein sequence databases (47). For phosphopeptide identification, variable modifications on the amino acid residues serine, threonine, and tyrosine (+80 atomic mass units (a.m.u.)) are applied prior to database searching. This step generates a database in which proteins with possible phosphorylation on serine, threonine, and tyrosine residues are included. A phosphopeptide is identified when a match is found between an MS/MS spectrum and an *in silico* spectrum of a phosphopeptide from the database. In addition, the analysis of phosphopeptides by MS can be facilitated by specific data acquisition methods. These MS methods take advantage of chemical lability of the phosphoester bonds in phosphoserine, phosphothreonine, and, to a lesser degree, phosphotyrosine. The phosphoester bonds can be induced to fragment in a collision cell or the ion source of an MS instrument, resulting in loss of phosphoric acid from the peptide. This characteristic fragmentation pattern can be used to detect phosphopeptides among nonphosphorylated peptides using a number of phosphopeptide-specific scans.

2.3.1 Neutral loss and precursor ion scanning

Neutral loss and precursor ion scanning were traditionally performed using triple quadrupole instruments. Now they are also increasingly carried out on higher-resolution instruments such as quadrupole time-of-flight (QTOF) mass spectrometers. Phosphorylated serine and threonine residues exhibit a loss of phosphoric acid due to a gas-phase β-elimination reaction, which can be monitored in a neutral loss scanning experiment (48). In these experiments, neutral H_3PO_4 loss scanning for different charge states of phosphopeptides has been applied to obtain phosphopeptide candidate m/z values, which were then used to initiate product ion scans. The neutral loss scan selectively detects phosphopeptides among nonphosphorylated peptides. It can be used alone or combined with phosphopeptide wet-lab enrichment for the characterization of phosphorylated peptides. The presence of the phosphate can be confirmed in subsequent fragment ion species analysis by the presence of dehydroalanine or dehydroamino-2-butyric acid, respectively.

Serine, threonine, and tyrosine phosphopeptides also generate a diagnostic PO_3^- ion ($m/z = -79$) when subjected to fragmentation by collision-induced dissociation (CID) in the negative-ion mode. This has been used to detect phosphopeptides among nonphosphorylated peptides using a precursor ion scan in the negative-ion mode (49). The actual sequencing of the phosphopeptide, however, usually requires fragmentation of positive ions. The detection and identification of phosphopeptides using this approach therefore requires special data acquisition protocols with quick switching between positive- and negative-ion modes (50). Another precursor ion experiment to specifically detect phosphotyrosine-containing peptides has been described in which the production of a diagnostic immonium ion of phosphotyrosine residues ($m/z = 216.04$) was monitored (51). This method allows the detection and sequencing of phosphotyrosine-containing peptides in the positive-ion mode on a high-resolution instrument such as a QTOF mass spectrometer (52).

2.3.2 MS^n and electron transfer dissociation

The predominant loss of phosphoric acid in the collision cell of the mass spectrometer creates a problem for sequencing serine/threonine phosphate-containing peptides because the resulting spectra that are dominated by the eliminated phosphate are the dephosphorylated peptide (28). The lack of informative fragmentation at the peptide backbone severely reduces the ability of database searching algorithms to identify the phosphopeptide unambiguously. Furthermore, when a phosphopeptide is identified, it is often not possible to assign the site of phosphorylation to a particular serine or threonine residue.

To obtain conclusive sequence information of serine/threonine-phosphorylated peptide, the dominant fragment ion resulting from the neutral loss can be subjected to an MS/MS/MS (MS^3) experiment (see *Fig. 4* for example). The systematic use of this strategy in an ion trap mass spectrometer resulted in the identification of a number of phosphopeptides that were not identified by MS^2 experiments (46). The disadvantage of MS^3 experiments is the significant loss of trapping ions over several stages of MS. As a result, MS^3 experiments were mainly successful for phosphopeptides with relatively high signal intensity. The introduction of MS instruments with high trapping capacity such as quadrupole linear ion traps facilitates MS^3 experiments on phosphopeptides. Correspondingly, a specific data acquisition method, named data-dependent MS^3-NL scanning, was developed. Using the MS^3-NL scan, an MS^3 spectrum is automatically collected by iso-

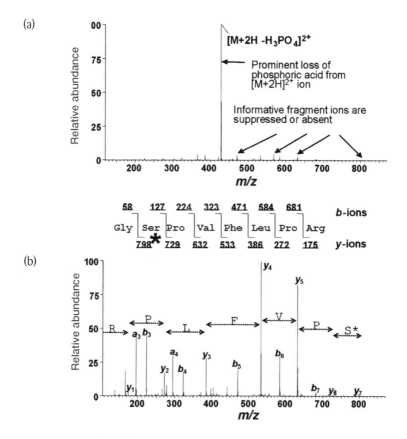

Figure 4. Examples of MS^n spectra for the identification of phosphopeptides.
(a) Example of an MS^2 spectrum of a phosphopeptide showing a typical extensive neutral loss of phosphoric acid. (b) An MS^3 spectrum of the neutral loss precursor ion from (a). Abundant peptide bond fragmentation permitted the unambiguous identification of this peptide from the protein cell division cycle 2-related protein kinase 7, with a phosphorylated serine residue marked by an asterisk (from 46).

lating and fragmenting the neutral loss fragment ion from the MS/MS spectrum if a significant loss of phosphoric acid upon fragmentation and the neutral loss event is one of the most intense ions detected in the MS^2 spectrum.

Recently, a new dissociation method designated electron transfer dissociation (ETD) (53, 54) has emerged as a powerful method for the study of PTMs. The method uses chemical ionization to generate a radical anion, which transfers electrons to multiply charged peptide cations trapped in the same mass analyzer. Electron capture by a protonated peptide is exothermic and causes the peptide backbone to fragment by a nonergodic process, e.g. one that does not involve intramolecular vibrational energy redistribution. ETD induces more extensive fragmentation of the peptide backbone than CID, providing greater sequence coverage. ETD has been used successfully for analysis of phosphorylated peptides. As shown in *Fig. 5*, in contrast to conventional CID, no loss of phosphoric acid, phosphate, or water from the

176 ■ CHAPTER 9: CHARACTERIZATION OF POST-TRANSLATIONAL MODIFICATIONS

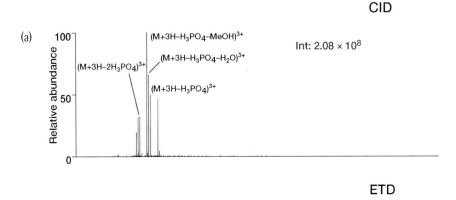

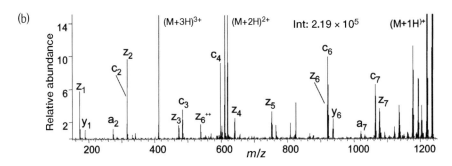

Figure 5. Comparison of single-scan CID and ETD mass spectra.
The mass spectra were recorded during data-dependent analyses (microcapillary HPLC–nanoESI-MS/MS) of a phosphopeptide (ERpSLpSRER) generated in a tryptic digest of human nuclear proteins. (*a*) CID spectrum dominated by fragment ions corresponding to the loss of phosphoric acid and either methanol or water. (*b*) ETD spectrum containing 13 out of 14 possible c- and z-type product ions (from 53).

parental peptide or the fragments is seen when ETD-based sequencing of phosphopeptides is performed. This allows direct assignment of phosphorylation sites from the MS^2 spectra. The broad application of ETD technology in proteomics in the future, in particular in PTM analysis, depends on two factors:

1. The success of identifying chemical reagents that can efficiently transfer electrons to multiply charged peptides with minimum proton transfer reaction as the competing reaction.
2. The ease of modifying the existing MS instruments to incorporate the ETD method.

2.4 Quantitative phosphoproteomics

Quantitative determination of phosphorylation is of particular biological significance. Kinases and their substrates are likely biomarkers in a variety of diseases and the determination of the absolute and relative expression levels of these proteins can be used for diagnosis or molecular targeting. In addition, one phosphoprotein might be involved in more than one signaling pathway, inducing overlapping patterns of phosphorylation if the response to different stimuli is being studied. Likewise, the ratio of phosphorylation of a protein on multiple residues might be crucial for its function. The techniques that have traditionally been used for quantitation of phosphorylation are phosphoamino acid analysis and Edman degradation. Recently, MS-based approaches have been rapidly evolving as new techniques for quantitative analysis.

MS has been used increasingly for quantitative analyses in proteomics. This is commonly achieved by comparing the intensity of peptides of the same sequence but differential stable isotope composition (55). The labeling step can be done *in vitro* by chemical derivatization (56) or enzymatic reaction (57), or *in vivo* by the incorporation of isotopes by metabolic labeling (58). In theory, quantitative phosphoproteomics can be achieved by the combination of any existing stable isotope labeling method with phosphoprotein/phosphopeptide enrichment. However, special caution has to be taken for quantitative phosphorylation analysis. As only a small fraction of proteins are phosphorylated, quantitative information at the level of proteins does not necessarily reflect the phosphorylation status. Instead, quantitative measurement has to be made at the level of the phosphopeptide. In addition, stable isotope labeling, in particular chemical derivatization, usually introduces extra steps into the sample preparation, which could have a negative effect on the analysis of phosphoproteins that are already of low abundance. Finally, many enrichment methods in phosphoproteome analysis are affinity based and therefore may not be quantitative.

A few years ago, we introduced an *in vitro* labeling method to quantitate proteins by labeling cysteine residues, a strategy designated isotope-coded affinity tagging (ICAT) (56). The same commercially available ICAT reagents were used to introduce a biotin tag into phosphoserine and phosphothreonine residues by β-elimination and Michael addition, enabling enrichment and simultaneous quantitation of phosphoproteins (59). Several global labeling methods (e.g. N-terminal derivatization and methylation of carboxylic groups in peptides, combined with the enrichment of phosphopeptides) have also been reported for quantitative phosphorylation (15, 33).

All proteins in the whole cell can be labeled *in vivo* by growing them in ^{15}N-labeled medium without any further manipulation (60). This can also be used to quantify the extent of phosphorylation. An alternative strategy to label proteins *in vivo*, designated stable isotope labeling by amino acids in cell culture (SILAC) (61–63), which uses amino acids containing a stable isotope, has been increasingly used. For example, a human cell line was grown in medium containing normal or $^{13}C_6$-containing lysine and arginine (this increases the mass of every lysine- or arginine-containing peptide by 6 Da) (62). The resulting mass spectrum clearly showed the differences in intensity between two triply charged peptide ions separated by 6 Da. The advantage of using stable-isotope-containing amino acids over medium containing ^{15}N is that it can be used in cases where the sequence is not known. Nevertheless, the *in vivo* labeling strategies are mostly restricted to cells that can be grown in culture.

Protocol 8

Analysis of phosphopeptides by micro-liquid chromatography (μLC)-MS^n

Equipment and Reagents
- Phosphopeptide sample (from *Protocol 7*)
- Solvent A: 0.1% formic acid in HPLC-grade water
- Solvent B: 0.1% formic acid in acetonitrile
- Microcentrifuge
- Agilent 1100 HPLC system
- Finnigan LTQ (Thermo Electron)
- SEQUEST software (Thermo Electron)
- Trans-Proteomics Pipeline (TPP)[a] software (Institute for Systems Biology)

Method

1. Reconstitute the phosphopeptide sample in 8 μl of solvent A. Spin in a microcentrifuge at the highest speed for 10 min. Transfer the clear solution into an autosampler vial.

2. Load the sample onto the Agilent pre-column using the autosampler. Wash the pre-column with solvent A for 5 min.

3. Start the HPLC gradient. Set up a linear gradient of 10–35% of solvent B over different time periods depending on the complexity of sample. Control the flow rate at ~200 nl/min. Most peptides will elute off from the reverse-phase column between 10 and 35% of solvent B.

4. Analyze phosphopeptides that elute from the column using a Finnigan LTQ instrument. Acquire data-dependent MS spectra by switching among normal MS, MS^2, and MS^3-NL modes.

5. Convert the MS data into mzXML format, a common data format compatible with most proteomics software (64).

6. Submit the mzXML files to SEQUEST software to search against protein databases (65). When searching MS^2 data, specify the phosphorylation of interest (e.g. serine, threonine, or tyrosine) as a variable modification (+79.966 a.m.u.) on these residues.

7. Run TPP software[a] to filter the data and display the results using PEPTIDEPROPHET (66) and PROTEINPROPHET (67). For quantitative analysis, run the ASAPRATIO software (68).

Note

[a]TPP is a set of software provided as open-source proteomics tools developed by the Institute for Systems Biology (ISB), Seattle, Washington, USA, available at http://www.systemsbiology.org.

3. TROUBLESHOOTING

- Best results are obtained when starting with proteins free of contamination from oligonucleotides and phospholipids. Oligonucleotides and phospholipids contain multiple phosphate groups, which are likely to interfere with most affinity- or reactivity-based enrichment methods.
- Many signaling proteins are membrane proteins. Detergents such as SDS are commonly used to solubilize these proteins. It is important to remove deter-

gents before MS analysis, as they are incompatible with the analysis and will significantly diminish the sensitivity of MS analysis.
- During sample preparation, the phosphate groups on proteins and peptides are subjected to chemical or enzymatic dephosphorylation. Therefore, it is important to have a positive control in all procedures to ensure that phosphorylation has been preserved. Western blotting using anti-phospho antibodies is an efficient and sensitive tool for overall protein phosphorylation status, whilst spiking trace amounts of isotopically labeled peptides (^{32}P) can be used to monitor peptide phosphorylation.
- It is useful to apply a few standard peptides including phosphopeptides and nonphosphopeptides to ensure that the purification procedure is efficient. Standard peptides can be monitored using a simple MALDI-TOF instrument in each step.

4. REFERENCES

1. Huber LA, Pfaller K & Vietor I (2003) *Circ. Res.* **92**, 962–968.
2. Patterson SD (2004) *Curr. Proteomics*, **1**, 3–12.
★ 3. Mann M & Jensen ON (2003) *Nat. Biotechnol.* **21**, 255–261. – *A review of the analysis of PTMs by MS.*
4. Aebersold R & Goodlett DR (2001) *Chem. Rev.* **101**, 269–295.
5. Mann M, Ong SE, Gronborg M, Steen H, Jensen ON & Pandey A (2002) *Trends Biotechnol.* **20**, 261–268.
6. Hunter T (2000) *Cell*, **100**, 113–127.
7. Patterson SD & Aebersold RH (2003) *Nat. Genet.* **33** (Suppl.), 311–323.
★ 8. Machida K, Mayer BJ & Nollau P (2003) *Mol. Cell. Proteomics*, **2**, 215–233. – *A review of profiling of global tyrosine phosphorylation based on SH2 and phosphotyrosine-binding domains.*
9. Kalume DE, Molina H & Pandey A (2003) *Curr. Opin. Chem. Biol.* **7**, 64–69.
10. Sickmann A & Meyer HE (2001) *Proteomics*, **1**, 200–206.
11. Gold MR, Yungwirth T, Sutherland CL, *et al.* (1994) *Electrophoresis*, **15**, 441–453.
12. Pandey A, Podtelejnikov AV, Blagoev B, Bustelo XR, Mann M & Lodish HF (2000) *Proc. Natl. Acad. Sci. U.S.A.* **97**, 179–184.
13. Maguire PB, Wynne KJ, Harney DF, O'Donoghue NM, Stephens G & Fitzgerald DJ (2002) *Proteomics*, **2**, 642–648.
14. Blagoev B, Ong SE, Kratchmarova I & Mann M (2004) *Nat. Biotechnol.* **22**, 1139–1145.
★★ 15. Tao WA, Wollscheid B, O'Brien R, *et al.* (2005) *Nat. Methods*, **2**, 591–598. – *First report on the use of a soluble polymer support for the isolation of phosphopeptides.*
16. Peirce MJ, Begum S, Saklatvala J, Cope AP & Wait R (2005) *Proteomics*, **5**, 2417–2421.
17. Blagoev B, Kratchmarova I, Ong SE, Nielsen M, Foster LJ & Mann M (2003) *Nat. Biotechnol.* **21**, 315–318.
18. Jin J, Smith FD, Stark C, *et al.* (2004) *Curr. Biol.* **14**, 1436–1450.
19. Gronborg M, Kristiansen TZ, Stensballe A, *et al.* (2002) *Mol. Cell. Proteomics*, **1**, 517–527.
20. Kane S, Sano H, Liu SC, *et al.* (2002) *J. Biol. Chem.* **277**, 22115–22118.
21. Shi X, Belton RJ Jr, Burkin HR, Vieira AP & Miller DJ (2004) *Anal. Biochem.* **329**, 289–292.
22. Dubrovska A & Souchelnytskyi S (2005) *Proteomics*, **5**, 4678–4683.
23. Bieber AL, Tubbs KA & Nelson RW (2004) *Mol. Cell. Proteomics*, **3**, 266–272.
24. Bodenmiller B, Mueller LN, Mueller M, Domon B & Aebersold R (2007) *Nat. Methods*, **4**, 231–237.
25. Rush J, Moritz A, Lee KA, *et al.* (2005) *Nat. Biotechnol.* **23**, 94–101.
26. Zhang Y, Wolf-Yadlin A, Ross PL, *et al.* (2005) *Mol. Cell. Proteomics*, **4**, 1240–1250.

27. Brill LM, Salomon AR, Ficarro SB, Mukherji M, Stettler-Gill M & Peters EC (2004) *Anal. Chem.* **76**, 2763–2772.
★★ 28. Ficarro SB, McCleland ML, Stukenberg PT, *et al.* (2002) *Nat. Biotechnol.* **20**, 301–305. – *Optimization of IMAC via methylation of carboxylic groups in peptides. The first report on large-scale phosphoproteomics analysis.*
29. Posewitz MC & Tempst P (1999) *Anal. Chem.* **71**, 2883–2892.
30. Pandey A & Andersen JS & Mann M (2000) *Sci STKE* **2000**, PL1.
31. Salomon AR, Ficarro SB, Brill LM, *et al.* (2003) *Proc. Natl. Acad. Sci. U.S.A.* **100**, 443–448.
32. Shu H, Chen S, Bi Q, Mumby M & Brekken DL (2004) *Mol. Cell. Proteomics*, **3**, 279–286.
33. Riggs L, Seeley EH & Regnier FE (2005) *J. Chromatogr. B Analyt. Technol. Biomed. Life Sci.* **817**, 89–96.
34. Haydon CE, Eyers PA, Aveline-Wolf LD, Resing KA, Maller JL & Ahn NG (2003) *Mol. Cell. Proteomics*, **2**, 1055–1067.
35. Larsen MR, Thingholm TE, Jensen ON, Roepstorff P & Jorgensen TJ (2005) *Mol. Cell. Proteomics*, **4**, 873–886.
★★ 36. Olsen JV, Blagoev B, Gnad F, *et al.* (2006) *Cell*, **127**, 635–648. – *A comprehensive study of phosphorylation using TiO_2 and SILAC.*
37. Pocsfalvi G, Cuccurullo M, Schlosser G, Scacco S, Papa S & Malorni A (2007) *Mol. Cell. Proteomics*, **6**, 231–237.
38. Cantin GT, Shock TR, Park SK, Madhani HD & Iii JR (2007) *Anal. Chem.* **79**, 4666–4673.
39. Chen CT & Chen YC (2005) *Anal. Chem.* **77**, 5912–5919.
40. Thingholm TE, Jorgensen TJ, Jensen ON & Larsen MR (2006) *Nat. Protoc.* **1**, 1929–1935.
41. Kweon HK & Hakansson K (2006) *Anal. Chem.* **78**, 1743–1749.
42. Oda Y, Nagasu T & Chait BT (2001) *Nat. Biotechnol.* **19**, 379–382.
★★ 43. McLachlin DT & Chait BT (2003) *Anal. Chem.* **75**, 6826–6836. – *A detailed study on the application of β-elimination reactions for the isolation of phosphoserine and phosphothreonine peptides. Both pros and cons are addressed.*
44. Zhou H, Watts JD & Aebersold R (2001) *Nat. Biotechnol.* **19**, 375–378.
45. Chu B, Wahl GM & Orgel LE (1983) *Nucleic Acids Res.* **11**, 6513–6529.
★ 46. Beausoleil SA, Jedrychowski M, Schwartz D, *et al.* (2004) *Proc. Natl. Acad. Sci. U.S.A.* **101**, 12130–12135. – *Enrichment of phosphopeptides based on SCX.*
47. Kapp EA, Schutz F, Connolly LM, *et al.* (2005) *Proteomics*, **5**, 3475–3490.
48. Schlosser A, Pipkorn R, Bossemeyer D & Lehmann WD (2001) *Anal. Chem.* **73**, 170–176.
49. Neubauer G & Mann M (1999) *Anal. Chem.* **71**, 235–242.
50. Le Blanc JC, Hager JW, Ilisiu AM, Hunter C, Zhong F & Chu I (2003) *Proteomics*, **3**, 859–869.
51. Steen H, Kuster B, Fernandez M, Pandey A & Mann M (2001) *Anal. Chem.* **73**, 1440–1448.
52. Bateman RH, Carruthers R, Hoyes JB, *et al.* (2002) *J. Am. Soc. Mass Spectrom.* **13**, 792–803.
★★ 53. Syka JE, Coon JJ, Schroeder MJ, Shabanowitz J & Hunt DF (2004) *Proc. Natl. Acad. Sci. U.S.A.* **101**, 9528–9533. – *Original publication describing ETD and a comparison between CID and ETD of phosphopeptides.*
54. Coon JJ, Ueberheide B, Syka JE, *et al.* (2005) *Proc. Natl. Acad. Sci. U.S.A.* **102**, 9463–9468.
55. Tao WA & Aebersold R (2003) *Curr. Opin. Biotechnol.* **14**, 110–118.
56. Gygi SP, Rist B, Gerber SA, Turecek F, Gelb MH & Aebersold R (1999) *Nat. Biotechnol.* **17**, 994–999.
57. Yao X, Freas A, Ramirez J, Demirev PA & Fenselau C (2001) *Anal. Chem.* **73**, 2836–2842.
58. Ong SE, Blagoev B, Kratchmarova I, *et al.* (2002) *Mol. Cell. Proteomics*, **1**, 376–386.
59. Goshe MB, Veenstra TD, Panisko EA, Conrads TP, Angell NH & Smith RD (2002) *Anal. Chem.* **74**, 607–616.
60. Conrads TP, Alving K, Veenstra TD, *et al.* (2001) *Anal. Chem.* **73**, 2132–2139.
61. Ibarrola N, Kalume DE, Gronborg M, Iwahori A & Pandey A (2003) *Anal. Chem.* **75**, 6043–6049.

62. Gruhler A, Olsen JV, Mohammed S, *et al.* (2005) *Mol. Cell. Proteomics*, 4, 310–327.
63. Oda Y, Huang K, Cross FR, Cowburn D & Chait BT (1999) *Proc. Natl. Acad. Sci. U.S.A.* 96, 6591–6596.
64. Pedrioli PG, Eng JK, Hubley R, *et al.* (2004) *Nat. Biotechnol.* 22, 1459–1466.
65. Eng JK, McCormack AL & Yate JR (1994) *J. Am. Soc. Mass Spectrom.* 5, 976–989.
66. Keller A, Nesvizhskii AI, Kolker E & Aebersold R (2002) *Anal. Chem.* 74, 5383–5392.
67. Nesvizhskii AI, Keller A, Kolker E & Aebersold R (2003) *Anal. Chem.* 75, 4646–4658.
68. Li X, Zhang H, Ranish JA & Aebersold R (2003) *Anal. Chem.* 75, 6648–6657.

CHAPTER 10
Protein microarray technologies
Chien-Sheng Chen, Sheng-Ce Tao, and Heng Zhu

1. INTRODUCTION

The basic concepts behind the design of protein microarrays are the same as those for DNA microarrays (1–4): miniaturization, parallelism, and automation to achieve high-throughput screening. The general procedures for the fabrication of DNA and protein chips can be divided into similar steps, including probe (protein/DNA) preparation, array substrate selection, probe (protein/DNA) delivery, sample probing, and signal detection. Many tools developed for DNA microarrays can be adapted directly to protein microarrays, such as microarray printing robots, laser-based microarray scanners, and the software needed for data analysis. In the case of DNA microarray construction, probe preparation is simple: probes can either be synthesized (e.g. *in situ*-synthesized oligonucleotide microarrays from Affymetrix or NimbleGen Systems) or PCR-amplified (e.g. cDNA microarrays). However, the same step is much more complicated in the case of protein chip fabrication, and especially for the so-called functional protein chips (see below), as the proteins have to be translated from mRNAs, necessitating sequential steps of gene cloning, protein expression, and protein purification in most cases. Another significant difference is the selection of microarray substrate: unlike oligonucleotides/DNA fragments that share the same biochemical properties, proteins are broadly heterogeneous in their biochemical properties and/or activity, as well as in their size, shape, and stability. Therefore, the fabrication for protein microarray is much more challenging. As proteins are the major driving force in carrying out cellular functions, it is clear that studies performed at the protein level can provide more direct information than DNA to elucidate the molecular mechanisms behind various pathways and networks.

Protein microarrays can be classified into two types, depending on their application: analytical protein microarrays and functional protein microarrays. Applications of both analytical and functional protein microarrays are shown in *Fig. 1*. In the following sections, we will provide examples of various applications of both types of microarray, focusing mainly on functional protein microarrays. We will also address the development of some new protein microarray technologies.

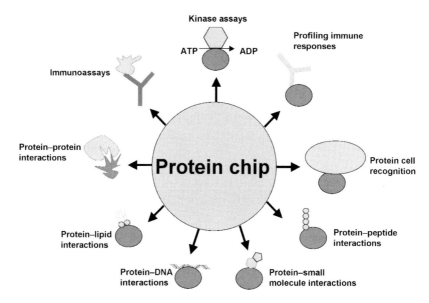

Figure 1. Applications of analytical and functional protein microarrays. Antibody arrays can be used for clinical diagnosis or environmental/food safety analysis. Functional protein arrays are used mainly to study various types of protein activity, including protein–protein, protein–lipid, protein–DNA, protein–drug, protein–peptide, and protein–cell interactions; to identify enzyme substrates; and to profile immune responses.

2. PROTEIN MICROARRAY FABRICATION

Protein microarrays, also known as protein chips, are miniaturized, parallel assay systems that contain small amounts of purified proteins at high density on a solid surface (5, 6). They allow simultaneous screenings with a variety of analytes from small amounts of samples in a single experiment. Recent years have witnessed a flourish of studies on surface materials, protein printing, assay platforms, and detection methods.

2.1 Surface materials

As glass slides have become the preferred solid supports capable of immobilizing 1600 spots/cm^2 (7), protein microarrays are typically prepared by spotting proteins onto a microscopic slide using a standard contact (5, 8) or noncontact (9–11) microarrayer. Glass also has great durability, good optical properties, and compatibility with the platforms already established for DNA microarrays. To achieve specific and strong protein attachment, glass slides are also activated with coupling groups, such as aldehyde, epoxy, carboxylic esters, and mercaptopropyl trimethoxysilane, to covalently crosslink to protein through amines (12, 13). In addition to covalent immobilization, nitrocellulose (14, 15) or gel (16, 17) is coated for attachment through diffusion and adsorption/absorption. These surfaces prevent both rapid evaporation and the close contact of the protein with the surface, thereby preserving the three-dimensional (3D) structures of immobilized biomolecules. However, in addition to the sophisticated processes required to create such 3D matrixes, the 3D chips involve higher levels of difficulty in changing buffers and recovering trapped molecules from the matrix microarrays (18). It has also been reported that reproducibility is poorer with hydrogel slides (16).

Immobilization via affinity capture is also attractive because of the uniformity of the immobilized proteins on the surface. For example, nickel-coated slides for noncovalent affinity attachment of His_6-tagged proteins have been used and are reported to provide tenfold better sensitivity than has been obtained with other random attachment methods (5). Proteins fused to a high-affinity tag at their C or N terminus are linked to the surface of the chip through this tag; hence, all of the attached proteins should be oriented uniformly away from the surface (5, 19–21). When this method is used, immobilized proteins are more likely to remain in their native conformation, and the analytes have easier access to the active sites of the proteins.

The special advantage of using a gold-coated surface resides in its ability to integrate surface plasmon resonance (22, 23) or mass spectrometry (MS) (24) as the detection system. In general, a self-assembled monolayer can be formed by treating the gold surface with a bifunctional thio-alkylene, which carries an SH group that reacts with the gold surface and a free end that reacts with the biorecognition molecules. This approach offers the opportunity to study the dynamics of biochemical reactions in a high-throughput fashion and has great potential for use in drug and drug-target discovery and in biomedical research.

2.2 Protein printing

The standard protein delivery system used for protein microarrays is the same as that used for DNA microarrays: a high-throughput printing robot (4), which can deliver a tiny amount of protein onto the appropriate substrate in an addressable fashion. In order to maintain the native conformation and activity of the printed protein, a relatively high humidity is required and low temperature during printing is preferred.

The printing robotic systems can be divided into two major classes: the contact and noncontact microarrayers. A contact microarrrayer delivers subnanoliter sample volumes directly to the surface by using pins with tiny tips, with or without capillary quills. It is particularly convenient and simple, requiring no extraordinary apparatus or skill, and yields spots of approximately 150–200 µm in diameter (1600 spots/cm^2) (8, 25). The TeleChem Stealth steel pin is quite popular because of its flexibility and accuracy. However, there are several disadvantages associated with robotic systems equipped with these pins: carry-over contamination is still an issue, and handling of the pins can be painful because they are very fragile. Also, it is not easy for contact printing to align pins to the prefabricated structures and they may cause damage to the substrates, especially on gel- and nitrocellulose-coated slides (18). To improve the quality of the contact printing, the so-called Silicon Microarray technology has been developed. Instead of the pins being fabricated individually, the silicon pins are fabricated in parallel on a single crystal silicon wafer using high-precision micromachining. This process improves pin-to-pin uniformity, eliminates pre-spotting, and reduces dead volume, carryover, and 'missing spot' phenomena.

Noncontact microarrayers, which use ink-jet technology, have been used to deposit nanoliter- to picoliter-sized droplets onto glass slides, polyacrylamide gel packets (26), and nanowells (27). Ink-jet printers do not contact the printing substrate, thereby avoiding the possible damage caused by the contact (28). Also, it is not restricted in terms of the surface structure and is well suited for more complicated biochemical assays. However, the ink-jet microarrayer is slow when spotting many different samples and the shearing force during drop formation may cause damage to samples (29). In addition, a larger protein volume is required, and the size of the spots usually is bigger than those produced by a contact printer.

Piezoelectric dispensers are the main type of ink-jet printer used for protein microarrays as they allow recovery of the portion of the sample that is not dispensed (10) and there is no change in temperature involved in the printing process. These printers are equipped with borosilicate glass capillaries surrounded by a piezoelectric-element collar. The sample is dispensed by the application of a voltage to the piezoelectric collar, typically resulting in the release of a droplet of less than 1 nl (10). They are generally believed to yield the lowest spot-to-spot variability in the amount of antibody deposited (30).

2.3 Assay platforms

Typically, a chip is either probed directly with a labeled molecule or probed in two steps, either by first using a tagged probe (e.g. biotin) that can then be detected in a second step using a labeled affinity reagent (e.g. streptavidin) or by using specific labeled antibodies. Fluorescence is preferred for labeling, as fluorescent tags are simple, safe, and sensitive, and offer high resolution (31). Various enzymes have also been used for labeling (32, 33). Radioisotope labeling has been used on many functional protein microarrays; Ge (34), Zhu *et al.* (5), and Ptacek *et al.* (35) have respectively applied radioisotopes to study protein–protein, protein–DNA, and protein–drug interactions on nitrocellulose-coated arrays, as well as kinase–substrate interactions.

Usually, a cover slip is used to distribute sample evenly over the entire microarray and to avoid evaporation; this method requires only approximately 50 µl of sample. However, this approach lacks the possibility of simultaneous analysis of different samples on one chip, as the entire chip usually can only be exposed to one sample at a time. This does not allow a direct, accurate comparison of different samples without deviations in the results caused by interchip variations, which can range from 12 to 60% depending on the coating of the microarrays (16). An accurate comparison between samples and standards is important for quantitative assays, which typically require approximately ten samples for standard curves, negative controls, and several serial sample dilutions. To increase sample throughput, Jones et al. (36) fabricated 96 arrays on a glass substrate to match the spacing of a microtiter plate. After printing, the glass substrate was attached to a bottomless 96-well plate using an intervening silicone gasket. A 16-well pad exactly fitted to a typical microscope glass slide is now commercially available. It is also an excellent tool for increasing sample throughput.

A microarray platform at each well of a microtiter plate has been used to provide both array and sample throughput. A microprinter for glass slides is adapted to print microspots by adsorption in the wells of a regular polystyrene microtiter plate (37), a silanized glass plate (38), or an *N*-hydroxysuccinimide-activated glass plate (39). More than 100 spots displayed in a single well have been reported (30, 39). A charge-coupled device (CCD) camera is generally used to image the arrays quantitatively, although a microtiter-well scanner is now commercially available. However, it is more difficult to print biomolecules on a microtiter plate than on a glass slide.

After incubation with samples, the entire chip is washed with shaking. The wash conditions vary depending on the assay and probes used. Typically, we wash with 150 mM NaCl in Tris-buffered saline for three 10 min incubations at room temperature. Sometimes, we wash with 500 mM NaCl in Tris-buffered saline at a higher temperature or even wash with 0.5% sodium dodecyl sulfate (SDS). On the other hand, for some assays such as cell probing, one can wash only once and very gently. Before measuring the signal, residual liquid is usually removed from the chip by centrifugation.

2.4 Detection methods

The laser scanner and the CCD imaging detector are the two main detection systems for the analysis of fluorescence-based microarrays (30). The laser scanner provides more sensitive detection than a CCD, but is slower than a CCD, as it scans every spot on the chip individually. A typical protein microarray image achieved by a laser scanner is showed in *Fig. 2* (also available in the color section). Evanescent wave excitation of a planar waveguide has been integrated with a microchip and a CCD camera to identify signals simultaneously across the entire area of the planar waveguide (40–42). The evanescent wave, an electromagnetic component of the light guided down the microscope slide, is used for excitation and extends out from the surface of the microscope slide into the lower refractive index medium, decaying exponentially with distance from the surface. As an evanescent field extends only a few hundred nanometers into the solution, only the surface-bound fluorophores are excited, thereby greatly eliminating nonspe-

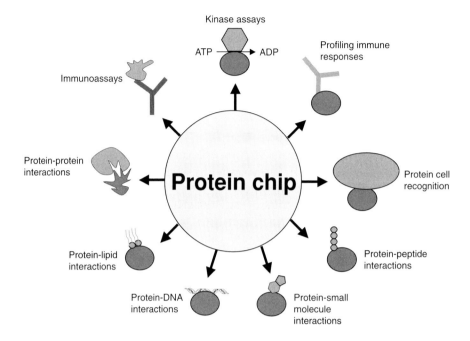

Figure 2. Image of a typical protein microarray (see page xxviii for color version). A yeast protein microarray was probed with an anti-GST antibody, followed by a Cy3-conjugated secondary antibody to visualize the immobilized proteins. An enlarged image of one of the 48 blocks is shown below the protein chip.

cific signals (43). The planar waveguide technology has been applied successfully to detect 1 pg/ml of interleukin-6 (44).

A radioisotope-based microarray is usually detected using X-ray film. When using ^{32}P or ^{33}P as label, the microarray is usually exposed to the film for only a few minutes. However, ^{14}C or ^{3}H has to be exposed to the film for several weeks. In comparison with fluorescence-based microarrays, radioisotope-based microarrays also have safety issues and it is more difficult to score the arrays. However, they have proved useful for some assays such as kinase–substrate interactions.

As labeling reagents sometimes affect protein activity and their use is also limited by the available detection channels, label-free detection has advantages as a direct detection method for protein microarrays. Furthermore, nonlabel methods use simpler protocols and can provide real-time measurement. Surface plasmon resonance (SPR) (45), oblique-incidence optical reflectivity difference microscope (46), MS (47, 48), atomic force microscopy (49), and microcantilevers (50) have all been used for label-free detection on microarrays.

SPR biosensors use an evanescent field to quantify interactions between analytes and surface-immobilized ligands by changes in surface refractive index, thus providing real-time measurement of biomolecular interactions without labeling and washing steps (22, 23). The feature of not requiring washing is especially important for low-affinity antibody–antigen interactions that would not be stable if washed before analysis. This makes SPR a versatile detection tool for studying the kinetics of receptor–ligand interactions with a wide range of molecular masses, affinities, and binding rates (51–53). Efforts have been made in both academic and industrial settings to explore the potential application of SPR to the field of protein microarrays. Myszka and Rich (54) have described a sensor surface with 64 individual immobilization sites in a single-flow cell. Biacore has two SPR systems available for array users: the Flexchip and the A100. The Flexchip platform allows 400 probes to be spotted on a slide and then screened against a single sample. In contrast, the A100 system is limited to 20 immobilized probes. When combined with a parallel-flow system, however, it can perform up to four simultaneous screens with large numbers of samples.

Alternatively, Sapsford et al. (55) have developed an antibody array biosensor to study the kinetics of antigen binding using a planar waveguide as the detection method. More importantly, they have demonstrated that significant signal intensity can be obtained from spots as small as 200 µm in diameter. In addition, the oblique-incidence optical reflectivity difference microscope has gained some heat in recent years as an alternative real-time, label-free detection system as demonstrated by Zhu and colleagues (46). One of the major advantages of this new system is its flexibility with surface chemistry – any type of transparent surface can be used. As both systems can detect signals from protein spots smaller than 300 µm, they are well suited for high-throughput, real-time kinetics studies.

MS can be used to determine the structural features of bound proteins. Surface-enhanced laser desorption/ionization (SELDI)-MS has been used to detect low-density captured proteins in an array on a metal surface. With SELDI, protein arrays act as a surface to which the sample binds uniformly, and the matrix is placed on the microchip after the proteins have been attached. The captured proteins are vaporized using a laser beam, followed by the analysis of the mass spectra to reveal the identities of the proteins (24). In addition to being label-free, another major advantage of the combined MS/protein microarray approach is that a mixture of samples can be probed on the slide at the same time. Also, MS can quantitatively decode the bound molecules one by one, a decided advantage for the study of protein–small molecule interactions, including the process of drug screening (56).

The atomic force microscope method detects changes in surface topography with a force probe to identify proteins captured in an antibody array. The atomically sharp probe is scanned over a surface with feedback mechanisms that enable the piezoelectric scanners to maintain the probe at a constant force (to obtain height information) or height above the sample surface (to obtain force information). The detection system does not measure force or height directly. It senses the deflection of the cantilever with the probe at its end. Generally, a light beam is reflected from the mirrored surface on the back side of the cantilever onto a position-sensitive photodetector. A small deflection of the cantilever will tilt the reflected beam and change the position of the beam on the photodetector. This approach relies on the change in the height or force that results from ligand–receptor binding and therefore does not require the use of labeled receptors. Height changes of 3–4 nm have been observed as a consequence of adsorption of antigenic IgG to a gold or SiO_2 surface, followed by an additional increase upon antibody–antigen binding (11, 57).

Another relatively new technological approach is illustrated by the arrays based on microcantilevers that have been developed by Protiveris. Microcantilevers are extremely sensitive nano-mechanical surface stress and mass sensors, and respond to minute changes in mass and stress through their deflection and resonance behaviors. They can be used to detect molecular-scale stresses caused by molecular recognition of biomolecules or the adsorption of chemicals, as well as the change in mass resulting from these binding events.

3. ANALYTICAL PROTEIN MICROARRAYS

Perhaps the most representative class of analytical microarray is the antibody microarray, in which antibodies are arrayed at a high density on glass surfaces. The biggest challenge associated with antibody microarrays is the production of antibodies that are able to identify the proteins of interest with high specificity and affinity in a high-throughput fashion. As the traditional method for generating monoclonal antibodies is time-consuming and laborious, researchers have recently sought alternative approaches. For example, phage antibody display, ribosome display, systematic evolution of ligands by exponential enrichment, mRNA display, and affibody display have been developed to expedite the production of antibodies with high specificity (29, 31, 58, 59). Each of these methods involves the construction of a large library of viable regions with potential binding activity, which can then be selected by multiple rounds of affinity purification. The binding affinity of the resulting candidate clones can be further improved using maturation strategies. However, the ideal selection system has yet to be fully developed: one that is not only fast, robust, sensitive, and low cost, but also automated and miniaturized.

Despite the challenges involved in obtaining specific antibodies, direct (60), sandwich (30, 61), and competitive (23, 62) multiplexed immunoassays have all been used in protein chips. Some examples are given below.

Schweitzer *et al.* (61) used sandwich immunoassays to measure 75 cytokines on two separate arrays using rolling-circle amplification to label antibodies with DNA. Once the detection antibodies are localized to the antigens on the protein array, their DNA sequences are extended by DNA polymerase *in situ* to form long DNA polymers of defined sequence that are tethered to the detection antibody. After this polymerization step, the extended DNA sequence is hybridized to fluorescently labeled DNA of complementary sequence. As the extended DNA polymers are very long, multiple copies of the fluorescently labeled DNA are attached to each detection antibody, thereby amplifying the signal. Detection limits as low as 0.5 pg/ml and a dynamic range of three decades have been achieved. However, the additional procedures required for signal amplification increase the assay time and experimental difficulties.

Sreekumar et al. (63) created antibody arrays with 146 distinct antibodies against proteins involved in the stress response, cell-cycle progression, and apoptosis on poly-L-lysine-coated or superaldehyde-modified glass slides. Microcontact printing was used to monitor the alterations in protein levels in LoVo colon carcinoma cells that had been treated with ionizing radiation. The reference standards and samples were labeled separately using either Cy5 or Cy3 dye. The slides were incubated with a labeled protein mixture, washed with buffer, and the signals were detected using a confocal microarray scanner. Differential expression profiles were observed with radiation-induced upregulation of apoptotic regulators such as p53, DNA fragmentation factors, and tumor necrosis factor-related ligands.

Delehanty and Ligler (64) developed an antibody microchip for the rapid detection of proteins and bacterial analytes. A piezoelectric noncontact dispenser was used to immobilize biotinylated capture antibodies on the surface of an avidin-coated glass slide. The assay was carried out using a microfluidic six-channel flow module for sequential introduction of samples, detection antibodies, and wash buffers. The signals on the microchip were measured using a scanning confocal microscope. Assays could be completed in 15 min, and cholera toxin, staphylococcal enterotoxin B, and ricin, as well as *Bacillus globigii*, were detected at concentrations as low as 8, 4, and 10 ng/ml and 6.2×10^4 c.f.u./ml, respectively.

Taitt et al. (43) patterned biotinylated antibodies in microfluidic networks (three channels) on avidin-coated glass slides and conducted the assay with microfluidic devices (three channels), forming a 3×3 array for nine analytes. Evanescent wave excitation of a planar waveguide was used and a CCD camera was employed for visualization and quantification of the spots. Staphylococcal enterotoxin B, ricin, cholera toxin, *Bacillus anthracis* Sterne, *Bacillus globigii*, *Francisella tularensis* LVS, *Yersinia pestis* F1 antigen, MS2 coliphage and *Salmonella typhimurium* were detectable at concentrations of 100, 200, and 100 ng/ml; 1.5×10^4, 1×10^5, and 9×10^6 c.f.u./ml; 100 ng/ml; 1×10^9 p.f.u./ml and 5×10^6 c.f.u./ml, respectively.

Knecht et al. (23) reported the simultaneous detection of ten antibiotics in milk using an automated microarray system. They chose an indirect competitive immunoassay format of immobilizing haptens on glass slides modified with (3-glycidyloxypropyl)trimethoxysilane by a noncontact piezoelectric arrayer. Antibody binding was detected using a second antibody labeled with horseradish peroxidase, generating enhanced chemiluminescence, which was recorded with a CCD camera. As all liquid handling was fully automated, analysis took only 5 min. The detection limits for all antibiotics ranged from 0.12 to 32 µg/l, which were far below the maximum residue limits, except for penicillin G.

4. FUNCTIONAL PROTEIN MICROARRAYS

A variety of types of functional microarray have been developed, including human (65), yeast (5, 21), plant (66), bacteria (67) and virus (68). These microarrays, in particular yeast proteome microarrays, have recently been applied to many aspects of discovery-based biology, including protein–protein, protein–lipid, protein–DNA, protein–drug, and protein–peptide interactions.

4.1 Expression-ready open reading frame collections and high-throughput production of proteins

The first two steps in fabricating a proteome microarray are the construction of proteome-wide, expression-ready open reading frame (ORF) collections and the development of high-throughput strategies for the production of a large number of proteins. In recent years, many research groups and companies have put tremendous effort into these fundamental steps.

Phizicky and colleagues (69) first attempted to clone all 6000 yeast ORFs into an expression vector using a yeast recombination cloning strategy and then applied this collection to screening of novel biochemical activities in yeast. This first ORFeome collection in yeast, however, was later found to be unstable after repetitive replications, as the recombinant clones were not rescued and thus each expression yeast strain was not clonal. Furthermore, the identity and extent of the sequence error in the recombinant clones were not checked. In 2001, Snyder and colleagues (5) applied the same cloning strategy to clone 5800 yeast ORFs into an expression vector that overproduces N-terminal glutathione S-transferase (GST)::His$_6$-tagged fusion proteins in yeast upon induction with galactose. Every recombinant clone was rescued into bacteria and sequenced from one end to confirm its identity and to determine whether any frame-shift mutations had occurred. The verified recombinant clones were then reintroduced into yeast for protein purification. Although this process was more laborious, it guaranteed a high-quality ORFeome collection that could be stably stored for many generations. Later, the Snyder and Phizicky groups together constructed a C-terminal tandem affinity purification tag collection in yeast using the same strategy (21).

Based on the genomic sequence data of the *E. coli* K-12 strain, Mori and coworkers (70, 71) constructed two sets of 4364 cloned individual genes encoding His-tagged proteins, with or without fused green fluorescent protein (GFP), for use in functional genomic analysis. Each clone encodes a protein product of the predicted ORF attached to a His tag at the N-terminal end and/or GFP at the C-terminal end. The expression of the cloned ORF is under the control of an isopropyl-β-D-thiogalactopyranoside-inducible promoter.

The total number of annotated human ORFs is expected to be about 22 000 (including 2188 predicted; 72), and the expected number of transcripts, including splice variants, is expected to be 1.54 times that number. Based on these figures, the total number of full-length cDNA clones required to cover the entire human genome is about 33 000. Several projects are underway in both industrial and academic spheres, with an ultimate goal of cloning all of the human genes. OriGene's TrueClone Collection covers over 70% of all known loci and contains more than 24 000 unique human full-length cDNA clones. All TrueClones are in a uniform mammalian expression vector (pCMV) with 5′- and 3′-untranslated regions. Invitrogen's Ultimate Human ORF collection represents 16 240 human

ORFs cloned in the Gateway Entry vector, which makes it easy to shuttle the inserts into any Gateway Destination vectors for expression and functional analysis of the target protein in a variety of hosts, including *E. coli*, yeast, baculovirus, CHO cells, and mammalian cell lines, as well as cell-free transcription and translation coupling systems, so that the collection is suitable for a wide variety of different applications. In addition, GeneCopeia's ORFEXPRES Gateway Shuttle Clone product line currently contains over 16 000 unique clones derived from sequence-validated, full-length cDNA clones or human tissue cDNA libraries. In academia, the collection from the Mammalian Gene Collection contains 24 759 full-length ORF human clones, encompassing 14 534 nonredundant ORFs. Vidal and colleagues recently created a collection of ORF-only clones (full-length cDNAs without untranslated regions) in the Gateway system; at present, this collection contains 12 212 ORFs representing 10 214 unique genes (73, 74), with future updates being planned.

After the successful construction of these expression-ready collections, the next crucial step in protein chip fabrication is to express and purify these proteins simultaneously in a high-throughput manner while retaining their biological activity. Affinity purification was chosen for the high-throughput production of fusion proteins based on the fact that the combination of recombinant proteins and affinity purification has been widely used to express and purify proteins from various hosts, including *E. coli*, yeast, insects, and human cell lines (5, 33, 75–77), plus the advantage that this combination offers almost one-step purification and a high purity of the resulting proteins. Affinity tags were fused to either the C-terminal or N-terminal ends of each of the collections described above (His_6 (70, 71), GST (5), and ZZ domains (21), respectively, as tags) and the corresponding matrices were used for the affinity purification (Ni-NTA–Sepharose, glutathioine–Sepharose, and IgG–Sepharose). By combining affinity purification with a 96-well format that handled all of the steps from culture to protein elution, and with the aid of several semi-automatic devices, the throughput of protein purification reached 1000 proteins per day (5). Labaer and colleagues (75) have created a system (FLEXP) that carries out all of the steps from cDNA cloning to protein production (from *E. coli*) in a fully automated fashion. In a test case, ~80% of 336 random cDNA clones could be successfully purified as full-length fusion proteins. Because the protocol was automated, at least 1000 proteins could be purified per day. However, eukaryotic proteins expressed in *E. coli* lack their native post-translational modifications. As a result, the *in vivo* activity of these proteins may be lost. In contrast, the proteins in the yeast collections described above were expressed in their own (homologous) system, and, because the proteins were in their native environment, they were subject to the usual post-translational modifications and were able to interact with their natural partners. Therefore, these proteins are well suited for biochemical analysis on protein microarrays (5, 21, 78).

Although we have attempted to describe all of the major developments and applications of functional protein microarrays, it is impossible to cover all the instances in which they have been explored. Therefore, we will mainly focus here on examples of yeast proteome microarrays (see *Fig. 3*, also available in the color section).

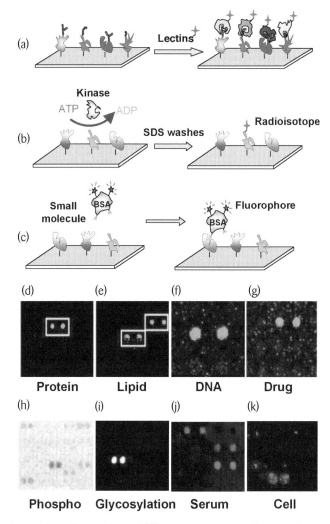

Figure 3. Protocols and examples of different assays using functional protein microarrays (see page xxix for color version).
Experimental protocols for three assays for screening glycosylated proteins using a lectin probe (a), in vitro kinase substrates (b), and drug targets (c) are illustrated. Chip images of various biochemical assays for protein–protein (d), protein–lipid (e), protein–DNA (f), protein–drug (g) and protein–cell (k) interactions are also shown. In addition, the protein chip approach may be applied to monitor immune responses in patients (j) or post-translational modifications of proteins, such as phosphorylation (h) and glycosylation (i).

4.2 Protein–protein and protein–lipid interactions

Zhu et al. (5) reported the construction and application of the first proteome microarrays, which contained >5800 individually purified yeast proteins, or 85% of the yeast proteome. These proteome chips were first used to study protein–protein interactions, with the chips being incubated with biotinylated calmodulin to identify new binding partners. In addition, protein microarrays have also been used to examine interactions with various phospholipids that are known to act as secondary messengers. When biotinylated liposomes containing various phospholipids of interest were used as probes, more than 150 phospholipid-binding proteins were identified, and a wide range of proteins was found to bind to the lipid vesicles; over 50% of these were previously known to be associated with membranes.

4.3 Protein–DNA interactions

The identification of DNA-binding proteins and the DNA elements to which they bind is a fundamental problem in biology. Such protein–DNA binding information is necessary for uncovering gene regulatory networks that control cellular and developmental processes. In a later report, Snyder's group also used yeast proteome chips to screen for novel DNA-binding activities using fluorescently labeled yeast genomic DNA (78). A total of 200 proteins that reproducibly bound DNA were identified. Half of these had not previously been shown to have DNA-binding activity; these new proteins fell into a wide variety of functional categories. The most surprising discovery in this category was the identification of ARG5,6 as a DNA-binding protein. The *ARG5,6* gene encodes two mitochondrial enzymes that mediate two key steps in the biosynthesis of ornithine (a precursor to arginine). Follow-up experiments revealed that these enzymes associate with specific mitochondrial loci *in vivo*, and this information was used to define a DNA-binding motif for these proteins. Thus, a novel DNA-binding activity was found to be associated with a well-characterized protein, thereby identifying a novel function for that protein.

Recently, Snyder and colleagues have constructed a protein chip with 282 known and potential yeast transcriptional factors (79). Forty Cy3-end-labeled dsDNA oligonucleotides, containing 75 DNA motifs previously identified on the basis of their evolutionary conservation were probed with the protein chips. In all, 211 specific DNA–protein interactions were identified, 131 of which were previously unknown. Yjl103, a previously uncharacterized DNA-binding protein, was characterized and its binding site on DNA was identified as two direct repeats of CGG separated by eight or nine nucleotides. Among the 22 genes tested *in vivo*, Yjl103 was found to be able to bind 19 of them upstream, and all 22 contained the $CGGN_8CGG$ motif. In many cases, the known or predicted functions of the proteins encoded by these genes was primarily associated with the metabolism of carbon compounds and carbohydrates, a finding that is consistent with the proposed role of Yjl103 in energy utilization.

4.4 Protein–drug interactions

Protein microarrays also have great potential for use in drug discovery and the identification of drug targets. As the binding profile of a drug of interest can be simultaneously obtained across an entire proteome using this approach, the specificity or 'off-target effects' of a drug can be monitored. This information can also provide important clues about how to improve drug design (80). To demonstrate that protein microarrays can be used to identify drug targets, Huang et al. (81) probed yeast proteome chips with biotinylated small-molecule inhibitors of rapamycin (SMIRs) to find genetic modifiers of the target of rapamycin (TOR) signaling network. They identified candidate drug targets of the SMIRs and validated a previously unknown protein as the target of the SMIRs. Interestingly, in an independent study, De Virgilio and colleagues (82) also identified the same protein in the TOR signaling pathway using a different approach.

4.5 Protein–peptide interactions

In a recent report, Jones and colleagues (36) demonstrated that they could measure quantitative interactions between proteins and peptides in an array format. They cloned, expressed, and purified almost all of the human Src homology 2 (SH2)- and phosphotyrosine-binding domains. They then printed the 159 domains on aldehyde-modified glass substrates and incubated 61 peptides representing physiological sites of tyrosine phosphorylation on the four ErbB receptors with the protein chips. For quantitative measurement, eight concentrations of each peptide, ranging from 10 nM to 5 µM, were used in the assay, allowing the binding affinity of each peptide to be measured. With this microarray, 43 of the 65 previously reported interactions were detected and 116 new interactions were identified. In addition, ErbB1 and ErbB2 were found to become more promiscuous with increasing concentration, whereas ErbB3 did not. As ErbB1 and ErbB2 are overexpressed in many human cancers, the authors suggested that this potential for increased promiscuity might contribute to the oncogenic potential of receptor tyrosine kinases.

4.6 Protein–cell interactions

Cell-surface-specific protein microarray microarrays have the potential to elucidate the behavior of both normal and abnormal cells by allowing researchers to characterize cells from a complex mixture at the molecular level (see *Fig. 4*, also available in the color section).

Soen et al. (83) have fabricated analytical microarrays using peptide–MHC complexes to detect and characterize antigen-specific T-cell populations, an area of knowledge that is critical for our understanding of the development and physiology of the immune system and its responses in health and disease. Phycoerythrin-labeled peptide–MHC tetramers with the MHC class I (H-2k^b) murine antigen OVA and MHC class II (I-E^k) murine antigen moth cytochrome c (MCC) were printed onto polyacrylamide-film-coated glass slides using a piezoelectric noncontact arrayer. The microarrays were then probed directly with T cells, and after incubation and washing steps, binding between the antibody and the MHC tetramer spots was observed. To define the cell-binding specificity further, OVA-specific OT-1 and MCC-specific 5c.c7 lymphocytes (6×10^5) were mixed at a 1 : 1 ratio and added to the preprinted peptide–MHC array: whilst anti-CD3 and anti-CD28 mAb spots captured both lymphocyte populations, the OVA or

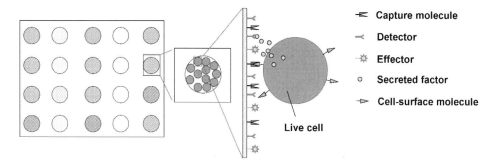

Figure 4. Analytical protein microarrays for live-cell studies (see page xxix for color version). A schematic of an analytical protein microarray for profiling of live cells is shown. On such types of chip, affinity reagents, such as capture molecules, detectors, and effectors, can be immobilized, with the ability to recognize cell-surface molecules, such as surface glycans, antigens, and receptors, or molecules secreted by cells.

MCC tetramers captured only the OVA-specific or MCC-specific cells, respectively. The authors also showed that this peptide–MHC microarray could be used to detect particular antigen-specific or rare T cells within a complex population. This study is the first to report the successful development of cellular microarrays with single-cell resolution.

By co-spotting mixtures of capture molecules, detector probes, and effectors in a microarray format, Chen et al. (84) developed a new analytical protein microarray. Blood, lymph nodes, and surgical biopsies from patients with malignancies were dissociated into single-cell suspensions and tested on this microarray. In addition to offering rapid molecular and functional profiling, this analytical microarray was also able to detect the presence of unexpected therapeutic targets, such as overexpression of c-K*it*, on the surface of tumor cells from some of the clinical samples.

Kuschel and co-workers (85) have generated an extracellular matrix (ECM) protein microarray by using a piezoelectric microarray printer to print ECM protein solutions onto nitrocellulose-coated glass slides. Five thousand or fewer cells could be applied to an array of 4 × 4 mm, consisting of 64 microspots. By using this ECM microarray, differences in the adhesive properties of three cell lines to 14 different ECM proteins were identified.

Glycosylation of bacterial cell surfaces is a critical factor in symbiosis, pathogenesis, cell–cell interactions, and immune evasion. Hsu et al. (86) fabricated a lectin chip with 21 lectins for use in profiling the surface lipopolysaccharides in bacterial cells. When labeled *E. coli* cells were incubated on the lectin chips, the lectins were able to capture the bacterial cells onto the chip surfaces via the interactions with lipopolysaccharides attached to the repeating sugar units, the O-antigens, on the bacteria. Two closely related *E. coli* strains, JM109 and HB101, could be distinguished by their differences in cell-surface glycosylation. In addition, dynamic changes in the surface glycosylation of the neonatal meningitis-associated *E. coli* strain RS218 were observed.

4.7 Identification of kinase substrates on protein chips

As phosphorylation is known to be involved in almost every aspect of cell processes, identification of the downstream substrates of protein kinases is a critical step in understanding the effects of phosphorylation on protein functions. In order to demonstrate that the protein chip approach is suitable for such investigations, Zhu et al. (27) first analyzed the substrate specificity of 119 yeast kinases on 17 different substrates using nanowell protein chips. Recently, as an extension of the same idea but on a much larger scale, the so-called 'Phosphorylome Project' was accomplished using yeast proteome microarrays (35). The goal was to identify all of the potential protein substrates of each yeast kinase. In vitro kinase reactions were carried out on the yeast proteome chips using 87 individually purified kinases/kinase complexes in the presence of [^{33}P]ATP. In all, 4129 phosphorylation events involving 1325 different proteins were identified. To ensure that the signals resulted from phosphorylation events, 5% SDS was used to denature the proteins on the chips to remove the signals from binding of kinase proteins or [^{33}P]ATP. These phosphorylation results have been assembled into a first-generation, global kinase signaling network in yeast.

4.8 Protein glycosylation analysis

Glycosylation is one of the most important protein post-translational modifications and is highly related to the regulation and function of many proteins. Although 20–50% of eukaryotic proteins are predicted to be glycosylated, there are only a few proteins with known carbohydrate linkages. To evaluate the feasibility of using protein microarrays to study protein glycosylation on the proteome scale, Gelperin and co-workers (21) constructed the so-called 'MORF' yeast proteome microarray. This microarray consisted of 5573 C-terminally tagged purified yeast proteins. A polyclonal antibody that recognized yeast glycans was incubated with the proteome chip, resulting in the identification of 599 putative glycosylated proteins. The chip results were validated by gel-shift assays using proteins treated with the deglycosylation enzymes endoglycosidase H and *N*-glycosidase. Of the 49 known glycoproteins, 33 exhibited a mobility shift after deglycosylation treatment, including 21 of the 25 known *N*-linked glycoproteins. Among 344 candidates, 109 new glycoproteins were confirmed, nearly twice the number of previously identified *N*-linked glycoproteins in yeast.

4.9 Profiling immune responses

The microarray-based identification of the autoantigens targeted by autoantibodies during the immune response has considerable potential for use in diagnosis, classification, and prognosis (65). Robinson and colleagues (87) published the first report of the simultaneous analysis of multiple human disease sera via protein microarray. They arrayed 196 distinct biomolecules involved in eight distinct human autoimmune diseases, including proteins, peptides, enzyme complexes, ribonucleoprotein complexes, DNA, and post-translationally modified antigens, onto glass slides to form the autoantigen microarrays. These arrays were incubated with patient serum samples as a means of defining the pathogenesis of autoantibody responses in human autoimmune diseases. Recently, Cahill and colleagues (88) constructed a protein array consisting of polypeptides translated from 37 200 random human cDNA clones in *E. coli* and used this array to identify potential autoantigens involved in the pathogenesis of alopecia areata. Eight autoantigens were identified and successfully confirmed by Western blotting. Likewise, the human protein chips from the same group (65) were incubated with sera from alopecia and arthritis patients; sera from clinically unremarkable persons were used as controls. High intensities for sera from diffuse alopecia and arthritis were obtained for autoantigen p69. The sera of patients with rheumatoid arthritis recognized a protein of the RAS association domain family 1 and diglycerol kinase-ζ. The alopecia serum recognized two additional proteins, the tumor suppressor p33 ING homolog and a protein with unknown function.

Zhu *et al.* (68) have also fabricated protein chips that allowed them rapidly and sensitively to distinguish the immune responses of severe acute respiratory syndrome coronavirus (SARS-CoV)-infected and healthy people. These protein chips harbored all of the SARS-CoV proteins, as well as proteins from five additional coronaviruses that can infect humans (human coronaviruses 229E and OC43), cows (bovine coronavirus), cats (feline infectious peritonitis virus), and mice (murine hepatitis virus A59). The presence of human IgG and IgM antibodies against SARS-CoV was detected on the chips with labeled anti-human IgG and IgM antibodies. Sera from patients could quickly be clustered as SARS-positive or SARS-negative on the basis of the serum-probing signals. Comparison with other methods such as enzyme-linked immunosorbent assay (ELISA) and immunofluorescent assay (IFA) indicated that the origin of 94% of the sera could correctly be predicted on the microarrays. The chip-based assay was at least 100-fold more sensitive than ELISA or IFA and required a smaller amount of sample.

5. DEVELOPMENT OF NEW PROTEIN MICROARRAY TECHNOLOGIES

As indicated above, tremendous efforts have been made during the past few years to develop a variety of different analytical and functional protein microarrays, as well as to explore their applications. In addition to these, there are important protein microarray technologies that cannot simply be classified as either analytical or functional. Here, we describe two interesting examples: the so-called 'reverse-phase protein microarray', which can be used to analyze hundreds to thousands of samples simultaneously on a single chip (73), and the new protein microarray fabrication technologies that make it possible to generate protein microarrays directly on the glass slides at a low cost in a highly efficient fashion (89–91).

5.1 Reverse-phase protein microarrays

In contrast to previous protein arrays that immobilize the probe, Liotta and co-workers (92) have developed a so-called 'reverse-phase' protein array that relies on immobilized lysates that represent the state of individual tissue cell populations undergoing disease transitions. Using the reverse-phase protein microarray, they analyzed the pro-survival checkpoint proteins in patient-matched samples at the microscopic transition stage from histologically normal prostate epithelium to prostate intraepithelial neoplasia and then to invasive prostate cancer. Cancer progression was associated with an increased phosphorylation of Akt, suppression of apoptosis pathways, and decreased phosphorylation of ERK. At the transition from histologically normal epithelium to prostate intraepithelial neoplasia, a significant increase in phosphorylated Akt and a concomitant suppression of downstream apoptosis pathways were observed, preceding the transition into invasive carcinoma. Use of the same strategy revealed two promising diagnostic markers (villin and moesin) for distinguishing colon from ovarian adenocarcinomas (93). Similarly, the same group constructed another reverse-phase protein microarray (94) on which lysates representing 60 human cancer cell lines (NCI-60) were printed. Fifty-two mouse monoclonal antibodies were probed individually to test the expression of 52 proteins on the microarrays. By using this reverse-phase protein microarray, the authors found that cell-structure-related proteins almost invariably showed a high correlation between mRNA and protein levels across the NCI-60 cell lines, whilst non-cell-structure-related proteins showed a poor correlation.

5.2 Making protein microarrays without large-scale cloning and expressions

Protein microarray technology has not been widely adopted, mainly because it is time-consuming and labor-intensive to purify the thousands of proteins required (5, 95). The general process of protein chip fabrication currently requires large-scale gene cloning and transformation, protein induction, and high-throughput protein purification. In recent years, several strategies have been introduced in an attempt to overcome the tedious protein preparation procedure. One strategy has taken advantage of cell-free systems that have been developed to synthesize proteins *in vitro*, including the *E. coli*, wheat germ, and rabbit reticulocyte systems. For example, an *E. coli* cell-free protein synthesis system has been used to synthesize proteins in a 96-well format (96). However, after protein synthesis, a further protein purification step is still required.

One of the most important developments in protein chip fabrication is the so-called nucleic acid programmable protein array (91). LaBaer and colleagues immobilized plasmids that could express target proteins fused to GST using an *in vitro* transcription/translation system. The *de novo*-synthesized proteins were captured locally through binding of the GST moiety to anti-GST antibodies spotted together with the plasmid DNAs. Prior to *in vitro* transcription/translation, such chips can easily be stored because DNA and antibodies are very stable. As the proteins are translated and immobilized locally, no complicated purification step is needed. This technology can easily be applied to expression-ready cDNA libraries as long as they express tagged fusion proteins. However, this strategy requires the cloning of cDNAs in such expression-ready vectors, and because transcription/

translation reactions occur on the chip, both the nascent mRNAs and proteins are free in solution and are therefore sensitive to vibration, which can cause diffusion among the adjacent spots. Therefore, to prevent cross-contamination, the adjacent spots must be kept at a certain distance from each other and this requirement limits the density of the features on these chips.

An alternative approach to synthesizing proteins is to use the so-called PROFusion technology, which is based on the mRNA display strategy (97–99). A poly(dA)$_{27}$CCA oligonucleotide labeled with a puromycin molecule at its 3′ end is ligated to the 3′ end of a mRNA molecule, which carries a ribosome-binding site at its 5′ end. Upon incubation with an *in vitro* translation system, a ribosome translates the encoded protein until it stalls at the RNA–DNA junction. As a result, the puromycin molecule tethered at the 3′ end of the RNA–DNA hybrid has sufficient time to enter the A site of the ribosome and form a covalent link to the nascent peptide before the dissociation of the ribosome–mRNA complex. Therefore, the synthesized peptide is tagged with its coding sequence. Because of this unique feature, a large number of such peptides can be screened for a particular activity in a mixture, and the selected peptides can later be decoded and retrieved based on the tagged mRNA sequences using a reverse transcriptase polymerase chain reaction (RT-PCR). Such selection can continue through multiple rounds until the desired activity is obtained (100). Furthermore, the decoding process can be performed on an addressable microarray containing the complementary oligonucleotides to form a high-density peptide/protein microarray, the PROFusion microarray. This strategy obviates the necessity for a purification procedure, and different proteins can be translated in a single tube. However, before protein immobilization, a ligation step is required to form an RNA–DNA hybrid in order to incorporate puromycin molecules individually for every mRNA species to be translated. As it is cumbersome to ligate single-stranded RNA and DNA molecules and the yield is somewhat low, this approach is hard to adapt to high-throughput peptide/protein production. Furthermore, in order to enhance the quantities of the proteins immobilized to a glass surface, the untranslated RNA–DNA hybrids have to be removed to eliminate any competition with RNA-tagged peptides/proteins.

Building upon these earlier approaches, our group (89) has developed two new strategies that greatly simplify the process of making peptide/protein chips using *in vitro* translation systems. Specifically, we have developed new methods to stall ribosomes during protein translation to allow the capture of the nascent polypeptides by puromycin, which is grafted at the end of an oligonucleotide immobilized on a solid surface. As proof of concept, a peptide chip with three commonly used epitopes/peptides, namely His$_6$, FLAG, and StrepTag II, was constructed. On the peptide chip, His$_6$ and FLAG could sensitively and specifically be detected by their corresponding antibodies, whilst StrepTag II could only be detected by streptavidin. To demonstrate that high-density protein chips can be fabricated, a second strategy was applied to produce several proteins of various families with a wide range of molecular mass from 27 to 57 kDa, including GST, two protein kinases (Tpk1 and Hrr25), two transcription factors (Mcm1 and Cbf1), and one autofluorescent protein (m-RFP). As revealed by antibody detection, all of these proteins could be translated and captured at full length at high density. Protein–DNA motif interactions have been carried out successfully on the chip, strongly suggesting that protein chips of this kind can be used to analyze protein activity.

6. OUTLOOK

Protein microarray technology has been shown to be a useful tool for multiplexed detection and proteomics studies. Femtomolar sensitivity has been achieved in analytical protein microarrays, and the number of applications of functional protein microarrays has grown dramatically. Novel applications utilizing protein microarrays and new protein microarray technologies are continually emerging. However, there are still several issues that need to be resolved before protein chip technology can be widely applied. First, the traditional cloning/expression/purification/printing approach is still the gold standard procedure for making protein microarrays, especially for proteome microarrays. Because of the sophisticated expertise required and the high cost of production, it is almost impossible for most the laboratories to make their own microarrays, but the price for the commercial protein chips is unacceptably high. A variety of promising strategies has already been tested to bypass the traditional procedure. However, none of these is close to being used for the large-scale fabrication of protein microarrays. Thus, to make protein microarray technology more applicable, a simpler and more powerful strategy is needed. Secondly, there is no widely accepted experimental standardization protocol. Fortunately, this issue is now being investigated by the Human Proteome Organization (HUPO), which is developing guidelines for experimental design and data annotation. Thirdly, at present most protein microarray results are only semi-quantitative. In order to reach the goal of accurate quantification, new technology is needed and is eagerly anticipated.

Although still in their infancy, protein microarrays will no doubt prove to be one of the most powerful tools in both basic science and clinical research. Improvements in our ability to generate large sets of high-quality proteins and antibodies or their mimetics will play a key role in quantitative analysis and promote the extension of this technology to many model organisms.

Acknowledgements

We thank the NIH (U54RR020839) for financial support and Dr Deborah McClellan for editorial assistance.

7. REFERENCES

1. Fodor SP, Read JL, Pirrung MC, Stryer L, Lu AT & Solas D (1991) *Science*, **251**, 767–773.
2. Schena M, Shalon D, Heller R, Chai A, Brown PO & Davis RW (1996) *Proc. Natl. Acad. Sci. U.S.A.* **93**, 10614–10619.
3. DeRisi J, Penland L, Brown PO, *et al.* (1996) *Nat. Genet.* **14**, 457–460.
4. Schena M, Shalon D, Davis RW & Brown PO (1995) *Science*, **270**, 467–470.

★ 5. Zhu H, Bilgin M, Bangham R, et al. (2001) *Science*, **293**, 2101–2105. – *To analyze the yeast proteome, these authors cloned 94% of the yeast ORFs in an expression vector, purified 80% of the encoded proteins as full-length fusions, and spotted them on microscopic slides to form the first proteome microarray.*
6. Chen CS & Zhu H (2006) *Biotechniques*, **40**, 423, 425, 427.
7. Afanassiev V, Hanemann V & Wolfl S (2000) *Nucleic Acids Res.* **28**, E66.
★ 8. MacBeath G & Schreiber SL (2000) *Science*, **289**, 1760–1763. – *Demonstration that proteins could be spotted on a glass slide at high-density to analyze for protein–protein, protein–small molecule, and enzyme–substrate interactions by using three test systems.*
9. Delehanty JB & Ligler FS (2003) *Biotechniques*, **34**, 380–385.
10. Delehanty JB (2004) *Methods Mol. Biol.* **264**, 135–143.
11. Jones VW, Kenseth JR, Porter MD, Mosher CL & Henderson E (1998) *Anal. Chem.* **70**, 1233–1241.
12. Kusnezow W, Jacob A, Walijew A, Diehl F & Hoheisel JD (2003) *Proteomics*, **3**, 254–264.
13. Kusnezow W & Hoheisel JD (2003) *J. Mol. Recognit.* **16**, 165–176.
14. Kramer A, Feilner T, Possling A, et al. (2004) *Phytochemistry*, **65**, 1777–1784.
15. Stillman BA & Tonkinson JL (2000) *Biotechniques*, **29**, 630–635.
16. Angenendt P, Glokler J, Murphy D, Lehrach H & Cahill DJ (2002) *Anal. Biochem.* **309**, 253–260.
17. Charles PT, Goldman ER, Rangasammy JG, Schauer CL, Chen MS & Taitt CR (2004) *Biosens. Bioelectron.* **20**, 753–764.
18. Zhu H & Snyder M (2003) *Curr. Opin. Chem. Biol.* **7**, 55–63.
19. Bieri C, Ernst OP, Heyse S, Hofmann KP & Vogel H (1999) *Nat. Biotechnol.* **17**, 1105–1108.
20. Houseman BT, Huh JH, Kron SJ & Mrksich M (2002) *Nat. Biotechnol.* **20**, 270–274.
21. Gelperin DM, White MA, Wilkinson ML, et al. (2005) *Genes Dev.* **19**, 2816–2826.
22. Yuk JS & Ha KS (2005) *Exp. Mol. Med.* **37**, 1–10.
23. Knecht BG, Strasser A, Dietrich R, Martlbauer E, Niessner R & Weller MG (2004) *Anal. Chem.* **76**, 646–654.
24. Davies H, Lomas L & Austen B (1999) *Biotechniques*, **27**, 1258–1261.
25. Bernard A, Fitzli D, Sonderegger P, et al. (2001) *Nat. Biotechnol.* **19**, 866–869.
26. Arenkov P, Kukhtin A, Gemmell A, Voloshchuk S, Chupeeva V & Mirzabekov A (2000) *Anal. Biochem.* **278**, 123–131.
★ 27. Zhu H, Klemic JF, Chang S, et al. (2000) *Nat. Genet.* **26**, 283–289. – *The first large-scale screening for specific kinase–substrate relationships. This study analysed almost all yeast protein kinases for their substrate specificities.*
28. Roda A, Guardigli M, Russo C, Pasini P & Baraldini M (2000) *Biotechniques*, **28**, 492–496.
29. Haab BB (2001) *Curr. Opin. Drug Discov. Devel.* **4**, 116–123.
30. Nielsen UB & Geierstanger BH (2004) *J. Immunol. Methods*, **290**, 107–120.
31. Templin MF, Stoll D, Schrenk M, Traub PC, Vohringer CF & Joos TO (2002) *Trends Biotechnol.* **20**, 160–166.
32. Joos TO, Schrenk M, Hopfl P, et al. (2000) *Electrophoresis*, **21**, 2641–2650.
33. Bussow K, Cahill D, Nietfeld W, et al. (1998) *Nucleic Acids Res.* **26**, 5007–5008.
34. Ge H (2000) *Nucleic Acids Res.* **28**, e3.
★ 35. Ptacek J, Devgan G, Michaud G, et al. (2005) *Nature*, **438**, 679–684. – *Mapping of a first-generation, global kinase signaling network in yeast in an attempt to identify all of the potential protein substrates of each yeast kinase using yeast proteome chips.*
★ 36. Jones RB, Gordus A, Krall JA & Macbeath G (2005) *Nature*, **439**, 168–174. – *The first report to quantify protein–peptide interactions on protein chips.*
37. Moody MD, Van Arsdell SW, Murphy KP, Orencole SF & Burns C (2001) *Biotechniques*, **31**, 186–184.
38. Wiese R, Belosludtsev Y, Powdrill T, Thompson P & Hogan M (2001) *Clin. Chem.* **47**, 1451–1457.
39. Mendoza LG, McQuary P, Mongan A, Gangadharan R, Brignac S & Eggers M (1999) *Biotechniques*, **27**, 778–780, 782–786, 788.
40. Rowe CA, Scruggs SB, Feldstein MJ, Golden JP & Ligler FS (1999) *Anal. Chem.* **71**, 433–439.

41. Rowe CA, Tender LM, Feldstein MJ, *et al.* (1999) *Anal. Chem.* **71**, 3846–3852.
42. Rowe-Taitt CA, Hazzard JW, Hoffman KE, Cras JJ, Golden JP & Ligler FS (2000) *Biosens. Bioelectron.* **15**, 579–589.
43. Taitt CR, Anderson GP, Lingerfelt BM, Feldstein MJ & Ligler FS (2002) *Anal. Chem.* **74**, 6114–6120.
44. Weinberger SR, Morris TS & Pawlak M (2000) *Pharmacogenomics*, **1**, 395–416.
45. Rich RL, Day YSN, Morton TA & Myszka DG (2001) *Anal. Biochem.* **296**, 197–207.
46. Landry JP, Zhu XD & Gregg JP (2004) *Opt. Lett.* **29**, 581–583.
47. Hu S, Zhang S, Hu Z, Xing Z & Zhang X (2007) *Anal. Chem.* **79**, 923–929.
48. Kersten B, Possling A, Blaesing F, Mirgorodskaya E, Gobom J & Seitz H (2004) *Anal. Biochem.* **331**, 303–313.
49. Soultani-Vigneron S, Dugas V, Rouillat MH, *et al.* (2005) *J. Chromatogr. B Analyt. Technol. Biomed. Life Sci.* **822**, 304–310.
50. Huber F, Hegner M, Gerber C, Guntherodt HJ & Lang HP (2006) *Biosens. Bioelectron.* **21**, 1599–1605.
51. McDonnell JM (2001) *Curr. Opin. Chem. Biol.* **5**, 572–577.
52. Salamon Z, Brown MF & Tollin G (1999) *Trends Biochem. Sci.* **24**, 213–219.
53. Nieba L, Nieba-Axmann SE, Persson A, *et al.* (1997) *Anal. Biochem.* **252**, 217–228.
54. Myszka DG & Rich RL (2000) *Pharm. Sci Technol. Today*, **3**, 310–317.
55. Sapsford KE, Liron Z, Shubin YS & Ligler FS (2001) *Anal. Chem.* **73**, 5518–5524.
56. Finnskog D, Ressine A, Laurell T & Marko-Varga G (2004) *J. Proteome Res.* **3**, 988–994.
57. Joos TO, Stoll D & Templin MF (2002) *Curr. Opin. Chem. Biol.* **6**, 76–80.
58. Stoll D, Templin MF, Schrenk M, Traub PC, Vohringer CF & Joos TO (2002) *Front. Biosci.* **7**, c13–c32.
59. Cahill DJ (2001) *J. Immunol. Methods*, **250**, 81–91.
60. Espina V, Woodhouse EC, Wulfkuhle J, Asmussen HD, Petricoin EF, III & Liotta LA (2004) *J. Immunol. Methods*, **290**, 121–133.
61. Schweitzer B, Roberts S, Grimwade B, *et al.* (2002) *Nat. Biotechnol.* **20**, 359–365.
62. Barry R, Diggle T, Terrett J & Soloviev M (2003) *J. Biomol. Screen.* **8**, 257–263.
63. Sreekumar A, Nyati MK, Varambally S, *et al.* (2001) *Cancer Res.* **61**, 7585–7593.
64. Delehanty JB & Ligler FS (2002) *Anal. Chem.* **74**, 5681–5687.
65. Lueking A, Possling A, Huber O, *et al.* (2003) *Mol. Cell. Proteomics*, **2**, 1342–1349.
66. Popescu SC, Popescu GV, Bachan S, *et al.* (2007) *Proc. Natl. Acad. Sci. U.S.A* **104**, 4730–4735.
67. Eyles JE, Unal B, Hartley MG, *et al.* (2007) *Proteomics*, **7**, 2172–2183.
68. Zhu H, Hu S, Jona G, *et al.* (2006) *Proc. Natl. Acad. Sci. U.S.A* **103**, 4011–4016.
69. Martzen MR, McCraith SM, Spinelli SL, *et al.* (1999) *Science*, **286**, 1153–1155.
70. Kitagawa M, Ara T, Arifuzzaman M, *et al.* (2005) *DNA Res.* **12**, 291–299.
71. Arifuzzaman M, Maeda M, Itoh A, *et al.* (2006) *Genome Res.* **16**, 686–691.
72. International Human Genome Sequencing Consortium (2004) *Nature*, **431**, 931–945.
73. Rual JF, Venkatesan K, Hao T, *et al.* (2005) *Nature*, **437**, 1173–1178.
74. Rual JF, Hirozane-Kishikawa T, Hao T, *et al.* (2004) *Genome Res.* **14**, 2128–2135.
75. Braun P, Hu Y, Shen B, *et al.* (2002) *Proc. Natl. Acad. Sci. U.S.A.* **99**, 2654–2659.
76. Hall DA, Ptacek J & Snyder M (2006) *Mech. Ageing Dev.* **128**, 161–167.
77. Albala JS, Franke K, McConnell IR, *et al.* (2000) *J. Cell. Biochem.* **80**, 187–191.
★ 78. Hall DA, Zhu H, Zhu X, Royce T, Gerstein M & Snyder M (2004) *Science*, **306**, 482–484.
 – The first report to study protein–DNA interactions using proteome chips.
79. Ho SW, Jona G, Chen CT, Johnston M & Snyder M (2006) *Proc. Natl. Acad. Sci. U.S.A.* **103**, 9940–9945.
★ 80. Huang YH, Li D, Winoto A & Robey EA (2004) *Proc. Natl. Acad. Sci. U.S.A.* **101**, 4936–4941. – The first report to demonstrate that protein microarrays can be used to identify drug targets.
81. Huang J, Zhu H, Haggarty SJ, *et al.* (2004) *Proc. Natl. Acad. Sci. U.S.A.* **101**, 16594–16599.
82. Dubouloz F, Deloche O, Wanke V, Cameroni E & De Virgilio C (2005) *Mol. Cell*, **19**, 15–26.
83. Soen Y, Chen DS, Kraft DL, Davis MM & Brown PO (2003) *PLoS Biol.* **1**, E65.

84. Chen DS, Soen Y, Davis MM & Brown PO (2004) *J. Clin.Oncol.* **22**, 9507.
85. Kuschel C, Steuer H, Maurer AN, Kanzok B, Stoop R & Angres B (2006) *Biotechniques,* **40**, 523–531.
86. Hsu KL, Pilobello KT & Mahal LK (2006) *Nat. Chem. Biol.* **2**, 153–157.
87. Robinson WH, DiGennaro C, Hueber W, *et al.* (2002) *Nat. Med.* **8**, 295–301.
88. Lueking A, Huber O, Wirths C, *et al.* (2005) *Mol. Cell. Proteomics,* **4**, 1382–1390.
★ 89. Tao SC & Zhu H (2006) *Nat. Biotechnol.* **24**, 1253–1254. – *Report of two similar strategies that greatly simplify the process of fabricating peptide/protein chips. Specifically, puromycin was used to capture the nascent polypeptides during translation onto a solid surface. Using these new strategies, efficiently fabrication of both high-density peptide and protein chips was possible, further demonstrating that such protein chips can be used for analysis of protein activity.*
90. Angenendt P, Kreutzberger J, Glokler J & Hoheisel JD (2006) *Mol. Cell. Proteomics,* **5**, 1658–1666.
91. Ramachandran N, Hainsworth E, Bhullar B, *et al.* (2004) *Science,* **305**, 86–90.
92. Paweletz CP, Charboneau L, Bichsel VE, *et al.* (2001) *Oncogene,* **20**, 1981–1989.
93. Nishizuka S, Chen ST, Gwadry FG, *et al.* (2003) *Cancer Res.* **63**, 5243–5250.
94. Nishizuka S, Charboneau L, Young L, *et al.* (2003) *Proc. Natl. Acad. Sci. U.S.A.* **100**, 14229–14234.
95. Zhu H, Bilgin M & Snyder M (2003) *Annu. Rev. Biochem.* **72**, 783–812.
96. Murthy TV, Wu W, Qiu QQ, Shi Z, LaBaer J & Brizuela L (2004) *Protein Expr. Purif.* **36**, 217–225.
97. Liu R, Barrick JE, Szostak JW & Roberts RW (2000) *Methods Enzymol.* **318**, 268–293.
98. Roberts RW & Szostak JW (1997) *Proc. Natl. Acad. Sci. U.S.A.* **94**, 12297–12302.
99. Weng S, Gu K, Hammond PW, *et al.* (2002) *Proteomics,* **2**, 48–57.
100. Wilson DS, Keefe AD & Szostak JW (2001) *Proc. Natl. Acad. Sci. U.S.A.* **98**, 3750–3755.

CHAPTER 11

Intelligent mining of complex data: challenging the proteomic bottleneck

Dan Bach Kristensen and Alexandre Podtelejnikov

1. INTRODUCTION

During the last decade or so, the field of proteomics – broadly defined as the characterization of the protein complement of the genome (1) – has undergone incredible technological developments, and it is now broadly applied in both basic and applied sciences in academia and the industry (2). One of the key driving forces behind the proteomic revolution has been technological development in the field of mass spectrometry (MS). This fact was reflected in the 2002 Nobel Prize in Chemistry, which was awarded in part to John B. Fenn and Koichi Tanaka for 'their development of soft desorption ionization methods for mass spectrometric analyses of biological macromolecules' (http://nobelprize.org/chemistry/laureates/2002/index.html). During the 1990s, the MS technology for biological macromolecules matured to a stage where it became possible to interface MS and liquid chromatography (LC) instruments, and so LC-MS for proteomics came of age. The technology has established itself as one of the primary workhorses in proteomics, due to the significant throughput and the ability of modern MS instruments to analyze peptides and proteins on the fly as they elute from the LC instrument.

Shotgun proteomics via LC tandem MS (LC-MS/MS) is a particularly well-established technology in proteomics and a typical shotgun proteomics workflow is shown in *Fig. 1*. In shotgun proteomics, a protein mixture is digested with a protease to produce a mixture of peptides, which is desirable from an analytical point of view, as peptides are easier to analyze by MS than intact proteins. The peptide mixture is subjected to LC-MS/MS, during which individual peptides are isolated and fragmented to produce an MS/MS spectrum (3). Using a search engine, such as Sequest (4) or Mascot (5), peptides are finally identified by matching the associated MS/MS spectrum against a theoretical fragmentation pattern, which can be predicted from nucleotide or protein sequence databases. Thousands of peptides can be analyzed during a single LC-MS/MS experiment, and a single biological sample may produce tens or even hundreds of thousands of MS/MS spectra. For instance, the research we conducted on plasma membrane-enriched samples

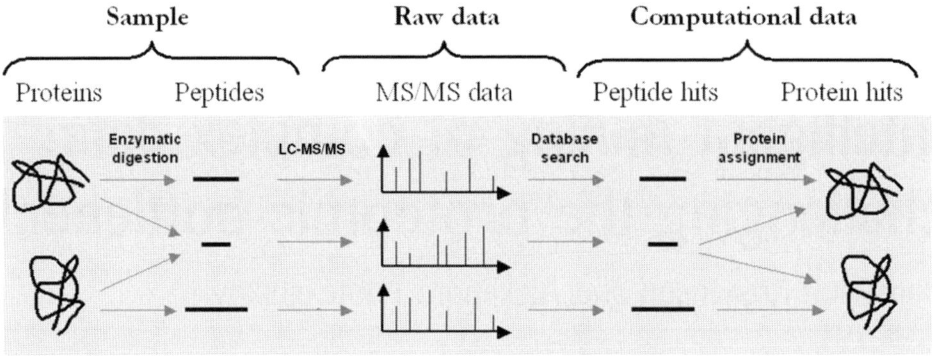

Figure 1. A shotgun proteomics workflow.
A protein sample is digested with a protease and the resulting peptides are subsequently analyzed by LC-MS/MS. Using a search engine, such as MASCOT, the acquired MS/MS spectra are matched against theoretical fragmentation patterns derived from a sequence database. The search engine identifies the most likely peptide hits for the experimental MS/MS spectra, and the peptide hits are finally associated with the proteins from which they are derived.

from cancer cells typically produced 25 000 MS/MS spectra per sample preparation, and samples were typically run in triplicate, producing a total of 75 000 MS/MS spectra per sample (data not shown). The literature contains examples with much higher numbers, and it is becoming increasingly clear that proteomic researchers are facing a new bottleneck: the ability to produce data has outpaced the ability to mine the data intelligently.

Whilst the ability to generate data has exploded, many researchers are still relying on lists of protein hits – usually the direct output from the search engine – to evaluate their shotgun proteomics experiments. This may be tedious and time-consuming at best, but a more dire consequence may be that the interpretation of the data is compromised, for example if the protein evidence is not presented correctly (see next section). Rather than considering the result output of a shotgun proteomics experiment as a fixed list of protein hits, a more dynamic and interactive data mining approach is required to extract relevant answers to biological and analytical questions from what otherwise may appear to be an ocean of data.

In addition to having to deal with large quantities of data in shotgun proteomics, another challenge lies in interpretation of the data. Peptides are the experimental evidence in shotgun proteomics, but researchers are mainly interested in knowing the proteins that are present in the sample. Inferring protein identity from a peptide hit is not always straightforward, as a peptide can be derived from multiple proteins, include splicing variants, homologous members of a gene family, and variants derived from sequence polymorphism. Furthermore, a peptide may point towards multiple entries in a database due to sequence redundancy derived from the presence of sequence errors and partial sequences. Consequently, in most cases, a peptide hit will point towards a group of proteins rather than a single protein (6). Protein evidence in shotgun proteomics should therefore be presented as protein groups rather than individual proteins, although this is currently not the standard.

This chapter deals with data mining strategies in shotgun proteomics. However, one should always bear in mind that the quality of a proteomic study will ultimately be determined by the weakest link in the chain of the workflow. Consequently, a brief coverage of the other components of a typical shotgun proteomic workflow (i.e. instruments, search engines, and databases) is given in section 2. For more in-depth coverage, particularly with respect to database search engines and data mining strategies, the reader may refer to recent reviews in the literature (6–9). Here a novel, peptide-centric database concept called EPICENTER (Experimental Peptide Identification Center) (10) is presented for the mining of shotgun proteomics data. The driving force behind the development of EPICENTER was the need for:

- Complete flexibility in data organization and selection (data evaluation should be possible on any combination of experiments or subsets thereof);
- Comparative functionality (to show similarities or differences between selected datasets);
- Dynamic, real-time filters (for extraction of concise, question-driven information);
- Multiple results view for data evaluation at different levels (protein groups, peptide assignments, statistical summaries at the peptide and protein level).

The speed of data evaluation can be increased significantly by combining these features in a single platform accessible through a standard web browser. Furthermore, the complete flexibility in terms of data organization, filtering, and comparison allows the researcher to address biological and analytical questions in new ways. The concept and functionality of EPICENTER is demonstrated in section 3 after covering the components – including other data mining approaches – of a shotgun proteomics workflow in section 2.

2. METHODS AND APPROACHES

Shotgun proteomics has become one of the methods of choice when performing comprehensive protein identification studies. The success of the technology is derived from its robustness, high throughput, and amenability to automation, which allows 24/7 operations. Shotgun proteomics is also capable of achieving significantly higher proteome coverage than two-dimensional electrophoresis, another widely used technology in proteomics. One reason for the higher coverage is that loss of low-solubility proteins (e.g. with protein transmembrane domains) can be avoided, and a broader protein molecular mass and pI coverage can also be obtained. At the same time, current shotgun analysis has significant limitations, which will be addressed in this chapter, together with possible ways of resolving the challenges.

The quality of the final output of a shotgun proteomics experiment will always depend on the weakest link in the workflow chain. Consequently, a brief coverage of the four main components of a shotgun proteomic workflow (i.e. instrumentation, databases, search engines, and data mining strategies) is given here. Finally, in section 3 we present our approach to the data mining step.

2.1 Instrumentation

A broad range of mass spectrometers is available today, including ion traps, quadrupole time-of-flight (QTOF), tandem TOF/TOF, linear-ion trap Fourier transform ion cyclotron resonance (FTICR), and Orbitrap mass spectrometers. The price and quality (i.e. sensitivity, dynamic range, speed, mass accuracy, and resolution) of these instruments vary considerably. However, from an analytical point of view, the data quality produced by the mass spectrometer will have a large impact on the final quality of a proteomics study: basically noise in equals noise out! For example, data obtained from low-resolution ion trap instruments can generate significant problems during data validation due to their inability to distinguish the charge state of multiple charged ions. Mass accuracy of precursor ions also plays a critical role in shotgun proteomics: low mass accuracy data must be accompanied by a broad mass tolerance in the database search, and this may result in an explosion in the number of theoretical peptides with an acceptable mass (11). For instance, Olsen and Mann obtained a 100-fold increase in peptide confidence using the high mass accuracy of an FTICR instrument compared with a typical ion trap instrument (12). Furthermore, ultrahigh confidence in peptide hits can be achieved by using the high mass accuracy of an FTICR in MS mode combined with MS^2 and MS^3 (13). Instrument quality is thus a crucial parameter in shotgun proteomics, where researchers have to deal with complex samples resulting in massive datasets that have to be matched against large and constantly increasing sequence databases. Whenever possible, we therefore recommend employing instruments with isotopic resolution of peptides and mass accuracy better than 50 p.p.m. in shotgun proteomics. This will allow charge-state determination and a relatively small mass window for precursor ions during database searches, both of which contribute to a significantly narrower database search. For nonconfident hits, additional fragmentation information can be obtained with alternative fragmentation approaches, including electron-capture dissociation (14), electron-transfer dissociation (15), and combinations of thereof, such as collisionally activated dissociation-based MS^2 and MS^3 (13) and collisionally activated dissociation combined with electron-capture dissociation (16). By exploiting such advanced technology, one can not only increase the confidence in peptide hits, but also find and characterize previously undescribed post-translational modifications.

With modern mass spectrometers, data generation is no longer a bottleneck in proteomic studies. However, tandem mass spectrometers are generally capable of detecting more precursor ions than they can successfully subject to MS/MS during an LC-MS run. Consequently, for complex samples, only the intense precursor ions are selected for MS/MS. For instance, a typical acquisition cycle on a QTOF instrument may last for 7 s and be composed of a 1 s MS survey followed by three MS/MS acquisitions of 2 s each. Thus, the maximum throughput in this example would be roughly 25 MS/MS acquisitions/min. However, the QTOF is easily capable of detecting more than 100 peptides in a single MS survey, and consequently only a fraction of these would be subjected to MS/MS, assuming the peptides elute in approximately 1 min.

There are multiple ways of addressing this MS/MS bottleneck, including: (i) employment of a longer LC gradient, (ii) reduction of sample complexity prior to LC MS/MS, and (iii) multiple LC MS/MS analyses of the same sample.

Longer peptide elution times could be achieved by using a longer LC gradient, but there is always a balance to be achieved: longer gradients mean more time for MS/MS but also less signal intensity; shorter run times mean less time for MS/MS but also higher signal intensity. Alternatively, reduction of sample complexity can be achieved by splitting the sample into more off-line fractions prior to LC-MS/MS analysis. In the end, however, the best experimental set-up is the one that produces the best data with minimal manual intervention, whilst fully exploiting the potential of the employed LC-MS/MS unit.

We have discovered that multiple analyses of the same sample using an exclusion list approach is an efficient way of addressing the MS/MS bottleneck during LC-MS/MS. In this strategy, a sample is analyzed multiple times (e.g. three times) and precursor ions are placed on an exclusion list after being subjected to MS/MS so that they are analyzed only once. This allows the precursor ions missed in one run to be analyzed in subsequent runs and, as a result, more precursor ions, including those of lower intensity, are subjected to MS/MS. Besides effectively addressing the MS/MS bottleneck, this approach relies only on additional work from the LC-MS unit, so it is easily automated. We have shown that both the number of identified proteins and the average sequence coverage per protein increases with this approach (17).

In addition to increasing both protein identification and the average sequence coverage, the multiple analyses approach adds a statistical component to shotgun proteomics. For instance, the reproducibility of peptide retention times and signal amplitude can be evaluated. The latter may add analytical confidence to quantitative shotgun studies where the signal intensity of, for example, light and heavy versions of isotope-labeled peptides is compared (18). Furthermore, multiple exclusion list analyses can be used to improve the signal-to-noise ratio by averaging the MS signal across the different runs, such as in the Rosetta Elucidator Protein Expression Data Analysis system (http://www.rosettabio.com/products/elucidator/default.htm). This strategy exploits multiple analyses to increase true signals, whilst reducing random noise.

The MS/MS bottleneck during LC-MS analyses can also be addressed using improved or different MS hardware. For instance, a new instrument with an increase in sensitivity and speed of MS/MS acquisition would, in principle, improve the situation, although a higher sensitivity would also result in the detection of more precursor ions. Alternatively, an LC matrix-assisted laser desorption ionization (MALDI) approach can be used. In this case, the LC run is split into many small fractions, which are deposited on a MALDI target and subsequently analyzed by MALDI-TOF/TOF. The key advantage of this approach is an indefinite amount of time – at least in principle – available for MS/MS. The disadvantage, however, is that MALDI almost exclusively produces single-charged precursor ions, and these are considerably more challenging to obtain good MS/MS data from using conventional fragmentation technology.

2.2 Database selection

There is number of databases available, ranging from well-annotated databases such as SwissProt (19) to massive expressed sequence tag (EST) nucleotide and genomic databases. Recently, we performed comparison studies of different protein databases using a mouse mitochondrial membrane proteome (data not shown). Database searches were performed using IPI mouse (20), SwissProt, RefSeq (21), and NCBI nrdb (22). In total, 927 protein groups were identified, with IPI, nrdb, RefSeq, and SwissProt producing 86, 20, 24, and 0 unique protein groups, respectively. This example illustrates the importance of selecting the correct database for extracting the maximum amount of high-quality information from the datasets. In our group, the IPI database is considered one of the most comprehensive, yet nonredundant, protein databases. However, currently it can only be used for a limited number of species, including human, mouse, chicken, rat, cow, zebrafish, and *Arabidopsis thaliana*.

In an ideal situation, a database search should be performed on the most comprehensive database, which should include information on all splicing variants, point mutations, and amino acid modifications. Some of this information is available from sources other than protein databases, for example, the scientific literature. A number of efforts have led to the creation of databases of post-translationally modified sequences identified from the literature, including PhosphoELM (23) and GlycoSuite DB (24). On the other hand, the databases have to be as concise and nonredundant as possible in order to minimize false results and requirements for computational power. For instance, currently a direct search of a shotgun proteomic dataset against EST databases can take several days, using a state-of-the-art, single CPU computer. Furthermore, validation of the results from a shotgun proteomics study matched against an EST database is extremely complicated.

To address this need for comprehensive results without sacrificing data quality, we developed an iterative database search algorithm (25). In this strategy, the proteomic dataset is searched initially against a protein database, typically IPI, and then all of the confident protein hits are extracted from the results into a temporary database. Next (i.e. the first iteration), the temporary database is searched for peptides with post-translational modifications (PTMs) using the MS/MS spectra that had no confident match in the initial search. As the temporary database comprises only a small set of high-confidence protein hits, the chance of correctly mapping a PTM increases significantly compared with a full database search, where random PTM matches are more frequent. We have successfully mapped several hundred PTMs from confident protein hits in shotgun studies with the iterative search approach (data not shown). Furthermore, the iterative database searches can be done in a significantly shorter time and with fewer computational resources than traditional searches, as both the number of entries in the temporary database and the number of MS/MS spectra are significantly reduced compared with standard database searches. The iterative database search can be expanded further to include information from single-nucleotide polymorphisms, EST, and alternative splicing variant databases, which will cover most sequence variations derived from alternative splicing and sequence polymorphism. The iterative database search strategy is thus an effective way of extracting comprehensive, high-quality data in shotgun proteomics.

2.3 Selection of a search engine

The success of a proteomics study depends to a large extent on the processing of the MS data. Data processing can produce highly varying results depending on the type of mass spectrometer, processing software, input parameters, and search engine. There are two major components in data processing:

1. Raw data processing.
2. Database searching.

The processing of raw data into a flat file format (e.g. a peak list, including a list of peptide masses and associated fragment ion masses in the case of MS^n), includes de-isotoping, centroiding, and peak picking. It is beyond the scope of this chapter to cover the area of raw MS data processing. However, it should be noted that many hardware vendors have their own raw data processing tools and data standards, and the processing results may vary considerably between the different solutions available. Consequently, it is important to point out that several efforts are currently under way to unify data processing according to a universal standard, independent of the MS platform employed. One such effort is the Proteome Standard Initiative Mass Spectrometry (PSI-MS) initiative, which is focused on providing vendor-neutral standard formats for representing MS experimental data and associated results (http://www.psidev.info/index.php?q=node/80). The PSI-MS file format, mzXML, supports the major MS vendors' software packages and facilitates the integration of large-scale datasets across different vendor platforms. Furthermore, several of the major instrument vendors are currently implementing mzXML support into their own software solutions. Taken together, these standardization efforts will be critical for free exchange and comparison of proteomic data across platforms and between study groups.

The second major component of data processing is the database search. A number of database search engines are currently available for the scientific community. Among the most popular commercial search engines are SEQUEST (4), which is based on a cross-correlation scoring algorithm, and MASCOT (5), which is based on probability scoring. In addition, there are a number of noncommercial search engines such as PROTEIN PROSPECTOR (26) and OMSSA (27). Most current search engines applied in shotgun proteomics work by matching experimental fragment ions against the theoretical fragmentation pattern extracted from the sequence database. Our group compared the results output from different search engines, including SEQUEST, MASCOT, SONAR, and PEPSEA, applied to low- and high-quality MS datasets derived from ion traps and QTOFs, respectively (28). As shown previously, the results output from the search engines depends to a certain extent on the instrument type and the associated mass accuracy. For example, SEQUEST provided the best results for the low mass accuracy dataset obtained from traditional ion trap instruments, whilst it produced more false-positive hits with high mass accuracy data acquired on QTOF instruments. In contrast, MASCOT, PEPSEA, and SONAR produced significantly more accurate results than SEQUEST for high-quality datasets. In the end, the best results were obtained by combining data from high-quality instruments with a probability-based search engine, such as Mascot.

The currently available database search engines share a common disadvantage. In order to produce a reliable hit, they require a significant number of experimental fragment ions to get a good match to the theoretical fragmentation pattern. Consequently, the majority of MS/MS spectra acquired during LC-MS/MS analysis are left unsearched, with typically only 10–30% of the spectra producing a peptide hit. To address this low MS/MS efficiency, efforts are currently under way to improve database search strategies, including the ability to combine results from multiple database searchers on the same dataset, or by exploiting multiple analyses of the same sample to improve overall statistics.

The use of combined results from multiple search engines is easy to implement, as it is based on existing technology, and encouraging results have been achieved. For instance, the SCAFFOLD software package from Proteome Software (http://www.proteomesoftware.com/Proteome_software_prod_Scaffold.html) has shown significant improvements in removing false-positive hits by combining SEQUEST and MASCOT results. However, whether using an algorithm based on cross-correlation or probability scoring, the search engines are fundamentally exploiting the same principle, i.e. they compare experimental and theoretical fragmentation patterns. Consequently, combined searches in some cases also produce false-positive results in situations where the different search engines all point to the same false-positive hit. For instance, the deamidated form of the peptide LSSPATLNSK is a commonly detected autolysis product of trypsin (m/z 523.79, 2+). We found that a number of search engines gave a false-positive hit to the tryptic plasminogen peptide, LSSPADITDK (data not shown). In this case, the combination of multiple search engines did not alleviate the problem.

Alternatively, the confidence of peptide hits could be increased by combining orthogonal search strategies. For instance, one possibility would be to combine the conventional search strategy with a sequence-tagging or *de novo* sequencing strategy (29). A sequence tag is highly specific; usually a 2–3 amino acid sequence tag is sufficient for unambiguous peptide identification, and it would consequently be highly complementary to a probability- or cross-correlation-based search strategy.

Among relatively new search engines are X! TANDEM (30) and PHENYX (http://www.phenyx-ms.com/). A unique feature of X! TANDEM is its ability to perform searches against a database composed of experimental MS/MS data, i.e. experimental MS/MS can be matched directly across different studies. This approach can significantly accelerate the database search itself and, furthermore, the direct comparison of MS/MS spectra between different experiments can be extremely useful for targeted analysis of biomarkers. As with other experimental databases, however, this approach may be limited for less-well-characterized species. Another advanced search engine is PHENYX, which is based on the OLAV algorithm (31). When performing database searches via the UniProt database, PHENYX takes annotated modifications from the database into consideration as variable modifications (i.e. they may be present or absent at the annotated position). This strategy provides a concise approach for detection of known PTMs, where many of the conventional permutation noise associated with allowance of multiple PTMs in a database search is eliminated. Of course, the approach only applies to known modifications found in well-annotated proteins.

Finally, it is worth mentioning that different strategies have been established for measuring the quality of shotgun proteomic data. One of the most common approaches is to compare results obtained from a database search against a normal database and the same database with the sequences reversed. The results derived from the reversed database are then compared with those of the normal database search and the percentage of 'false-positive' hits is calculated.

2.4 Data mining in shotgun proteomics

Despite the proven success of shotgun proteomics, there is currently a number of challenges to be addressed. For instance, most researchers are using search engines as stand-alone applications, without, for example, the necessary bioinformatic tools or support for processing of multiple datasets. MASCOT, for example, has an upper limit in terms of the number of MS/MS spectra that can be processed in a database search and, for large proteomic datasets, this means that the data have to be divided into multiple searches. Furthermore, for large datasets, the spectra processing can be extremely time-consuming. When establishing a shotgun proteomics platform, it is therefore critical to consider the data-mining component. As a minimum, we recommend the following features:

- Data mining across any number of experiments or datasets.
- Comparative functionality for observing similarities or differences among datasets.
- Filters (e.g. for quantitative data, PTMs, number of peptide hits, database annotations, etc.) for extracting question-driven information.
- Multiple result outputs (protein groups, peptide hits, statistics).

In-depth analysis via shotgun proteomics is inherently complex and, with the maturation of MS technology, the field of proteomics has entered an information explosion stage. Unfortunately, the 'classical' protein software tools are not ready to deal with the high data complexity of proteomics. At the same time, there are no simple solutions on how to resolve this data-mining challenge in a generic manner. Consequently, researchers currently tend to select different software tools depending on the task at hand, be that comparative analysis of proteomes, investigation of protein interaction networks, or quantitative proteomics. In the case of quantitative proteomics based on the metabolic labeling of proteins in cells, the MSQUANT software package is a powerful solution (32). MSQUANT automatically extracts quantitative information on peptides and proteins from MASCOT result files and associated raw files, with an option to perform manual validation of quantitative data for each spectrum, as well as protein identification assignment. MSQUANT has been applied successfully in a number of proteomic studies, including signal transduction studies (33).

One of the most comprehensive tools currently available for the proteomics community is the Trans-Proteomic Pipeline (TPP) (http://tools.proteomecenter.org/software.php). This package includes a set of different software applications covering the entire workflow in shotgun proteomics, including PEPTIDEPROPHET, which assigns probabilities to the peptide identifications returned by the search engines. Another component, PROTEINPROPHET, provides a statistical model for validation of peptide identifications at the protein level. The TPP provides a standard data analysis platform through the mzXML format (data converters are included for most MS vendors) and provides solid tools for statistical validation at the peptide and protein level. However, the TPP is not ideal in terms of combining datasets as desired, applying real-time filters on selected datasets, or when performing comparative studies across the selected datasets. These features should ideally be available in real time, thereby allowing the investigator to extract the relevant information from the relevant datasets on the fly during data mining. In the next section, we present data-mining examples using the EPICENTER technology, which allows real-time data selection, filtering, and comparison across any number of shotgun proteomic datasets.

3. EPICENTER

In EPICENTER, all peptide evidence derived from any number of LC-MS/MS experiments is stored in a structured fashion, i.e. the core data is a list of precursor ions to which peptides have been assigned by a search engine, such as MASCOT. Some of the key features of the platform include:

- A concise repository for LC-MS/MS-derived peptide evidence
- Peptide validation based on multiple experimental parameters
- Protein grouping based on shared peptide evidence
- Flexible data organization and selection via a user-definable data tree structure
- Powerful on-the-fly filtering for mining selected data
- Multiple result views for evaluating peptide or protein evidence
- Export of protein evidence in a generic text format

The combination of the above features makes EPICENTER a highly dynamic and interactive platform for real-time mining of LC-MS/MS data in shotgun proteomics. Application examples on how to exploit EPICENTER for data mining in shotgun proteomics are given below. To test the functionality described below (with the exception of data organization and import), a demonstration version of EPICENTER can be accessed at http://epicenter.proxeon.com/.

3.1 Data organization and import

EPICENTER has been designed specifically to process large LC-MS/MS datasets from proteomic studies. The software package is organized to support an unlimited number of experiments, and for each project the data have to be organized in a user-definable data structure in EPICENTER (see *Fig. 2*, also available in the color

(a)

(b)
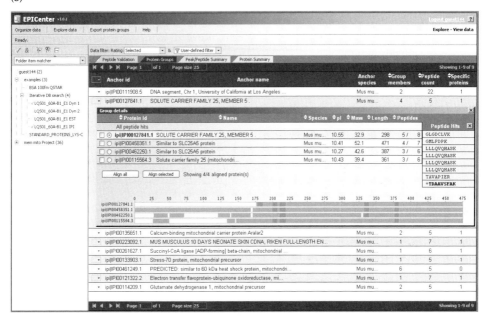

Figure 2. Data mining in EPICENTER (see page xxx for color version).
The user-definable data tree is shown to the left in (a) and (b). In this example, two data folders (LQS01_60A-B1_E1 IPI and LQS01_60A-B1_E1 Dyn1) have been selected in the path guest144/examples/Iterative DB search. To the right of the data tree, one of the four result views ('Peptide Validation', 'Protein Groups', 'Peak/Peptide Summary', and 'Protein Summary') is displayed, in this case the 'Protein Groups' view. Each line in the 'Protein Groups' view in (a) represents a protein group. To access the protein group, click on the anchor Id or the name, which will open a 'Group details' window (b). This window contains the accession number, protein name, species, pI, average mass, length, and number of peptides (number of unique peptides/number of matching precursor ions) for all of the protein members of the group. Unambiguously identified proteins are indicated with bold protein Ids, and the anchor protein is selected by clicking its radio button to the left of the protein Id. To the left in the group details views, the peptide hits are shown for the highlighted group member. Unique peptide hits (TDAAVSFAK) are shown in bold and modified peptides are indicated with an asterisk (TDAAVSFAK is N-terminally acetylated). In addition, the 'Group details' view shows a CLUSTALW alignment of all group members at the bottom of the window, with the sequenced peptides highlighted in red boxes.

section), similar to the way folders and files are organized and viewed in Windows Explorer. Data levels can be expanded/collapsed for viewing and selected/unselected for mining. There can be any number of data levels in a project, and *Fig. 2* shows an example where two data levels have been selected (LQS01_60A-B1_E1 IPI and LQS01_60A-B1_E1 Dyn 1) in the data tree path guest144/examples/Iterative DB search. The selected data levels are indicated with a √ sign next to the left of the data levels. The data should be organized in a way that makes subsequent data mining feasible. For instance, if the aim of a study is to determine the effect of iterative database searching (see example in *Fig. 2*), then a data level (Iterative DB search in *Fig. 2*) containing one data level for each of the search iterations should be established (four in the example shown in *Fig. 2*).

During EPICENTER installation, the location of the search engine or data storage has to be defined in order for EPICENTER to extract the relevant information from the MASCOT *.dat files. The import data process goes through several major stages, including selection and downloading of the MASCOT *.dat file, followed by association of the downloaded file to the corresponding tree/folder in EPICENTER, and finally automatic validation of the peptide hits.

3.2 Automatic and manual peptide validation

After importing LC-MS/MS data, EPICENTER performs an automated validation of the peptide hits returned by the search engine. This validation model combines the accumulated knowledge and evidence concerning a certain spectrum and attempts to assign the most probable peptide. Initially, a set of high-confidence peptides are identified based on a number of user-defined parameters, including:

- A minimum peptide length (i.e. number of amino acids)
- A minimum peptide hit score (from the search engine)
- A minimum delta score (score difference between peptides ranked 1 and 2)
- Minimum y-ion and b-ion scores (a measure of the number of consecutive fragment ions)

In order to create confident protein groups, we recommend using a minimum peptide length of 6 amino acids. The minimum peptide score derived from the MASCOT search engine depends on the protein database and type of mass spectrometer, but we typically use a minimum score equal to 28 when using a QTOF mass spectrometer. The delta score (i.e. the difference in scores between the first- and second-ranked peptides) is typically set to 5. For tryptic peptides analyzed on a QTOF instrument, the minimum y-ion and b-ion scores are set to 50 and 2, respectively. However, these values will depend on the instrument type and fragmentation technology. After identifying the confident peptide hits, less-confident peptide hits are identified, primarily based on the number of sibling peptides (the number of peptides matching the same protein entry), b- and y-ion scoring model, and the search engine score. After the automatic validation process, all of the precursors included in the validation process will have one of the following assignments:

- One confident *selected* peptide hit (all others are rejected)
- One or more *potential* peptide hits (all others are rejected)
- All *rejected* peptide hits

Typically, shotgun proteomics studies leave more than half of the acquired MS/MS spectra without any peptide assignments. The ability of EPICENTER to merge any number of datasets followed by peptide validation partially addresses this issue, as the overall statistics improve (e.g. sibling peptides can be identified from complete datasets). At the same time, a significant number of spectra may still require manual validation, and a simple and concise method for extracting uncertain peptide hits is therefore desirable. In EPICENTER, all potential peptide hits (i.e. the uncertain peptide hits) can be extracted simply by selecting 'Potential' in the rating filter. After the automatic validation, the 'Peptide Validation' view can then be used for manually inspecting and assigning the uncertain peptide hits (see *Fig. 3*, also available in the color section).

The 'Peptide Validation' view shows a list of precursor ions together with one or more suggested peptide hits, with a maximum of one selected hit per precur-

Peak ⇕350.224 / ⇕16.76	Database	Search	Rating	Δ mass	Quant ratio	Ions match/exp	y/b ion score	Proline score	Total NSP	Engine score
VVQPTR	IPI_mouse 60A MitoMem IPI L...			0.02 (2+)	-	6/46	41.6/1.0	40.0	0	24.53
VVQVVR	IPI_mouse 60A MitoMem IPI L...			0.01 (2+)	-	6/36	42.4/1.0	0.0	0	24.53

Peak ⇕350.224 / ⇕32.08	Database	Search	Rating	Δ mass	Quant ratio	Ions match/exp	y/b ion score	Proline score	Total NSP	Engine score
IQLLGR	IPI_mouse 60A MitoMem IPI L...			0.01 (2+)	-	9/38	98.7/9.4	0.0	9	33.54
LAGLIGR	IPI_mouse 60A MitoMem IPI L...			0.01 (2+)	-	8/36	75.3/8.0	0.0	0	31.57
LAGLLGR	IPI_mouse 60A MitoMem IPI L...			0.01 (2+)	-	8/36	75.3/8.0	0.0	0	31.57
IQLGIR	IPI_mouse 60A MitoMem IPI L...			0.01 (2+)	-	5/38	37.9/1.0	0.0	0	25.73
LQLVAR	IPI_mouse 60A MitoMem IPI L...			0.01 (2+)	-	5/38	37.9/1.0	0.0	0	25.73
IKLAVR	IPI_mouse 60A MitoMem IPI L...			0.05 (2+)	-	5/38	37.7/1.0	0.0	0	25.73
LKIGLR	IPI_mouse 60A MitoMem IPI L...			0.05 (2+)	-	5/38	37.7/1.0	0.0	0	25.73
IAGIGLR	IPI_mouse 60A MitoMem IPI L...			0.01 (2+)	-	5/36	11.0/1.0	0.0	0	25.73
LQLGLR	IPI_mouse 60A MitoMem IPI L...			0.01 (2+)	-	5/38	37.9/1.0	0.0	0	25.73
LKLGLR	IPI_mouse 60A MitoMem IPI L...			0.05 (2+)	-	5/38	37.7/1.0	0.0	0	25.73

Peak ⇕350.226 / ⇕25.17	Database	Search	Rating	Δ mass	Quant ratio	Ions match/exp	y/b ion score	Proline score	Total NSP	Engine score
LIGLQR	IPI_mouse 60A MitoMem IPI L...			0.01 (2+)	-	7/32	42.5/1.0	0.0	0	25.51
LLGLQR	IPI_mouse 60A MitoMem IPI L...			0.01 (2+)	-	7/32	42.5/1.0	0.0	0	25.51
ILGLKR	IPI_mouse 60A MitoMem IPI L...			0.04 (2+)	-	7/32	41.6/1.0	0.0	0	25.51
LLGLKR	IPI_mouse 60A MitoMem IPI L...			0.04 (2+)	-	7/32	41.6/1.0	0.0	0	25.51
LLLGKR	IPI_mouse 60A MitoMem IPI L...			0.04 (2+)	-	6/32	41.6/1.0	0.0	0	23.58
LLGKLR	IPI_mouse 60A MitoMem IPI L...			0.04 (2+)	-	6/34	41.6/1.0	0.0	0	22.42
*QNGAIGR	IPI_mouse 60A MitoMem IPI L...			0.10 (2+)	-	6/48	34.2/1.0	0.0	0	21.55
ILIANR	IPI_mouse 60A MitoMem IPI L...			0.01 (2+)	-	5/32	42.5/1.0	0.0	20	20.39

Figure 3. Peptide validation (see page xxxi for color version).
After the automatic validation, the 'Peptide Validation' view can then be used for manually inspecting and rating peptide hits. For instance, the rating filter can be set to 'Potential' to evaluate only precursor ions with uncertain peptide assignments. The 'Peptide Validation' view shows a list of precursor ions together with one or more suggested peptide hits, with a maximum of one selected hit per precursor ion. In this example, three precursor ions with 2, 10, and 8 suggested peptide hits, respectively, are shown. In the rating column, the peptide rating (selected, rejected, or potential) can be changed manually. In this case, a peptide hit has been assigned manually to the top precursor ion for illustrative purposes. Modified peptides are indicated in red and with an asterisk; in this case, a pyroglutamic acid (Q) is seen for a suggested peptide for the lower precursor ion.

sor ion. One example of the 'Peptide Validation' view is presented in *Fig. 3*, where three precursor ions with 2, 10, and 8 suggested peptide hits, respectively, are shown. For each precursor ion, the experimental peptide mass and retention time is shown, together with all of the potential peptide hits and associated database, the search name (user defined), the peptide hit rating, the difference between the experimental and theoretical mass, the quantitative ratio (if available), matched and expected ion counts, ion scores, the proline score (a measure of the relative intensity of the fragment ion with an N-terminal proline), the total number of sibling peptides (number of peptides matching the same protein entry), and finally the search engine score. In the rating column, the peptide rating (selected, rejected, potential) can be changed manually. For instance, once a peptide is 'Selected' manually, then the rating of all of the other peptides for that precursor ion is set to 'Rejected', as EPICENTER only allows one peptide hit per precursor ion. For validation purposes it should be noted that the MASCOT peptide view – containing the MS/MS spectrum and fragment ion information – can be accessed for any suggested peptide by clicking on the peptide sequence in the 'Peptide Validation' view.

3.3 Data mining – exploring datasets

The central working window of EPICENTER is the 'Protein Groups' window, which is shown in *Fig. 2*. When working in this window, the peptide-rating filter is usually set to 'Selected' in order to see only protein groups based on confirmed peptide hits. Each line in the 'Protein Groups' view (see *Fig. 2a*) represents a protein group and contains the following information:

- Accession number (Id) for the anchor protein (usually the protein with the highest number of peptide hits)
- Name of the anchor protein
- Species of the anchor protein
- Number of protein members in the group
- Number of (distinct) peptide sequences in the group
- Number of specific proteins in the group (proteins with one or more unique peptide hits compared with other group members)

To access the protein group, simply click on the anchor Id or the name, which will open a 'Group details' window, as shown in *Fig. 2(b)*. This window contains the accession number, protein name, species, pI, average mass, length, and number of peptides (number of unique peptides/number of matching precursor ions) for all of the protein members of the group. Unambiguously identified protein is indicated with bold protein Ids, and the anchor protein can be selected manually by clicking its radio button to the left of the protein Id.

To the right in the 'Group details' views, the peptide hits are shown for the highlighted group member and, by clicking the protein name, different group members can be highlighted. Unique peptide hits found only in the highlighted group member are shown in bold and modified peptides are indicated with an asterisk (the peptide TDAAVSFAK is both unique and modified in the current example). In addition, the 'Group details' view shows a CLUSTALW alignment of all group members at the bottom of the window, with the identified peptides highlighted in red boxes (moving the cursor over the box will display the corresponding peptide sequence). For groups with many members or long sequences, the CLUSTALW alignment is not done automatically, as this may require significant computing power. To align these groups, simply click on the 'Align all' button above the sequence plot. Alternatively, click 'Align selected' to align only the proteins selected manually using the tick box to the left in the group details window.

The example shown in *Fig. 4* (also available in the color section) demonstrates the potential complexity of shotgun proteomic data analysis. Five protein members of the solute carrier family 25 are found in the protein group, and EPICENTER automatically selected the anchor protein based on the maximum number of peptide hits (selected radio button next to the Protein Id). At the same time, there are two other protein entries (Protein Ids highlighted in bold) with unique peptide hits that are not found in the other group members. This clearly indicates that for the solute carrier family 25, multiple variants are present in the sample. The protein-grouping concept of EPICENTER thus gives a comprehensive but concise overview of the protein level evidence in shotgun proteomics, including the presence of protein variants.

A key strength of EPICENTER is that multiple, dynamic filters can be applied simultaneously to facilitate the analysis, including the number of sibling peptides, amino acid sequence, protein name, or even accession numbers corresponding to the proteins of interest. The filters can be accessed from a drop-down menu as shown in *Fig. 5*, and in this example three filters were selected:

- Three or more peptides hits per protein group
- One or peptide score(s) above 30
- One or acetyl modifications per protein group

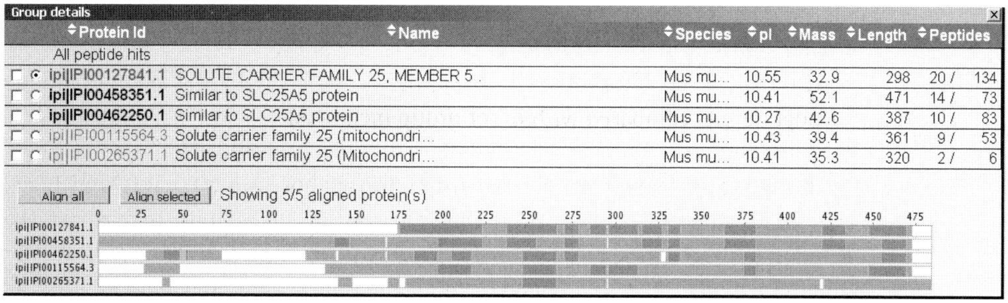

Figure 4. Detection of protein isoforms in the 'Group details' view (see page xxxii for color version).
Five protein members of the solute carrier family 25 are found in this protein group, and EPICENTER has automatically selected the anchor protein based on the maximum number of peptide hits (the radio button next to the protein Id is selected). At the same time, there are two other protein entries (protein Ids highlighted in bold) with unique peptide hits that are not found in the other group members. This clearly indicates the presence of three distinct variants of the solute carrier family 25. At the bottom, a CLUSTALW alignment is shown for the five group members.

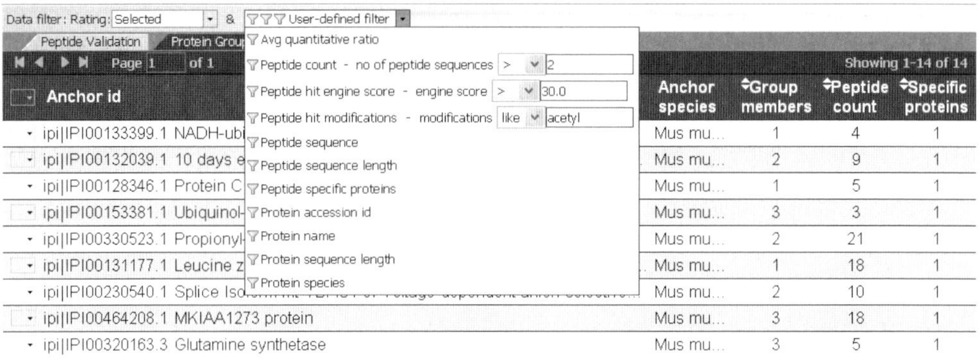

Figure. 5. Dynamic user-defined filters.
In EPICENTER, there is a set of dynamic user-defined filters, accessed from the drop-down menu. Common to all user-defined filters is that they work on the protein groups after the data has been pulled from the database and the groups have been built. The groups of protein identified by large-scale proteome study are presented according to the filter selection. In this example, the identified protein groups contain at least three peptide hits, the search engine score is higher than 30, and one or more acetylated peptides have been detected. To the left of the user-defined filters is the rating filter, for working with rejected, potential, or selected peptide hits.

Common to all of the filters is that they work on the protein groups after the data has been pulled from the database and the groups have been built. For instance, if you filter for a specific peptide sequence, then the protein group in which this sequence is found will be shown, including all other peptide sequences in that group.

In summary, EPICENTER allows the investigator to organize and select data, to filter peptide hits based on their rating (selected, potential, and rejected), and finally to apply dynamic user-defined filters in real time. This flexibility provides the investigator with powerful tools for intelligent mining of complex data.

3.4 Data mining – comparing datasets

Another EPICENTER functionality is the ability to compare datasets in real time, i.e. to find similarities or differences among two or more datasets from the data tree (see section 3.1). This comparative functionality can be exploited at many different levels, including biological applications (different samples, treated versus untreated samples, etc.), methodological applications (samples analyzed on different instruments, different MS acquisition methods, etc.), and data processing applications (different search engine parameters, iterative database searches, etc.). A simple example of a dataset comparison is shown in *Fig. 6*. Two samples were selected for comparison, one containing bovine serum albumin (BSA) only (dataset 1) and the other a protein standard mixture including BSA (dataset 2). The comparative functionality can extract the protein groups that are shared between the two datasets or unique to either of them. *Fig. 6(a)* shows the shared protein group(s) between datasets 1 and 2 (only BSA), whereas *Fig. 6(b)* shows

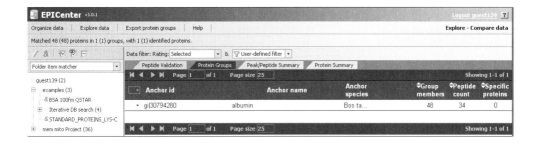

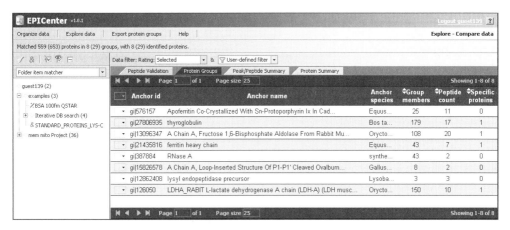

Figure 6. Dataset comparison.
EPICENTER also has a comparative functionality, which allows the investigator to determine differences or similarities among two or more datasets. In this example, two samples were selected for comparison, one containing bovine serum albumin (BSA) only (dataset 1) and the other a protein standard mixture including BSA (dataset 2). In (*a*), the protein group(s) shared between datasets 1 and 2 (only BSA) are shown, whilst (*b*) shows groups unique to dataset 2 (all standard proteins except BSA). In comparative mode, the selected data levels are indicated with '&' (inclusive) or '%' (exclusive) signs, as shown to the left in (*a*) and (*b*). The comparative functionality can be expanded to cover any number of datasets and can be combined with peptide rating and user-defined filters (see *Fig. 5*).

groups unique to dataset 2 (all standard proteins except BSA). Notice that, in comparative mode, the selected data levels are indicated with '&' (inclusive) or '%' (exclusive) signs, as shown to the left in *Fig. 6(a)* and (*b*). Dataset comparison can of course be expanded to a much higher level, allowing the comparison of hundreds of datasets or more in real time (processing time may take some seconds, depending on the size of the selected datasets and computer power). Again, in combination with the dynamic real-time filters mentioned in the previous section, comparative data mining can be performed in a highly interactive and targeted fashion, driven by the biological or analytical questions at hand.

3.5 Additional EPICENTER features

Among other important functionalities of EPICENTER, there are two statistical result views ('Peak/Peptide Summary' and 'Protein Summary', not shown here). The

'Peak/Peptide Summary' view is used to get a statistical overview of the precursor and peptide content in the selected dataset. For instance, the precursor rating filter should be set to 'Any' (selected, rejected, and potential hits all together) to see a summary of all the precursor ions selected in the data tree. In this way, the number of precursor ions in the search engine output, the number of precursor ions with associated hits in the database, and the number of precursor ions with a selected peptide assigned are presented in a summarized fashion in order to facilitate the statistical evaluation of the proteomics study. Similarly, the 'Protein Summary' view presents an overview of the proteins and protein groups found in the selected dataset, including the number of protein groups, the number of protein entries in all groups, and the number of unambiguously identified proteins, i.e. proteins with unique peptide hits.

In addition, EPICENTER creates reference libraries of MS/MS spectra. Using the 'Peptide sequence' filter, EPICENTER can extract all tandem mass spectra matching a given peptide, including modified peptides. This can, for instance, be used to extract statistical information on how often the peptide in question has been detected in different samples. Furthermore, it allows the investigator quickly to compare the different MS/MS spectra.

3.6 Conclusions

The data-mining concept presented here provides investigators with new and alternative approaches for mining of shotgun proteomic data. The data-mining process can be performed in a concise and targeted manner due to a high degree of flexibility in terms of data organization and selection, dynamic data filtering in real time, comparative functionality, and multiple result views. These functionalities can be combined intelligently to extract concise biological or analytical information from complex proteomic data, thereby challenging the data-mining bottleneck facing the proteomics community today.

4. REFERENCES

★ 1. Wasinger VC, Cordwell SJ, Cerpa-Poljak A, *et al.* (1995) *Electrophoresis*, **16**, 1090–1094. – *The original introduction of the proteome concept.*
★ 2. Aebersold R & Mann M (2003) *Nature*, **422**, 198–207. – *A comprehensive review on the applications of MS in proteomics.*
3. Steen H & Mann M (2004) *Nat. Rev. Mol. Cell Biol.* **5**, 699–711.
4. Yates JR III, Eng JK, McCormack AL & Schieltz D (1995) *Anal. Chem.* **67**, 1426–1436.
5. Perkins DN, Pappin DJ, Creasy DM & Cottrell JS (1999) *Electrophoresis*, **20**, 3551–3567.
★★ 6. Nesvizhskii AI & Aebersold R (2005) *Mol. Cell. Proteomics*, **4**, 1419–1440. – *A comprehensive review addressing the protein inference challenge in shotgun proteomics.*
★★ 7. Nesvizhskii AI & Aebersold R (2004) *Drug Discov. Today*, **9**, 173–181. – *An insightful review on analysis, statistical validation, and dissemination of large-scale proteomics datasets.*
8. MacCoss MJ (2005) *Curr. Opin. Chem. Biol.* **9**, 88–94.
9. Sadygov RG, Cociorva D & Yates JR III (2004) *Nat. Methods*, **1**, 195–202.
★ 10. Kristensen DB, Brond JC, Nielsen PA, *et al.* (2004) *Mol. Cell. Proteomics*, **3**, 1023–1038. – *The original paper describing the EPICENTER technology (previously called EPIR).*

11. Clauser KR, Baker P & Burlingame AL (1999) *Anal. Chem.* **71**, 2871–2882.
12. Olsen JV, Ong SE & Mann M (2004) *Mol. Cell. Proteomics*, **3**, 608–614.
13. Olsen JV & Mann M (2004) *Proc. Natl. Acad. Sci. U.S.A.* **101**, 13417–13422.
14. Zubarev RA (2004) *Curr. Opin. Biotechnol.* **15**, 12–16.
15. Syka JE, Coon JJ, Schroeder MJ, Shabanowitz J & Hunt DF (2004) *Proc. Natl. Acad. Sci. U.S.A.* **101**, 9528–9533.
16. Nielsen ML, Savitski MM & Zubarev RA (2005) *Mol. Cell. Proteomics*, **4**, 835–845.
17. Kristensen DB, Podtelejnikov AV, Brond JC, *et al.* (2003) Multiple LCMS exclusion list analyses: a tool to enhance protein identification from complex biological samples. Poster MPX 454. 51st Conference on Mass Spectrometry and Allied Topics, American Society for Mass Spectrometry, Montréal, Québec.
18. Ong SE, Foster LJ & Mann M. (2003) *Methods*, **29**, 124–130.
19. Boeckmann B, Bairoch A, Apweiler R, *et al.* (2003) *Nucleic Acids Res.* **31**, 365–370.
20. Kersey PJ, Duarte J, Williams A, Karavidopoulou Y, Birney E & Apweiler R (2004) *Proteomics*, **4**, 1985–1988.
21. Pruitt KD, Tatusova T & Maglott DR (2005) *Nucleic Acids Res.* **33**, D501–D504.
22. Wheeler DL, Barrett T, Benson DA, *et al.* (2005) *Nucleic Acids Res.* **33**, D39–D45.
23. Diella F, Cameron S, Gemund C, *et al.* (2004) *BMC Bioinformatics*, **5**, 79.
24. Cooper CA, Harrison MJ, Wilkins MR & Packer NH (2001) *Nucleic Acids Res.* **29**, 332–335.
25. Bennett KL, Kristensen DB, Schandorff S, *et al.* (2004) Iterative database searches: an approach to enhance the functional information obtained from LC-MS data. Poster TPA 020. 52nd Conference on Mass Spectrometry and Allied Topics, American Society for Mass Spectrometry, Nashville, Tennessee.
26. Clauser KR, Baker P & Burlingame AL (1999) *Anal. Chem.* **71**, 2871–2882.
27. Geer LY, Markey SP, Kowalak JA, *et al.* (2004) *J. Proteome Res*, **3**, 958–964.
28. Boutilier K, Ross M, Podtelejnikov AV, *et al.* (2005) *Anal. Chim. Acta*, **534**, 11–20.
29. Mann M & Wilm M (1994) *Anal. Chem.* **66**, 4390–4399.
30. Craig R & Beavis RC (2004) *Bioinformatics*, **20**, 1466–1467.
31. Colinge J, Masselot A, Giron M, Dessingy T & Magnin J (2003) *Proteomics*, **3**, 1454–1463.
32. Schulze WX & Mann M (2004) *J. Biol. Chem.* **279**, 10756–10764.
33. Blagoev B, Kratchmarova I, Ong SE, Nielsen M, Foster LJ & Mann M (2003) *Nat. Biotechnol.* **21**, 315–318.

CHAPTER 12
Bioinformatic approaches in proteomics

Sandra Orchard and Henning Hermjakob

1. INTRODUCTION

The molecular biologist now has a wealth of data to hand describing entire genomes at the DNA level, but this is insufficient to describe fully the protein content of an individual cell. The expression level of any one gene will vary with cell type, differentiation stage, cell cycle, and exposure to external signals. The mRNA transcript may be alternatively spliced, use varying initiation sites or promoters, or be edited prior to translation. Similarly, the protein itself may be modified during or after translation. The focus of attention is therefore moving towards the use of proteomics techniques to examine and understand the protein complement of the cell, the interactions proteins make with other molecule types, and their role in cellular processes and pathways. The sheer volume of data generated by such experiments, which includes supporting metadata such as sample preparation and storage, raw mass spectra, and the subsequent protein identification steps, can no longer be dealt with via the traditional routes of the laboratory notebook, the published paper, and the occasional textbook. Data is stored and maintained in purpose-built databases, from where it can easily be accessed and shared with other workers in the field. Specialist tools have been developed to allow further analysis of large datasets, and the field of bioinformatics has expanded to encompass entire proteome analysis in addition to genome analysis.

However, much data is still lost because of the fragmented approach taken by research workers both to data analyses and subsequent storage of the data generated. Whilst this typifies the early years in any rapidly expanding field, it results in data loss whilst compromising data comparability. The world of proteomics has benefited greatly from the large number of instrumentation manufacturers and software producers who have contributed to the wide range of hardware and analytical programs available to the scientist. The corresponding downside to this has been the accompanying plethora of proprietary software, which has made the comparison of datasets generated in different laboratories almost impossible. This issue is now being addressed by the Human Proteome Organization Proteomics Standards Initiative (HUPO-PSI), which has tasked working groups with the development of common formats and interchange standards for the description of proteomic data (1). The work of this organization will be described in greater detail in section 2.2.

Proteomics: *Methods Express* (C.D. O'Connor and B.D. Hames, eds)
© Scion Publishing, 2008

2. METHODS AND APPROACHES

2.1 *In silico* characterization of proteins

2.1.1 The use of protein diagnostic signatures

The input of new sequences into the nucleotide sequence databases has increased exponentially over the last few years and from there the designated coding sequences are transferred into protein sequence repositories. Unfortunately, much of this data is generated from large-scale projects and is little more than raw sequence originating from a known organism. Whilst this is a potentially invaluable resource for the proteomics scientist to work with, the lack of accompanying annotation limits its usefulness – it is of little value to find that your peptide of interest has a homolog in the human proteome if that is all that is known of the human protein. For that reason, the need for *in silico* protein sequence tools was recognized, to identify proteins, or regions within proteins, using diagnostic signatures derived from a seed alignment of proteins with high-quality annotation such as that found in UniProtKB/Swiss-Prot (2).

To create protein signatures, an alignment of related protein sequences is used to identify either specific regions of high homology or a consensus for the entire family. Highly conserved regions within a protein family are often closely associated with the protein function, for example catalytic residues or protein- or nucleic acid-binding sites. The simplest signatures use regular expressions to identify patterns of conserved amino acid residues. This core pattern is then analyzed against an increasing number of known proteins and optimized until it hits only the correct sequences in the test set. These patterns identify domains with a high degree of specificity; however, the loss or mutation of a single amino acid residue is enough to cause a pattern to miss a protein sequence. A more complex method, but with a much higher degree of flexibility, is the generation of profiles built from multiple sequence alignments. Profiles consist of tables of position-specific amino acids and gap costs or matrices describing the probability of finding an amino acid at a given position in the sequence. For each set of sequences, a threshold score is calculated to determine whether a protein is a member of that set or not.

Originally derived from profiles, hidden Markov models are essentially a statistics profile based on probabilities rather than scores (3). Profiles and hidden Markov models compensate for the limitations of regular expressions in that they cover larger areas of the sequence and are capable of identifying more divergent family members, but are less useful for pinpointing changes in protein function due to a minor amino acid change, for example identifying inactive enzymes. A number of databases making such protein prediction signatures available now exist, such as Pfam (4), ProSite (5), PIRSF (6), and PANTHER (7).

2.1.2 Protein alignment with known sequences

A complementary method for adding detail to an unknown protein sequence is to find a homolog in a closely related taxonomic species for which structural or functional information is available and transfer the annotation to your protein. The BLAST and FASTA algorithms have been written to compare protein or DNA queries within protein or DNA databases. Such methods are also used by producers of search engines, who attempt to match the peptides identified in high-throughput proteomics data to known proteins and, again, give identity and function to an amino acid sequence. The results of such a search will very much depend upon the quality of data against which it was instituted; searching against a highly redundant database with little or no additional annotation is of limited value.

The quality of the underlying protein sequences may also vary greatly among different protein sequence databases. Although the overwhelming majority of these are sourced from the DDBJ/EMBL/GenBank nucleotide databases (www.insdc.org/), there are differing levels of effort directed at the subsequent checking and, if necessary, correction of the sequence data, and the recognition of splice variants and the transcript products of alternative promoters and alternative initiation sites. The identification of positions of protein modification, such as cleavage sites and phosphorylation, methylation, and acetylation sites, is also vital information to the users of sequence information, as is the accurate mapping of single nucleotide polymorphisms, when known. This level of detail requires manual curation of each individual sequence, and the only database that supplies such a granularity of information is UniProtKB/Swiss-Prot.

However, for many organisms, the complete proteome is not yet represented within the protein sequence databases. Even for those with completely sequenced genomes, a complete proteome relies heavily on gene prediction algorithms and there is no consensus among such programs on either gene number or on the identity and structure of each gene. Additional programs and databases, such as the International Protein Index (8), have had to be established to reconcile these conflicting outputs in higher organisms.

2.1.3 Protein interaction studies

Alignment studies, and the prediction of which family a protein belongs to or what domains it contains, will enable the annotation of many proteins, but there still remain many others with no close analogs and which are not members of any known protein family. For these proteins, other techniques need to be employed to give an indication of their function. The interactions an individual protein makes with other proteins, with nucleic acids, lipids or small molecules, can give information as to the physiological role of the molecule within a cell. Protein interaction data has been reported in the literature for many years and is now being generated in increasingly large volumes. Such data are now being collected in molecular interaction databases, such as IntAct (9), MINT (10), DIP (11), and Impact (MIPS) (12), and are publicly available for searching and data retrieval. All of these databases hold both the results of large high-throughput experiments, which seek to define the 'interactome' of a particular organism derived using techniques such as the yeast two-hybrid system or the pull-down of tagged proteins, and the results of low-throughput interactions, concentrating on a limited number of specific proteins.

If two proteins interact in a physiologically relevant cellular environment and share the same subcellular location during a common period of time within the cell cycle or differentiation stage, it is reasonable to predict that they are involved in the same biological process, although it should be noted that an interaction with a ribosome or proteosome protein may indicate only the biosynthesis or degradation of the protein of interest. By examining the interactions a protein has been shown to make, valuable information may be gained as to its function and this may be a useful additional rich source of protein annotation data that has yet to be fully mined.

2.1.4 Protein expression studies

All protein function, interaction, and pathway data need to studied in the context of the cell type in which the protein has been expressed and state of the cells at the moment at which it is being studied. A protein may possess the same intrinsic enzyme activity wherever it is studied, but the processes it is involved in may vary greatly depending on whether the protein is expressed in a terminally differentiated skin fibroblast or an actively dividing T-cell precursor under antigenic stimulation. Protein expression data have traditionally been taken from Northern or Western blot experiments, in which the expression level of a particular protein is compared across a limited number of cell types or cell lines. With the advent of high-throughput proteomic techniques, expression data in different tissues under different differentiation stage or disease conditions has proliferated in the literature. Until recently, there have been no central repositories in which such data can be stored, accessed, and cross-referenced to protein sequence databases or databases of molecular interaction. The development of such repositories is now ongoing and new possibilities are opening up in the field of protein annotation.

2.2 Data standardization

The establishment of public domain repositories in which data on protein interaction and expression can be stored is a major step forward in the study of protein science, but isolated stand-alone databases are in themselves of limited use. The community benefits greatly if such databases are capable of interchanging data such that a common pool of highly curated information is available from a single source. When data is scattered across a number of separate sources, and each maintains its own individual data format, the user is forced into time-consuming and difficult parsing exercises before they can generate a single nonredundant dataset.

In April 2002, the Protein Standards Initiative (PSI) committee was formed by the Human Proteome Organization (HUPO) (13) and tasked with standardizing data formats within the field of proteomics. Public domain databases could be established or adapted for data deposition using these formats and data could then be exchanged between and downloaded from existing databases using the formats as a common exchange mechanism. In the interests of reducing this task to one of manageable proportions, the PSI first decided to limit their activities to two fields, molecular interactions and mass spectrometry (MS), whilst also working to establish a single data model that would describe and encompass central aspects of a proteomics experiment. The General Proteomics Standards group is now working closely with the Functional Genomics Experiment (FuGE) efforts to define a general standard in which to encode data that will enable a systems biology approach to data analysis, using interchangeable formats and common ontologies across the data model (http://fuge.sourceforge.net/).

The Molecular Interactions (MI) Group published their Level 1 XML-MI interchange schema at the beginning of 2004 (14) and most of the major molecular interaction databases are already offering data in that format. The Level 1 format only allowed the exchange of protein interaction data. Level 2 expanded this to allow all forms of molecular interaction data to be exchanged and Level 2.5, published in 2005, has further refined the schema to allow the description of situations such as the hierarchical build-up of protein complexes (S. Kerrien et al., unpublished). The success of PSI-MI led to an agreement between five founder public domain databases (BIND (15), DIP, IntAct, MINT, and MIPS) to share curation effort and exchange data on a nightly basis, a model analogous to the nucleotide database collaboration. Common curation standards have been agreed, and data exchange is initiated at the protein–protein interaction level (http://imex.sf.net).

The PSI-MS mzData interchange format was written to allow both the exchange of experimental data from proteomics experiments involving MS and also, with co-operation from instrumentation and search engine manufacturers, to enable researchers to generate data in such a standard directly from their instrumentation. These data written by the vendor data system are usable directly by search engines, as well as third-party software tools such as spectral databases and other computational tools. Adoption of the standard by manufacturers was widespread, for example the standard is now supported by Mascot (Matrix Science), Proteome Systems Ltd, Proteios (Lund University), Phenix (GeneBio), X! Tandem, and GPM (Global Proteome Machine Organization). A number of accompanying tools have also been made available, for example mzDataConverter, which allows the user to both convert from MS text formats to PSI-MS XML format, and mzDataViewer, which can be used to view and browse stored data in PSI-MS XML format. However, a second interchange format was published at much the same time by the Institute for Systems Biology (16), causing some confusion as to which was the most appropriate to use. It was agreed that the community would be best served by the merger of these two XML formats into a single agreed standard, mzML, which will be published in 2007.

Ongoing work will broaden the specification to allow a full description of acquisition and to encompass both mass array and mass intensity. mzData will soon be accompanied by a spectral analysis output format, supporting a common syntax for peptide/protein identification and for protein modification description (analysisXML). The analysisXML standard has been designed to capture results from MS search engines and represent the input parameters for analysis algorithms, thus unifying results from different search engines. The requirements for analysisXML include the need to support the identification of both proteins and peptides, by accession number or sequence, and must include the ability to describe modifications. Small-molecule identification by either CML or SMILES must also be supported. The rapid implementation of such standards by databases such as PRIDE, which is already mzData-compatible, and Peptide Atlas will again make data exchange and common curation possible.

2.3 Methodology

2.3.1 InterPro Protein Signature Database

A number of different databases now exist that have developed protein signatures for protein families and domains, each of which has its own focus, criteria, and method for creating the signature. Consequently, each has its own strengths and limitations. InterPro was created by a consortium of these databases and acts as a single comprehensive resource that combines the strengths of all of the member databases and, by combining the individual foci of the contributing sources, eliminates many weaknesses in the areas of limited taxonomic coverage. InterPro has been used by bench biologists analyzing a single protein and by sequencing groups annotating an entire genome to add biological detail to a stretch of unknown sequence (17).

Within InterPro, signatures from the member databases, PROSITE, Prints (18), Pfam, ProDom (19), SMART (20), TIGRFAm (21), PIRSF, PANTHER, SUPERFAMILY (22), and Gene3D (23), that describe the same family, domain, repeat, active site, binding site, or post-translational modification are grouped into a single InterPro entry. Signatures that describe a related subset in that entry are grouped into a second InterPro entry linked to the first through a parent/child relationship. Protein families maintain a 'contains/found' in relationship to domains, regions, and sites. Each InterPro entry contains high-quality manual annotation containing information on the protein family, domain, etc. Mapping of InterPro entries to gene ontology (GO) terms, where possible, supplies further functional annotation. Each InterPro entry contains a list of pre-computed matches to UniProtKB, which can be accessed via several different tabular and graphical views.

InterProScan is a tool that combines the protein function recognition methods of the member databases of InterPro into one application, allowing the user to enter one or more protein or nucleic acid sequences and receive back a full analysis of the protein's potential function (24). InterProScan is available at the European Bioinformatics Institute as a web browser (http://www.ebi.ac.uk/InterProScan). In addition, a standalone Perl version and a SOAP web service are also available to the users. *Protocol 1* describes the use of this facility on a hypothetical sample sequence.

Protocol 1

Annotation of the sequence of an unknown protein

A sequence analysis has produced the following translation of a nucleotide sequence:

MEEPQSDPSVEPPLSQETFSDLWKLLPENNVLSPLPSQAMDDLMLSPDDIEQWFTEDPGPDEAPRMPEA
APPVAPAPAAPTPAAPAPAPSWPLSSSVPSQKTYQGSYGFRLGFLHSGTAKSVTCTYSPALNKMFCQLAKT
CPVQLWVDSTPPPGTRVRAMAIYKQSQHMTEVVRRCPHHERCSDSDGLAPPQHLIRVEGNLRVEYLDDR
NTFRHSVVVPYEPPEVGSDCTTIHYNYMCNSSCMGGMNRRPILTIITLEDSSGNLLGRNSFEVRVCACPGR
DRRTEEENLRKKGEPHHELPPGSTKRALPNNTSSSPQPKKKPLDGEYFTLQIRGRERFEMFRELNEALELKD
AQAGKEPGGSRAHSSHLKSKKGQSTSRHKKLMFKTEGPDSD

To perform an *in silico* analysis:
1. Access InterProScan (www.ebi.ac.uk/InterProScan/).
2. Paste the sequence into the given box in any format. (If a nucleotide sequence has been pasted into the box, the user may also specify the length of a minimum open reading frame for analysis.)
3. Turn any of the applications on or off, as required.
4. Add an appropriate e-mail address to which a copy of the results may be sent.
5. Press 'Submit Job'.

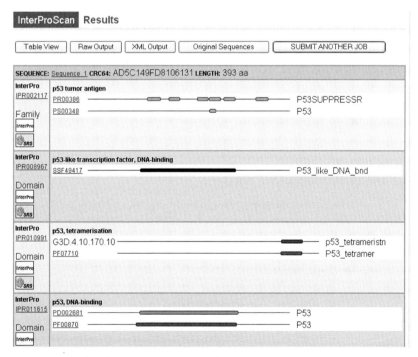

Figure 1. Example of an InterProScan result.
The InterProScan result on the unknown sequence analyzed in *Protocol 1*, identifying the protein as a member of the p53 tumor antigen family, containing both DNA-binding and tetramization domains.

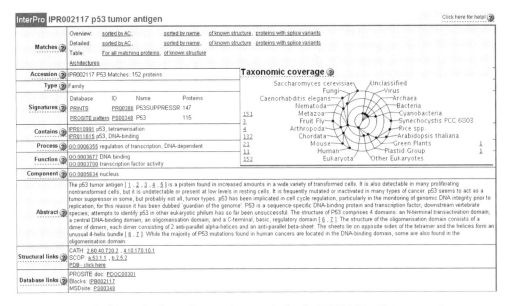

Figure 2. View of additional information on the protein family IPR02117 p53 tumor antigen.

The results of analyzing the protein sequence submitted in *Protocol 1* are shown in *Fig. 1*. As can be seen, the protein is hit by several diagnostic signatures from several of the InterPro member databases, suggesting that the protein is a member of the p53 family of tumor antigens. Using the hyperlink to the InterPro database entry for IPR002117 gives the reader additional information on the protein family and the functional domains contained within it. GO annotation (25) is appended to the entry and cross references to additional information, for example links to proteins on which structural studies have been performed (see *Fig. 2*). The taxonomic wheel allows the user to view the range of species in which this particular family or domain has been identified and access a set of species-specific entries, if required.

2.3.2 The UniProt sequence database

The UniProt database (www.uniprot.org) was created in 2002 from a union of the previously existing Swiss-Prot, TrEMBL, and PIR databases (2). The UniProt database consists of three database layers:

- The UniProt Archive (UniParc) provides a stable, comprehensive sequence collection without redundant sequences by storing the complete body of publicly available protein sequence data.
- The UniProt Knowledgebase (UniProtKB) is the central database of protein sequences with accurate, consistent, and rich sequence and functional annotation.
- The UniProt Reference Clusters (UniRef) databases provide nonredundant reference data collections based on UniProtKB in order to obtain complete coverage of sequence space at several resolutions.

UniProtKB is further divided into two sections – UniProtKB/Swiss-Prot and UniProtKB/TrEMBL. UniProtKB/Swiss-Prot is nonredundant, in that all reports for a given protein in a specific species are merged into a single-entry, manually annotated database. When an entry is entered into UniProtKB/Swiss-Prot, the sequence is checked and variants identified, as described above. The protein is given both a systematic protein and gene name and all known synonyms are recorded. Taxonomic data and citation information are checked and amended, if necessary. The database also strives to provide a high level of annotation through literature-based manual annotation. This includes descriptions of the properties of each individual protein such as function, post-translational modifications, domain and sites, secondary and quaternary structure, diseases associated with deficiency of that protein, tissues and developmental stage in which the protein is expressed, and pathways in which it is involved. Experimental evidence is supplemented by the results of sequence-analysis programs, such as those contained within InterProScan, and by sequence similarity comparison, which allows experimental evidence from species in which protein function has been experimentally verified to those closely related species in which the corresponding study has not been performed.

The production of a manually curated UniProtKB/Swiss-Prot entry is a highly labor-intensive process, and, as such, is the rate-limiting step in the growth of the database. To supplement this, and to address the ever-increasing amount of sequence information flowing from the nucleotide databases, UniProtKB/TrEMBL allows the automatic annotation of sequences prior to manual annotation into UniProtKB/Swiss-Prot. UniProtKB/TrEMBL consists of computer-annotated entries derived from the translation of all coding sequences in the nucleotide sequence databases, except for coding sequences already included in UniProtKB/Swiss-Prot. It also contains those protein sequences extracted from the literature or submitted directly by the user community that are not directly entered in UniProtKB/Swiss-Prot. UniProtKB/TrEMBL has a certain degree of sequence redundancy, namely that a single gene from an individual species may be represented by more than one entry. The UniProtKB/TrEMBL data content is enhanced by extensive automatic annotation procedures, based on a system of standardized transfer of annotation from well-characterized proteins in UniProtKB/Swiss-Prot to unannotated UniProtKB/TrEMBL entries belonging to defined groups. The identification of these groups is performed using the diagnostic signatures within InterPro.

UniProtKB may be regarded as a central hub of information, with the basic information on the protein, particularly with regard to sequence, given within the entry and cross-links to more than 90 data resources that extend out to many additional sources to expand the information summarized in the source record. These data sources include model organism, protein classification, structural, disease, molecular interaction, and pathway databases. Extensive referencing within a UniProtKB entry gives the user access to additional information on their protein.

2.3.3 The International Protein Index

As previously discussed, no one database currently covers the entire proteome content of the higher species, all of which depend heavily on prediction programs to describe all potential gene products of a genome. The International Protein

Index (IPI, www.ebi.ac.uk/ipi) attempts a semantic resolution of the different data in the various protein sequence and gene prediction databases, and its aim is to provide maximum coverage with minimal redundancy, stable identifiers to allow the tracking of sequences between releases, and inter-database cross-references (8). Each IPI entry represents a cluster of entries from the source databases believed to represent the same protein.

IPI merges the experimentally determined protein sequences held in the UniProtKB sequence database with the protein predictions of Ensembl (26) and both protein predictions and experimentally derived datasets provided by RefSeq (27), along with a number of species-specific datasets such as TAIR (The *Arabidopsis* Information Resource) (28), to provide a minimally redundant yet maximally complete set of human, mouse, rat, *Arabidopsis*, chicken, cow, and zebrafish proteins consisting of one sequence per transcript. All annotated splice variants are included in IPI as separate entries (unless their protein sequences are identical). IPI is produced automatically by mapping among the different datasets on the basis of protein similarity and maintains cross-references among the primary data sources.

IPI is updated monthly but maintains stable identifiers (with incremental versioning) to allow the tracking of sequences in IPI between IPI releases. When proteins disappear from source databases and a corresponding sequence cannot be identified, IPI identifiers are archived and can be traced by researchers who used the identifier in a particular dataset. Similarly, if two IPI entries are merged as a result of changing data within the source databases, a secondary identifier will be maintained to allow the original entry to be traced.

```
ID   IPI00025087.1         IPI;    PRT;    393 AA.
AC   IPI00025087; IPI00004351; IPI00030482; IPI00033269; IPI00014940;
AC   IPI00040121; IPI00034233; IPI00037402;
DT   01-OCT-2001 (IPI Human rel. 2.00, Created)
DT   01-OCT-2001 (IPI Human rel. 2.00, Last sequence update)
DE   SPLICE ISOFORM 1 OF CELLULAR TUMOR ANTIGEN P53.
OS   Homo sapiens (Human).
OC   Eukaryota; Metazoa; Chordata; Craniata; Vertebrata; Euteleostomi;
OC   Mammalia; Eutheria; Primates; Catarrhini; Hominidae; Homo.
OX   NCBI_TaxID=9606;
CC   -!- CHROMOSOME: 17.
CC   -!- START CO-ORDINATE: 7512464.
CC   -!- END CO-ORDINATE: 7531642.
DR   UniProtKB/Swiss-Prot; P04637-1; P53_HUMAN; M.
DR   UniProtKB/TrEMBL; Q3LRW3; Q3LRW3_HUMAN; -.
DR   UniProtKB/TrEMBL; Q3LRW4; Q3LRW4_HUMAN; -.
DR   UniProtKB/TrEMBL; Q5UOE4; Q5UOE4_HUMAN; -.
DR   REFSEQ_NP; NP_000537; GI:8400738; -.
DR   ENSEMBL; ENSP00000269305; ENSG00000141510; -.
DR   H-InvDB; HIT000031628; HIX0013510; -.
DR   UniParc; UPI000002ED67; -; -.
DR   HGNC; 11998; TP53; -.
DR   Entrez Gene; 7157; TP53; -.
DR   UniGene; Hs.408312; -; -.
DR   CCDS; CCDS11118.1; -; -.
DR   Transfac; G001075; -; -.
DR   EPD; EP11223; HS_TP53; Anti-oncogene.
DR   UTRdb; BB007450; U0015; 5'UTR.
DR   UTRdb; BB009158; U0015; 5'UTR.
DR   UTRdb; BB009159; U0015; 5'UTR.
DR   UTRdb; BB009675; -; 5'UTR.
DR   UTRdb; BB071175; -; 5'UTR.
DR   UTRdb; BB071176; -; 5'UTR.
```

Figure 3. Flat file view of the IPI entry IPI00025087.

As an example of the use of the IPI sequence database, the sequence given in *Protocol 1* is known to be human, so we may choose to match the sequence to the IPI sequence database to give maximum, nonredundant coverage of the human proteome. This procedure is described in *Protocol 2*.

Protocol 2

Accessing information on an unknown protein sequence using sequence similarity searching

1. A full toolbox of search engines can be accessed at www.ebi.ac.uk/services. For this example, a FASTA search will be performed.

2. Select FASTA, then select FASTA-protein in the Tools list, and on the query form, choose 'International Protein Index' from the list of available protein databases.

3. Select the version of FASTA you wish to run (in this example FASTA3 was selected) and paste in the sequence.

4. Click on 'Run Fasta3' to initiate the analysis. The sequence gives an exact match over its entire length to IPI00025087 splice isoform 1 of cellular tumor antigen p53 (see *Fig. 3* for flat file view). The entry gives a cross-reference to a UniProtKB/Swiss-Prot entry, giving direct access to a fully annotated record for this protein. The user may either access the entry directly via a hyperlink in the IPI entry or open UniProtKB (www.ebi.ac.uk/uniprot) and either type in the accession number (P04637) or the gene name 'TP53' and species 'human (or *Homo sapiens*)'. *Fig. 4* shows part of the detailed comments section for the human p53 entry. Examination of the full UniProt entry will give details of the function, co-factor requirements, the protein–protein interaction made by p53, and a detailed mapping of post-translational modifications, variants caused by single nucleotide polymorphisms, and splice variants to the given sequence.

SUBCELLULAR LOCATION	Cytoplasmic and nuclear.	
ALTERNATIVE PRODUCTS	Alternative splicing;2 named isoforms [Display all isoform sequences in Fasta format]	
	Name	1
	IsoformId	P04637-1
	Sequence	This is the isoform sequence displayed in this entry.
	Name	2
	Synonyms	I9RET
	IsoformId	P04637-2
	Sequence	VSP_006535, VSP_006536
	Note	Seems to be non-functional. Expressed in quiescent lymphocytes
DOMAIN	The nuclear export signal acts as a transcriptional repression domain.	
PTM	Acetylated. Acetylation of Lys-382 by CREBBP enhances transcriptional activity. Deacetylation of Lys-382 by SIRT1 impairs its ability to induce proapoptotic program and modulate cell senescence.	
PTM	Phosphorylated. Phosphorylation on Ser residues mediates transcriptional activation. Phosphorylated by HIPK1 (By similarity). Phosphorylated on Thr-18 by VRK1, which may prevent the interaction with MDM2. Phosphorylated on Thr-55 by TAF1, which promotes MDM2-mediated degradation. Phosphorylated on Ser-46 by HIPK2 upon UV irradiation. Phosphorylation on Ser-46 is required for acetylation by CREBBP.	
PTM	Dephosphorylated by PP2A. SV40 small T antigen inhibits the dephosphorylation by the AC form of PP2A.	
PTM	May be O-glycosylated in the C-terminal basic region. Studied in EB-1 cell line.	
DISEASE	TP53 is found in increased amounts in a wide variety of transformed cells. TP53 is frequently mutated or inactivated in about 60% of cancers.	
DISEASE	Defects in TP53 are involved in esophageal squamous cell carcinoma (ESCC) [MIM:133239]. ESCC is a tumor of the esophagus.	
DISEASE	Defects in TP53 are a cause of Li-Fraumeni syndrome (LFS) [MIM:151623]. LFS is an autosomal dominant familial cancer syndrome that in its classic form is defined by the existence of both a proband with a sarcoma and two other first-degree relatives with a cancer by age 45 years. In these families the affected relatives develop a diverse set of malignancies at unusually early ages. The spectrum of cancers in LFS includes breast carcinomas, soft-tissue sarcomas, brain tumors, osteosarcoma, leukemia and adreno-cortical carcinoma. Other possible component tumors of LFS are melanoma, gonadal cell tumors and carcinomas of the lung, pancreas and prostate.	
DISEASE	Defects in TP53 are found in Barrett metaplasia; also known as Barrett esophagus. It is a condition in which the normally stratified squamous epithelium of the lower esophagus is replaced by a metaplastic columnar epithelium. The condition develops as a complication in approximately 10% of patients with chronic gastroesophageal reflux disease and predisposes to the development of esophageal adenocarcinoma.	

Figure 4. A portion of the comments section of the UniProtKB/Swiss-Prot entry for human p53.

2.3.4 IntAct molecular interaction database

The IntAct molecular interaction (www.ebi.ac.uk/intact) database has been developed for the deposition and storage of protein–protein, protein–nucleic acid, and protein–small molecule information (9). IntAct also supports a team dedicated to the curation of experimental data already available in the literature. This effort is currently strongly biased towards the collection of protein–protein interaction data.

The IntAct data model has three main components: Experiment, Interaction, and Interactor. An Experiment groups a number of Interactions, usually from one publication, and classifies the experimental conditions in which these Interactions have been generated. An Experiment may have only a single interaction, or hundreds of interactions in the case of large-scale experiments. An Interactor is a biological entity participating in an Interaction, usually a protein, a nucleic acid sequence, or a small molecule. An Interaction contains one or more Interactors participating in the Interaction. The representation of interactions is not limited to binary interactions; data on multi-protein interactions (e.g. the results of tandem affinity purification experiments) can be represented as one interaction without artificially splitting them up into several binary interactions.

Molecule identifications, molecular binding features, and experimental techniques all need to be described in an unambiguous manner such that both the laboratory scientist and a computer program can search and access the data, understand the results, and be able to compare this with locally derived data or with data downloaded from other sources. IntAct achieves this by using extensive cross-referencing to key source databases and by the use of controlled vocabularies to describe any area of the database available to search.

Protein identification is dealt with by using the stable protein accession numbers generated by the UniProtKB database (2). In the rare cases where a protein is described in the absence of sequence information, for example a mouse antibody is used in a rat cell line but the rat protein has yet to be sequenced, a protein entry can be generated within IntAct until such time as the sequence becomes publicly available in UniProtKB. Similarly, nucleic acid identifiers will be annotated by DDBJ/EMBL/GenBank identifiers and small molecules using Chemical Entities of Biological Interest (ChEBI) descriptors and accession numbers (www.ebi.ac.uk/chebi). Interacting domains are identified by cross-referencing to InterPro, with residue numbering directly linked to that of the given sequence used by the corresponding UniProtKB entry, and residue renumbering is undertaken in line with sequence updates in the UniProtKB database.

All experimental detail and additional molecule feature information is described, wherever possible, by the use of controlled vocabularies. The use of ontologies and vocabularies is proving essential as biological data and the corresponding terms to describe such data proliferates. A single biological process can now be described by many different terms and their synonyms, which compromises the ability of search engines to identify complete data sets. For example, to the human eye, the phrases 'yeast-two-hybrid', 'yeast two hybrid' and 'Y2H' all describe the same process. To a text-mining program, they are very different and the program would require special instruction to group such data into a single set. By using a single term throughout the database to describe this experimental technique, and then making that term known to the user, data can be searched, sorted and filtered without suffering data loss due to poor use of terminology. Where possible, existing reference systems are used, such as the NCBI taxonomy database (26) to delineate taxonomy or GO (25) to describe complex function, involvement in biological processes, and subcellular location.

For a number of attributes specific to molecular interactions, new controlled vocabularies have been developed in the IntAct project, together with extensive definitions and cross-references. The vocabularies are used to describe many aspects of the experimental procedures and interaction details, for example 'participant type', 'interaction type', 'interaction detection method', 'participant detection method', 'feature type', and 'feature detection method'. In specific cases, these vocabularies are closely linked to, and cross reference, other specialist databases. For example, the post-translational modifications described in 'feature type' are all cross-referenced to the RESID database where such modifications are described in fuller detail (29). All of these controlled vocabularies have been made publicly available and have been incorporated into the PSI-MI standardization efforts.

The use of the IntAct molecular interaction database can be described by continuing the analysis of a protein sequence begun in *Protocol 1*. *Protocol 2*, using the IPI sequence database, has identified the IPI00025087 slice isoform I of human p53. *Protocol 3* shows how we can now use the IntAct database to examine molecular interactions made by this protein.

Protocol 3

Adding additional annotation by examining molecular interactions made by human p53

1. Open the molecular interaction database IntAct (www.ebi.ac.uk/intact) and enter either the accession number P04637 or the gene name TP53; as the latter is not species specific, this will result in a choice of entries being given and a further selection will have to be made.

2. The resulting table (see *Fig. 5*) gives a list of all proteins in the database that have been shown to interact with human p53, along with the number of separate experiments that have confirmed this observation.

3. A hyperlink from this number allows the viewer to see each set of fully annotated experimental detail.

4. Additional functionality allows a graphical representation of the data and the user may then highlight specific GO terms or InterPro domains within the proteins identified in a common cluster to categorize those proteins that share a common functionality, and potentially to add annotation to unknown sequences.

2.3.5 PRIDE: a proteomics identifications database

The PRIDE PRoteomics IDEntifications database (www.ebi.ac.uk/pride) is a centralized, standards-compliant, public data repository for proteomics data. It has been

p53_human	EBI-366083	94	P04637	TP53	Cellular tumor antigen p53
interacts with					
hd_human	EBI-466029	4	P42858	HD	Huntingtin
leu1_human	EBI-710057	1	O43261	DLEU1	Leukemia associated protein 1
btbd2_human	EBI-710091	2	Q9BX70	BTBD2	BTB/POZ domain containing protein 2
rsmn_human	EBI-712493	1	P63162	SNRPN	Small nuclear ribonucleoprotein associated protein N
pcda4_human	EBI-712273	1	Q9UN74	PCDHA4	Protocadherin alpha 4 precursor
gnl3_human	EBI-641642	1	Q9BVP2	GNL3	Guanine nucleotide binding protein-like 3
atf3_human	EBI-712767	1	P18847	ATF3	Cyclic AMP-dependent transcription factor ATF-3
q99786_human	EBI-711501	1	Q99786	ZNF581	Hypothetical protein
anxa3_human	EBI-712677	1	P12429	ANXA3	Annexin A3
dvl2_human	EBI-740850	1	O14641	DVL2	Segment polarity protein dishevelled homolog DVL-2
kith_human	EBI-712550	1	P04183	TK1	Thymidine kinase, cytosolic
p06748-2	EBI-354154	2	P06748-2	-	Nucleophosmin
hipk1_human	EBI-692891	2	Q86Z02	HIPK1	Homeodomain-interacting protein kinase 1
pa1b3_human	EBI-711522	1	Q15102	PAFAH1B3	Platelet-activating factor acetylhydrolase IB gamma subunit
cabl1_human	EBI-604615	1	Q8TDN4	CABLES1	CDK5 and ABL1 enzyme substrate 1
gstm4_human	EBI-713363	1	Q03013	GSTM4	Glutathione S-transferase Mu 4
znf24_human	EBI-707773	1	P17028	ZNF24	Zinc finger protein 24
cdn2c_human	EBI-711290	1	P42773	CDKN2C	Cyclin-dependent kinase 6 inhibitor

Figure 5. Table view of IntAct listing all proteins that have been shown to interact with human p53 tumor antigen.

developed to provide the proteomics community with a public repository for protein and peptide identification, together with the evidence supporting these identifications (30). The PRIDE project consists of a number of distinct parts. The XML format represents the basic data structure, whereas the relational database implementation is just one of the possible renderings of the hierarchical XML format in a relational schema. The PRIDE core libraries contain an object model of the PRIDE data structure and allow the programmer to interact seamlessly and effortlessly with the PRIDE XML format and reference database implementation. The PRIDE web libraries provide a web-based view on an underlying reference database and use the PRIDE core libraries for data access. Query results from the web can be sent in PRIDE XML format or in HTML after XML Stylesheet Language (XSL) transformation of the XML.

In PRIDE, one or more experiments are contained in the root tag 'ExperimentCollection'. The top-level structure of an experiment consists of seven conceptually distinct parts:

1. The experiment accession number, which is assigned after successful submission of experiment-associated data. The experiment accession number would be the data element of choice for inclusion in papers as the PRIDE reference because of its conciseness and because interested readers can use it easily for quick retrieval of all relevant data from the PRIDE web interface.
2. Meta-data about the experiment: a descriptive title, contact person and/or address, a short label, a description, and location information.
3. A description field and an attribute list of the sample being studied.

4. Protocol information: as well as a description and attribute list, this also holds one or more sections about the mass spectrometer(s) used, including manufacturer, model, source, and analyzer information, which can be supplemented further through an attribute list.
5. Details of information derived from the mass spectrometer. This section holds the MS coefficient (e.g. MS^2, MS^3), peak lists, optional raw data references, comments, and an attribute list.
6. The most intricate subsection of an experiment, dealing with the identifications obtained from the data specified in part 5. Identifications have been split into two different types: two-dimensional polyacrylamide gel electrophoresis (2D-PAGE)-based identifications and non-gel-based identifications. The shared elements are wrapped up in an abstract ancestor element called 'IdentificationType'. The additional information for 2D-PAGE-based identifications centers on protein-related data gathered during the gel separation phase, whereas the gel-free identifications typically require more information about the effective identification score and threshold (if available).
7. This part is not restricted to the experiment level but can be found in many of the smaller branches as well. This is the 'AttributeList', which represents a list of attributes, to be keyed from controlled vocabularies, allowing an extremely flexible way of integrating additional information into the core schema without sacrificing the structure of the whole. In fact, the PRIDE schema presents a minimum minimorum of information about protein identifications in present day proteomics.

Continuing with our analysis of human p53 in *Protocol 3*, we can use the PRIDE database to examine further the tissue specificity of this protein.

Protocol 4

Further examination of the tissue specificity of human p53 protein

1. Open the PRIDE PRoteomics IDEntifications database (www.ebi.ac.uk/pride) and search for TP53 by typing its IPI accession number (IPI00025087) into the Quick Search box and pressing 'Search'.
2. The search page gives a choice of result format. Many of these datasets are very large and attempting to view a full set of protein identifications alongside identifying spectra may not be a practical selection.
3. To view the example given in *Fig. 6*, select experimental Accession number 74 and press 'Download' to access the details. It is now possible to see a detailed experimental protocol, including sample preparation, experimental detail incorporating details of both the instrumentation and search engine used to identify the peptides, and full referencing to the experiment in the literature.
4. Additional information in the record is given by cross-referencing to external databases, the entries for which are hyperlinked within the web browser.

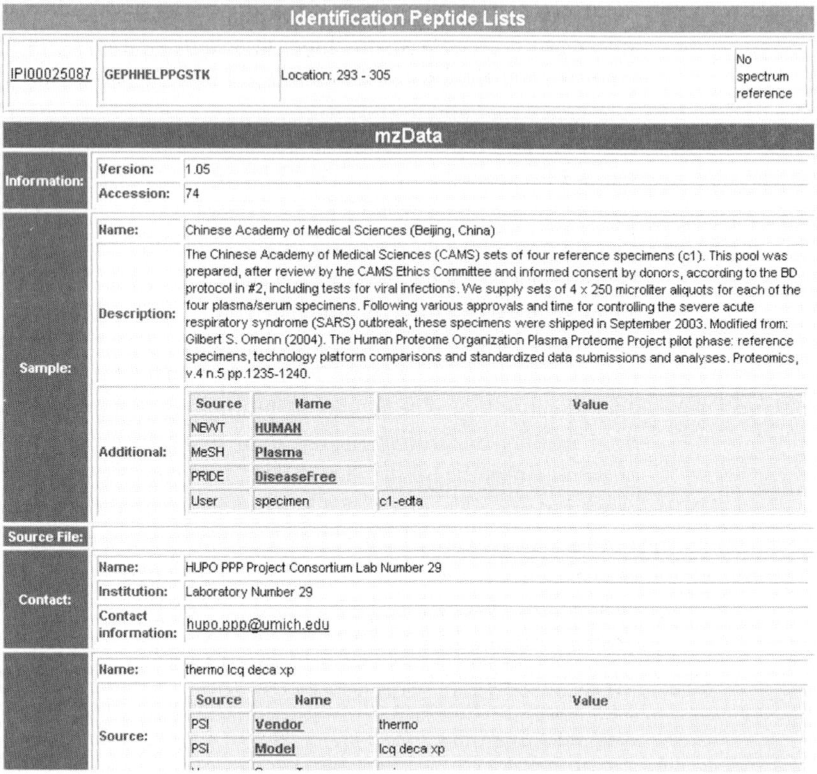

Figure 6. Snapshot of some of the experimental information held in the PRIDE database showing expression of human p53 tumor antigen in healthy human plasma.

3. TROUBLESHOOTING

- If your dataset is too large for the use of web-based tools to be practical, or you wish to manage the dataset in house without publishing it in the public domain, many of the tools and databases described in this chapter are completely open source and may be installed locally. Instructions on how to do so may be found on the application web site.
- Sequence accession numbers ideally should be versioned and archived such that the user may still access their original data, even if the sequence has since been updated or withdrawn. However, many users make the mistake of storing information by gene name, which may change with time, or using an identification number, which is lost should a sequence be withdrawn. In these cases, should the user have retained the sequence, it is possible to update the required information by performing an alignment against UniParc, the UniProt sequence archive, a nonredundant archive of protein sequences extracted from public databases. UniParc proteins are linked to their source

databases by database cross-references. Each cross-reference links one protein in UniParc to an accession number in a source database. The database cross-reference is active as long as the sequence identified by the source accession number remains unchanged. When the sequence is modified or removed in the source database, the cross-reference from UniParc becomes inactive. Active cross-references can be used directly to access the source databases but inactive cross-references can only be used to access sequence archives.
- Should there be problems with any of the databases described in this chapter, the help desks associated with the databases are extremely experienced at providing user support and should be contacted without hesitation.

4. REFERENCES

1. Orchard S, Taylor CF, Jones P, et al. (2007) Proteomics, 7, 337-339.
★★ 2. UniProt (2007) Nucleic Acids Res. 35, D193-197. – UniProt is the world's most comprehensive catalog of information on proteins.
3. Eddy SR (2004) Nat. Biotechnol. 22, 1315-1316.
4. Finn RD, Mistry J, Schuster-Bockler B, et al. (2006) Nucleic Acids Res. 34, D247-D251.
5. Hulo N, Bairoch A, Bulliard V, et al. (2006) Nucleic Acids Res. 34, D227-D230.
6. Wu C & Nebert DW (2004) Hum. Genomics, 1, 229-233.
7. Mi H, Lazareva-Ulitsky B, Loo R, et al. (2005) Nucleic Acids Res. 33, 284-288.
8. Kersey PJ, Duarte J, Williams A, Karavidopoulou Y, Birney E & Apweiler R (2004) Proteomics, 4, 1985-1988.
9. Kerrien S, Alam-Faruque Y, Aranda B, et al. (2007) Nucleic Acids Res. 35, D561-D565.
10. Zanzoni A, Montecchi-Palazzi L, Quondam M, Ausiello G, Helmer-Citterich M & Cesareni G (2002) FEBS Lett. 513, 135-140.
11. Salwinski L, Miller CS, Smith AJ, Pettit FK, Bowie JU & Eisenberg D (2004) Nucleic Acids Res. 32, 449-451.
12. Pagel P, Kovac S, Oesterheld M, et al. (2005) Bioinformatics, 21, 832-834.
13. Orchard S, Kersey P, Hermjakob H & Apweiler R (2003) Comp. Funct. Genomics, 4, 16-17.
★★ 14. Hermjakob H, Montecchi-Palazzi L, Bader G, et al. (2004) Nat. Biotechnol. 22, 177-183. – The HUPO-PSI molecular interaction XML interchange format.
15. Alfarano C, Andrade CE, Anthony K, et al. (2005) Nucleic Acids Res. 33, 418-424.
16. Pedrioli PG, Eng JK, Hubley R, et al. (2004) Nat. Biotechnol. 22, 1459-1466.
17. Mulder NJ, Apweiler R, Attwood TK, et al. (2005) Nucleic Acids Res. 33, 201-205.
18. Attwood TK, Bradley P, Flower DR, et al. (2003) Nucleic Acids Res. 31, 400-402.
19. Bru C, Courcelle E, Carrere S, Beausse Y, Dalmar S & Kahn D (2005) Nucleic Acids Res. 33, 212-215
20. Letunic I, Copley RR, Pils B, Pinkert S, Schultz J & Bork P (2006) Nucleic Acids Res. 34, D257-D260.
21. Haft DH, Selengut JD & White O (2003) Nucleic Acids Res. 31, 371-373.
22. Gough J, Karplus K, Hughey R & Chothia C (2001) J. Mol. Biol. 313, 903-919.
23. Yeats C, Maibaum M, Marsden R, et al. (2006) Nucleic Acids Res. 34, 281-284
24. Quevillon E, Silventoinen V, Pillai S, et al. (2005) Nucleic Acids Res. 33, 116-120.
25. Harris MA, Clark J, Ireland A, et al. (2004) Nucleic Acids Res. 32, 258-261.
26. Birney E, Andrews D, Caccamo M, et al. (2006) Nucleic Acids Res. 34, 556-561.
27. Wheeler DL, Barrett T, Benson DA, et al. (2006) Nucleic Acids Res. 34, 173-180.
28. Rhee SY, Beavis W, Berardini TZ, et al. (2003) Nucleic Acids Res. 31, 224-228.
29. Garavelli JS (2004) Proteomics, 4, 1527-1533.
30. Jones P, Cote RG, Martens L, et al. (2006) Nucleic Acids Res. 34, 659-663.

APPENDIX 1
List of suppliers

ABgene – www.abgene.com
Accurate Chemical and Scientific Corporation – www.accuratechemical.com
Agilent Technologies – www.agilent.com
Alexis Corporation – www.alexis-corp.com
Amersham Biosciences – www.amershambiosciences.com
Anachem Ltd – www.anachem.co.uk
Appleton Woods Ltd – www.appletonwoods.co.uk
Applied Biosystems – www.appliedbiosystems.com
ARC Laboratories – www.isotope.nl
AutoGen, Inc. – www.autogen.com
Axon Instruments – www.axon.com

Beckman Coulter, Inc. – www.beckman.com
Becton, Dickinson and Company – www.bd.com
Bio-Rad Laboratories, Inc. – www.bio-rad.com
BOC Group – www.boc.com
Brosch direct Ltd – www.broschdirect.com

Calbiochem – www.calbiochemicom
Cambrex Bio Science – www.cambrex.com
Cambridge Scientific Products – www.cambridgescientific.com
Carl Zeiss – www.zeiss.com
Chemicon International, Inc. – www.chemicon.com
Corning, Inc. – www.corning.com

DakoCytomation – www.dakocytomation.com
Difco Laboratories – www.difco.com
Dionex Corporation – www.dionex.com
DuPont – www.dupont.com

Elliot Scientific Ltd – www.elliotscientific.com
EMD Biosciences Inc. – www.emdbiosciences.com
Ethicon Inc. – www.ethicon.com
European Collection of Animal Cell Culture – www.ecacc.org.uk

Findel Education Ltd – www.fipd.co.uk
Fisher Scientific International – www.fishersci.com
Fluka – www.sigma-aldrich.com
Fluorochem – www.fluorochem.co.uk

Gibco BRL – www.invitrogen.com
Goodfellow Cambridge Ltd – www.goodfellow.com
Greiner Bio-One – www.gbo.com

Harlan – www.harlan.com
Hybaid – www.hybaid.com
HyClone Laboratories – www.hyclone.com

ICN Biomedicals, Inc. – www.icnbiomed.com
Insight Biotechnology – www.insightbio.com
Invitrogen Corporation – www.invitrogen.com

Jencons-PLS – www.jencons.co.uk

Kendro Laboratory Products – www.kendro.com
Kodak: Eastman Fine Chemicals – www.eastman.com

Lab-Plant Ltd – www.labplant.com
Lancaster – www.lancastersynthesis.com
Leica – www.leica.com
Life Technologies Inc. – www.lifetech.com
LOT-Oriel – www.lot-oriel.com

Mallinckrodt Baker – www.malbaker.com
Merck, Sharp and Dohme – www.msd.com
MetaChem – www.metachem.com
Millipore Corporation – www.millipore.com
Miltenyi Biotec – www.miltenyibiotec.com
MWG Biotech – www.mwg-biotech.com

National Diagnostics – www.nationaldiagnostics.com
New England BioLabs, Inc. – www.neb.com
Nikon Corporation – www.nikon.com

Olympus Corporation – www.olympus-global.com
Optivision Ltd – optivision.co.uk

Perbio Science – www.perbio.com
PerkinElmer, Inc. – www.perkinelmer.com
Pharmacia Biotech Europe – www.biochrom.co.uk
Photonic Solutions plc – www.psplc.com

Polysciences Inc. – www.polysciences.com
Promega Corporation – www.promega.com

Qiagen N.V. – www.qiagen.com

R&D Systems – www.rndsystems.com
Rathburn Chemical Ltd – www.rathburn.co.uk
Roche Diagnostics Corporation – www.roche-applied-science.com

Sanyo Gallenkamp – www.sanyogallenkamp.com
Sarstedt – www.sarstedt.com
Schleicher and Schuell Bioscience, Inc. – www.schleicher-schuell.com
Scientifica – www.scientifica.uk.com
Sefar America Inc. – www.afssociety.org
Serotec – www.serotec.com
Shandon Scientific Ltd – www.shandon.com
Sigma-Aldrich Company Ltd – www.sigma-aldrich.com
Sorvall – www.sorvall.com
Stratagene Corporation – www.stratagene.com

Thames Restek – www.thamesrestek.co.uk
Thermo Electron Corporation – www.thermo.com
Thistle Scientific – www.thistlescientific.co.uk
Thomas Scientific – www.thomassci.com

Upstate – www.upstate.com

Vector Laboratories – www.vectorlabs.com
Vedco Inc. – http://vedco.com
VWR International Ltd – www.bdh.com

Wolf Laboratories – www.wolflabs.co.uk
Wyeth – www.wyeth.com

York Glassware Services Ltd – www.ygs.net

Index

Page numbers in *italics* indicate illustrations; those followed by t indicate tables; those with roman numerals refer to illustrations in the color section.

α-defensins, 135, 144
Amino acids
   $^{13}C_6$-lysine and -arginine, 177
   deuterium-enriched, 18
   essential, 18
   in vivo labeling with, 17–32, 215
   ^{15}N-enriched, 18
Amino acid-coded mass tagging (AACT), *xix*, 17–32
   AACT/epitope dual-tagging strategy, *xx*, 27–29
Alzheimer's disease, 135
Angiotensin-converting enzyme (ACE), 3, 7, *8*, 9
Apolipoprotein A-II protein, 144
ASAPRATIO software, 178

7B2 protein, 144
b ions (*see* Mass spectrometry)
Band 4.1 protein, 3
β-actin, 3, 7, *8*, 9
β-amyloid protein, 135
β-defensins, 144
β-elimination/Michael addition, 167–169, 174, 177
   dehydroalanine formation, 174
   dehydroamino-2-butyric acid formation, 174
Biomarker(s), *see* Clinical proteomic profiling

C3a(desArg) protein, 144
Ca^{2+}-ATPase, 2
Calmodulin, 120, 194
   calmodulin-binding peptide, 120
Caveolin, 3, 7, *8*, 9, 13

Cells, 2
   astrocytes, 135
   binding to protein microarray, 194
   Caenorhabditis elegans, 120
   Candida albicans, 120
   culture, *in vivo* labeling of, 17–21
   Drosophila, 120
   endothelial, 2, 14
   Escherichia coli, 120, 197
   human, 120
   multipotent adult progenitor, 66
   yeast, 20, 120
Centrifugation
   Nycodenz, *4*, 6–7, 11
   sucrose density, 3, *4*, 9, 11
Cholera toxin-binding ganglioside G_{M1}, 3
Chromogranin B protein, 144
Clinical proteomic profiling, 135–160
   biological fluids for analysis, 139
      blood samples, 144–145
      cerebrospinal fluid, 139
      nipple aspirate, 139
      pancreatic juice, 139
      plasma, 139, 144–145
      serum, 139, 194
      urine, 139
   biomarkers, 1, 55, 62, 83, 103, *136*–137, 140, 145, 150, 151, 154, 176, 214
   data analysis and processing, *137*, 141–143, 150–156
      alignment of spectra, 150–151
      baseline subtraction, 150–151
      Bonferroni correction, 155
      classification of trees, 155
      coefficient of variation (CV), 152

Clinical proteomic profiling – *contd*
　data analysis and processing – *contd*
　　detecting differences between groups, 152–156
　　false discovery rate, 155
　　noise reduction, 151
　　normalization, 150–151
　　peak detection, 150–151
　　principal component analysis, *xxvi*, 152–*153*
　　quality control, 152
　　sample classification, 155
　　significance levels, 155–*156*
　immune responses, 198–199
　in lung tumors, *xxvi*, 136
　sample loading, 146
　stages in discovery, *137*
　study design, *137*, 140–141
　validation and translation, 137, 143–144
Cluster analysis, *xxvi*, 136
Coatomer (β-COP and ε-COP), 7, *8*, 9, 13
COFRADIC, *see* Reverse-phase diagonal chromatography
Cystatin C, 144
Cytochrome c, *8*, 13

Data
　bottleneck, 208
　dataset comparisons, 209
　different levels, 209
　dynamic real-time filters, 209
　metadata, 227, 240
　mining strategies, 207–225
　repositories, 216
　similarity searching, 237
　standardization, 230
Database(s)
　annotation of unknown proteins via InterProScan, 233
　BIND database, 231
　DDBJ/EMBL/GenBank nucleotide databases, 229, 238
　DIP database, 229
　EST databases, 212
　FASTA-formatted, 80, 237
　Gene3D, 232
　gene ontology (GO), 232
　genomes available, 96
　GlycoSuite, 212
　help desks, 243
　Impact (MIPS) database, 229
　IntAct database, 229, 238–239
　International Protein Index (IPI), 212, 229, 235–237
　InterPro, 232
　InterProScan, 232
　iterative database searching algorithm, 212
　MINT database, 229
　NCBI non-redundant database, *25*, *26*, 29, 212
　open source, 242
　PANTHER, 228, 232
　Pfam, 228, 232
　PhosphoELM, 212
　PIRSF, 228, 232
　PRIDE data repository, 232, 239–242
　　data structure, 240
　　identifications, 241
　　meta-data, 240
　　use to examine tissue specificity of a protein, 241
　　XML format, 240
　ProDom, 232
　ProSite, 228, 232
　RefSeq, 212
　searching, 94–97, 207
　　algorithms, 174
　　de novo correlation of MS/MS spectra, 95
　　direct correlation of MS/MS spectra, 95
　　for molecular interactions, 239
　　for sequence similarity, 237
　　influence of phosphorylation on, 162
　　searching parameters, 29, 80t
　selection, 212–213
　sequence accession numbers, 242
　SMART, 232
　SUPERFAMILY, 232
　TIGRFAm, 232
　UniProt, 212–213, 228–229, 232, 234–235, 238
　　UniParc sequence archive, 242
　use of hidden Markov models, 228
DBTOOLKIT, 80
DeCodon software, 50
Dendrimers, 169

Desorption electrospray ionization MS (DESI MS), *see* Mass spectrometry
Disease signatures, *see* Clinical proteomic profiling
Dynamin, 3

Early endosomal antigen 1 (EEA1), 7, *8*
Experimental Peptide Identification Center (EPICENTER) software, 209, 216–224
  data levels, 216–218
  data mining with, *217*, 220–223
  data organization and import, 216
  key features, 216
  peptide validation, 218–220
  reference libraries of MS/MS spectra, 224
  statistical result views, 224

Free-flow electrophoresis (FFE), 39–40

G-proteins, 13
  gas phase fractionation, *see* Mass spectrometry
Gel electrophoresis
  1-dimensional, 22, 126–127
    SDS-PAGE/LC-MS/MS, 24, 56
  2-dimensional, 14, 17, 22, 33–60, 61, 209
    carrier ampholytes, 35, 39, 41
    cyanine dyes, 42
    denaturants, 35
    detergents, 35
    difference in-gel electrophoresis (DIGE), 42–45
    electroendoosmotic effects, 45
    immobilised pH gradients (IPGs), 34–35, 40–41
    isoelectric focusing, 33, 41–42, 53
    reductants, 35, 45
    rehydration, 42
    second dimension, 46
  gel staining, 43, 47–49
    Coomassie blue, 47
    fluorescent, 48–49
    modification-specific, *xxi, 51*
      Pro-Q Diamond dye, 53
      Pro-Q Emerald dye, 53
    silver, 47–48, 127–128
    SYPRO dyes, 53
Genomes, 96

Gliomas, 135
Global Proteome Machine software, *see* Search engines
Golgi marker p58, *8*, 13
Gradiflow system, 39–40
GTP-binding nuclear protein Ran, 7, *8*

Haptoglobin protein, 144
HIP/PAP-1 protein, 144
HPLC, *see* Liquid chromatography
Human Proteome Organization (HUPO), 202, 227, 230

Immunoaffinity chromatography, 27–29, 52
Inositol 1, 4, 5-triphosphate receptor, 3
Isoelectric focusing, *see* Gel electrophoresis
Isotope-coded affinity tags (ICAT), 17, 26, 62, 177
Isotope-coding strategies, 1–32, 65, 177
  d_0- and d_3-labeling, 171
  $^{13}C_6$-lysine and -arginine labeling, 177
  ^{15}N uniform labeling, 18, 177
  ^{18}O labeling, 65, 67, 68–69

Liquid chromatography
  C18 columns, 67, 89
  capillary columns, *xxii*, 89
  construction of in-house columns, 90
  flow rates, 90
  flow splitting, 90
  HPLC, 24, 29, 56, 66, 78
  LC-ESI-MS/MS, *xxii, xxv*, 24, 29, 56, 66, 78, *89, 122*, 131, 144, 207, 210
  micro(capillary)-liquid chromatography (μ-LC), 90, 178
  nanocapillary-liquid chromatography (nano-LC), 90, 178
  packing columns, *xxii*, 89
  sample clean-up prior to LC-MS, 87–88, 130
  strong cation exchange, 172
  zero dead volume cross-connection, *xxii, 89*
Lung tumors, *136*
Lysosomal membrane glycoprotein 1 (Lamp1), 7, *8*, 13

Mass-coded abundance tagging (MCAT), 17
MASCOT software, see Search engines
Mass spectrometry
    acquisition of precursor ions, 90, 173, 174, 210
    b ions, *xxiii*, 92, 96t
    charge-state determination, 210–211
    collision-induced dissociation, 84, 89, 90, 174
    data
        formats, 178
        storage, 78
    desorption electrospray ionization (DESI), 99–118
        analysis of biological tissues, 103, 113–116
        analysis of intact proteins and oligopeptides, 105–106
        analysis of microtiter plate samples, 100
        coupling with separation methods, 110–112
        examples of spectra, *104, 106–108, 111, 113–115*
        geometry-independent DESI, *100*–101
        high throughput, *101*
        ion formation mechanisms, *102*–103
        ion source, *101*
        instrumentation, 100, *101*
        metabolite detection, 103
        reactive DESI-MS, 112–113
        thin layer chromatography/DESI-MS, *xxiv*, 110
        tryptic digest analysis, 106–109
    electron capture dissociation (ECD), 210
    electron transfer dissociation (ETD), 174–*176*, 207, 210
    electrospray ionization (ESI), 66, 83, 99
    fast atom bombardment, 99
    field desorption, 99
    Fourier transform ion cyclotron resonance (FTICR) MS, 210
    gas phase fractionation, *xxii, 84, 91,* 93–94
    high throughput for AACT analysis, 23–25
    imaging, 113–116, 135
    instrument resolution, 210
    intensity ratios, 23–24, 65
    ionization behavior, 162
    linear ion trap, 100, 174, 210, 213
    matrix-assisted laser desorption (MALDI), 66, 83, 99, 135, *138,* 207
        matrix solutions, 148
        time-of-flight (TOF)/TOF-MS, 66, 92, 210–211
    monoisotopic distribution patterns, 23
    MS/MS (tandem MS), *xxiii*, 26, 62, 66, *84*, 90–92, 131, 207, 211, 213
        data-dependent, 78, 90
        reference libraries of MS/MS spectra, 224
    MS^n experiments, 174–*175*, 210
    neutral loss, 173
    Orbitrap MS, 210
    plasma desorption, 99
    precursor ion scanning, 30, 90–91, 94–95, 97, 131, 173–175, 210–211, 216–217, 219–220, 224
    quadrupole time-of-flight (QTOF), 173–174, 210, 213
    quick-switching protocols, 174
    secondary ion, 99
    surface-enhanced laser desorption/ionization (SELDI), 135, *138*–139, 144, 189
        data analysis, 152
        data processing, 150–152
        Enterprise 4000 instrument, 139
        profile generation, 149–150
        ProteinChips, 135, *138*–139, 144, 146–148
        quality control, 152
    strategies for protein identification, 83–98
    undersampling, 62, 95
    use of exclusion lists, 211
    use of QSTAR XL for AACT analysis, *25*
    y ions, *xxiii*, 92, 96t
Membranes
    caveolae, 2, *4, 9, 10*, 13
    marker proteins, 3, 8, 13
    detergent extraction, 13–14
    endothelial, 2, *4, 8, 10*

lipid rafts, 13
microdomains, 1–16
plasma, 2, *8*, 13, 14, 124, 207
plasmalemma vesicles, 2, 13
purification of, 1–16
Micro-liquid chromatography (μ-LC), *see* Liquid chromatography
MSQUANT software, 215
Multi-compartment electrolysers, 39–40
Multi-dimensional protein identification technology (MuDPIT), 14, 24, *27*, 34
Myocardial tissue, 34

Nano-liquid chromatography (nano-LC), *see* Liquid chromatography
Nitric oxide synthase (NOS), 3, 7, *8*, 13
5' nucleotidase, *8*, 9

OLAV software, *see* Search engines
OMSSA software, *see* Search engines

p53 protein, *8*
Parvalbumin protein, 144
PEPSEA software, *see* Search engines
Peptides
    adducts, 87
    cysteinyl, 80
    desalting, 23, 130, 165
    extraction following SDS-PAGE and trypsinization, 24
    methionyl, 80
    phosphorylated, *see* Protein phosphorylation
    separation by μLC and MS/MS analysis, *xxii*, 24, 29, 56, 66, *89*
    separation by reverse-phase diagonal chromatography, 61–82
    stage tip purification, 130
    see also Proteins
PEPTIDEPROPHET software, *see* Search engines
PHENYX software, *see* Search engines
Polyacrylic acid (PAA), 5, 11
PROICAT software, *25*
PROID software, 29
Pro-Q Diamond dye, *see* Gel electrophoresis
Pro-Q Emerald dye, *see* Gel electrophoresis

Proteases
    tobacco etch virus (TEV), 120
    trypsin, 23, 69, *84*–86, 96, 106–109, 128–129
Proteins
    affinity purification of protein complexes, 119–134, 193
        biotin, 120
        calmodulin-binding peptide (CBP), 120
        c-myc, 120
        epitopes, 120
        FLAG tag, 120
        glutathione *S*-transferase (GST), 120, 193
        hemagglutinin, 120
        poly-His tags, 120, 193
        protein A, 120, 193
        tandem affinity purification, 119–134, 229
            cell extract preparation, 124
            example of purification, 132
            harvesting cells, 123–124
            overview, *xxv*, 122
            time line, 123
    alignment, 228–229
    alkaline, 56
    alkylation, 45
    basic, 54
    in blood, 139
    cytoskeletal, 14
    denaturants, 35, 87
    detergents for solubilization of, 35, 45, 87, 178
    epitope-tagging, *27*–29
    expression range, 1, 33, 94, 161, 227, 230
        expression versus abundance, 49
    extraction for MS, 22, 37
    glycosylation, 53, 56, 194, 197–198
        detection, 53
        sialidase treatment, 53
    high- and low-molecular mass, 33, 55
    hydrophobic, 17, 33, 38
    identification strategies, 83–98, 208, 212–220, 228
    interaction networks, *xxv*, 119, *122*, 229
    lipid-anchored, 14
    low abundance, 1, 26, 34, 55, 161

Proteins – contd
  markers of compartments, 3, 7, *8*, 13, 14
  membrane, 14, 26, 178
  phosphorylation, 29–30, 56, 161–182, 194
    acyl-phosphorylation, 162
    Akt phosphorylation, 200
    dephosphorylation, 179
    enrichment of phosphorylated proteins and peptides, 52, 62, 163–173
      anti-phosphotyrosine antibodies, 163–165
      charge-based, 172–173
      chemical derivatization, 167–171
      immobilized metal ion affinity chromatography (IMAC), 164, 166
      titanium dioxide, 166–167
    ERK phosphorylation, 200
    generic scheme for phosphoproteomic analysis, *163*
    immonium ion detection, 174
    mapping 'real-time' phosphorylation sites, 29
    neutral loss scanning in MS analysis, 173
    N-phosphorylation, 162
    occurrence, 162
    O-phosphorylation, 162
    quantitative phosphoproteomics, 176–177
    serine, 162, 174
    sorting of phosphorylated peptides, 76–77
    S-phosphorylation, 162
    stability, 162
    stoichiometry, 162
    threonine, 162, 174
    tyrosine, 162, 174
    see also Peptides
  physical properties, 39, *50*
  post-translational modification, 26, 41, 51, 161–182, 193, 212–213; see also Glycosylation; Phosphorylation
  prefractionation, 36, 140
  protein A, 120
  protein–cell interactions, 196–197
  protein–DNA interactions, 186, 194–195
  protein–drug interactions, 186, 194–195
  protein–lipid interactions, 194
  protein–protein interactions, 18, 119–134, 186, 194, 196, 229
    tandem affinity purification (TAP), see Proteins, affinity purification of complexes
    yeast two-hybrid system, 121, 229, 238
  recovery by precipitation, 12, 35
  reductants, 35
  separation, 22, 39
  staining, see Gel electrophoresis
  subcellular location, 39
  tandem affinity purification (TAP), see Proteins, affinity purification of complexes
  trypsin digestion, 23, 69, *84–86*, 96, 106–109, 128–129, 165
    missed cleavage sites, 97
PROTEINCHIP software, 150, 154
Protein disulphide isomerase A4 (ERP72), 7, *8*
Protein microarray technologies, 183–205
  analytical microarrays, 183, 190–191, *197*
  antibody arrays, 190–191
  applications, *184*
  assay platforms, 186–187
    direct probing with labeled molecule, 186
    fluorescence, 186
    radioisotopes, 186
    use of enzymes, 186
  detection methods, 187–190
    atomic force microscopy, 189
    CCD imaging, 187
    fluorescence, 187
    laser scanning, 187
    limits, 190, 202
    mass spectrometry, 189
    microcantilevers, 190
    planar waveguide technology, 187–188

surface plasmon resonance, 188
X-ray film, 187
functional microarrays, 183, 191–199
   de novo synthesis of proteins, 200–202
   expression-ready open reading frame collections, 192
   high throughput expression systems, 193
   microarray substrates, 183–185
      3-D matrices, 184
      glass, 184
      gold-coated, 184
      hydrogel, 184
      nickel-coated, 184
      self-assembled monolayers, 184
   kinase-substrate interactions, 186, 198
   printing, 185–186
   reverse-phase microarrays, 199
   yeast proteome microarrays, *xxix*, 191, 194
PROTEINPROPHET software, *see* Search engines
PROTEIN PROSPECTOR software, *see* Search engines
Protein sequence annotation via InterProScan, 233
PROTEIOS software, *see* Search engines
Proteome coverage, 17, 34, 79, 93, 209
Proteome/Protein Standard Initiative Mass Spectrometry (PSI-MS) initiative, 213, 227, 230
   analysisXML format, 231
   mzData interchange format, 231
   mzDataConverter, 231
   mzDataViewer, 231
   mzML, 231
   mzXML format, 178, 213
   small molecule identification support, 232
Ran protein, *see* GTP-binding nuclear protein Ran
Reverse-phase diagonal chromatography, 61–82
   combined fractional diagonal chromatography (COFRADIC), 63, *64*–68
   sorting of cysteinyl peptides, *64*, 71–74, 79

   sorting of methionyl peptides, *64*–65, 69–71
   sorting of N-terminal peptides, *64*, 66, 74–75, 79
   sorting of phosphorylated peptides, *64*, 76–77
Rosetta Elucidator protein expression data analysis system, 211
Rotofor system, 39–40

S100β protein, 135
Sample preparation, 1–16, 34–38, 69, 85, 87, 123–124, 227
SDS-polyacrylamide gel electrophoresis (SDS-PAGE), *see* Gel electrophoresis
Search engines
   Global Proteome Machine, 231
   MASCOT, 26, 29, 80, 132, 207–208, 213–216, 218, 220, 231
   OLAV, 213
   OMSSA, 213
   PEPSEA, 213
   PEPTIDEPROPHET, 178, 213
   PHENYX, 213
   PROTEINPROPHET, 178
   PROTEIN PROSPECTOR, 213
   PROTEIOS, 231
   raw data processing, 213
   SCAFFOLD, 213
   selection of, 213–215
   SEQUEST, 97, 178, 207, 213
   SONAR, 213
   Trans-Proteomics Pipeline (TPP), 178, 216
   X! TANDEM, 213
   see also Databases
SEQUEST software (*see* Search engines)
Serum Amyloid A protein, 144
Shotgun proteomics, *84–98*, 207–*208*
Silica particles, 3, 5, 14
SONAR, *see* Search engines
Stable isotope label (SIL) tags, 17
   in cell culture (SILAC), 18, 34, 65, 177
STATA software, 155
Suppliers, 245
SYPRO dyes (*see* Gel electrophoresis)
Systems biology, 83

Tandem affinity Purification (TAP), *see* Proteins, affinity purification of complexes
Transportin protein, *8*, 13
Trans-Proteomics Pipeline (TPP) software, *see* Search engines
Transthyretin protein, 144
Trypanosomiasis, 135

Vascular endothelium, 1–2
Vesicle-associated membrane protein (VAMP), 3, *8*
VGF protein, 144

y ions, *see* Mass spectrometry

X! TANDEM software, *see* Search engines